BIG IDEAS MATH®
Geometry

Phill

Student Journal

- Maintaining Mathematical Proficiency
- Exploration Journal
- Notetaking with Vocabulary
- Extra Practice

Erie, Pennsylvania

Photo Credits

Cover Image Allies Interactive/Shutterstock.com

97 Big Ideas Learning, LLC;
315 Mauro Rodrigues/Shutterstock.com;
320 *left* Elena Elisseeva/Shutterstock.com;
right ©Sam Beebe / CC-BY-2.0;
349 Mark Herreid/Shutterstock.com

Printed in the United States

ISBN 13: 978-1-60840-853-5
ISBN 10: 1-60840-853-1

9-VLP-18 17 16

Contents

Contents

Contents

Contents

Contents

Contents

Contents

About the Student Journal

Maintaining Mathematical Proficiency

The Maintaining Mathematical Proficiency corresponds to the Pupil Edition Chapter Opener. Here you have the opportunity to practice prior skills necessary to move forward.

Exploration Journal

The Exploration pages correspond to the Explorations and accompanying exercises in the Pupil Edition. Here you have room to show your work and record your answers.

Notetaking with Vocabulary

This student-friendly notetaking component is designed to be a reference for key vocabulary, properties, and core concepts from the lesson. There is room to add definitions in your words and take notes about the core concepts.

Extra Practice

Each section of the Pupil Edition has an additional Practice with room for you to show your work and record your answers.

Name____ Date__________

Chapter 1 Maintaining Mathematical Proficiency

Simplify the expression.

1. $\left|-3 + (-1)\right| =$

2. $\left|10 - 11\right| =$

3. $\left|-6 + 8\right| =$

4. $\left|9 - (-1)\right| =$

5. $\left|-12 - (-8)\right| =$

6. $\left|-15 - 7\right| =$

7. $\left|-12 + 3\right| =$

8. $\left|5 + (-15)\right| =$

9. $\left|1 - 12\right| =$

Find the area of the triangle.

10.

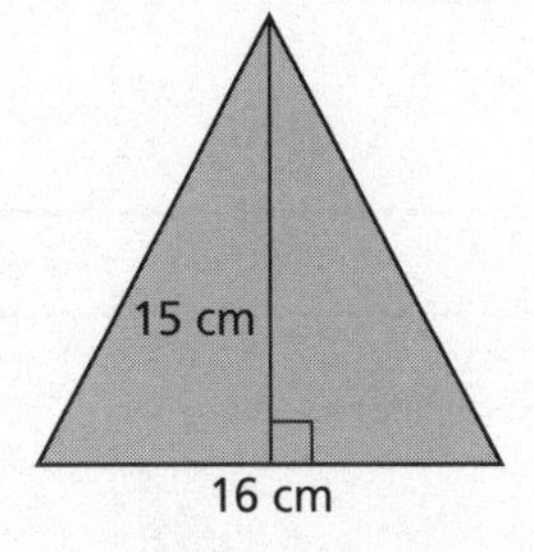

11.

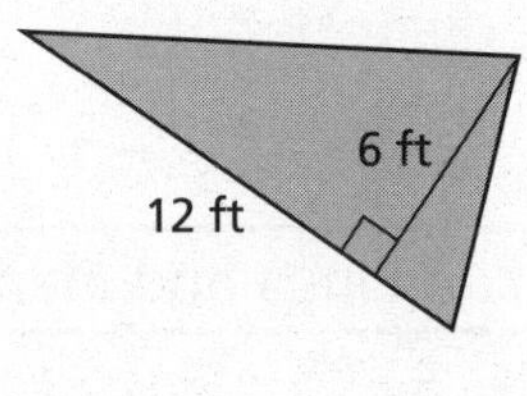

12.

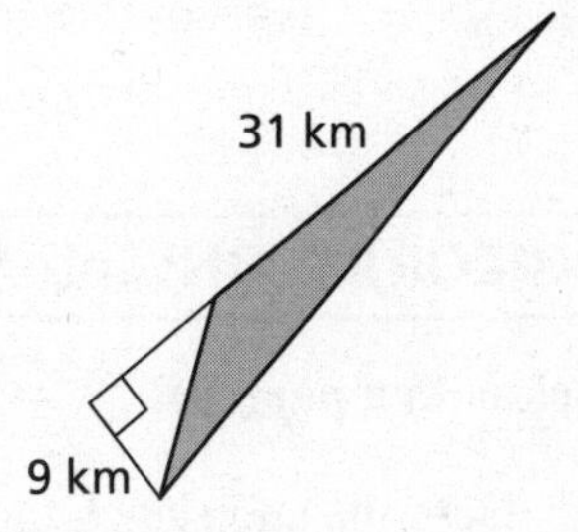

Name ____ Date __________

1.1 Points, Lines, and Planes

For use with Exploration 1.1

Essential Question How can you use dynamic geometry software to visualize geometric concepts?

1 EXPLORATION: Using Dynamic Geometry Software

Go to *BigIdeasMath.com* for an interactive tool to investigate this exploration.

Work with a partner. Use dynamic geometry software to draw several points. Also, draw some lines, line segments, and rays. What is the difference between a line, a line segment, and a ray?

Sample

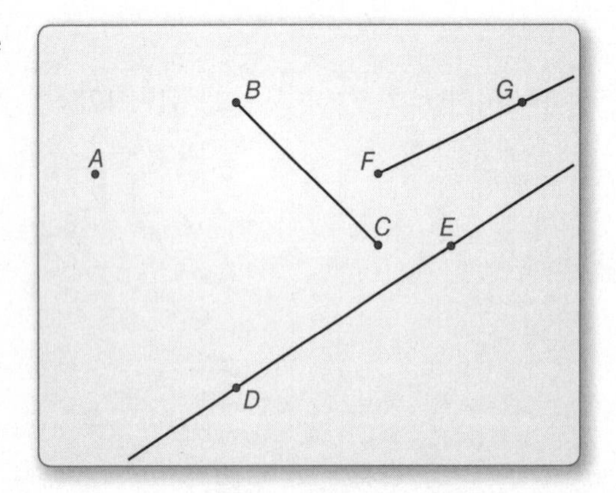

2 EXPLORATION: Intersections of Lines and Planes

Work with a partner.

a. Describe and sketch the ways in which two lines can intersect or not intersect. Give examples of each using the lines formed by the walls, floor, and ceiling in your classroom.

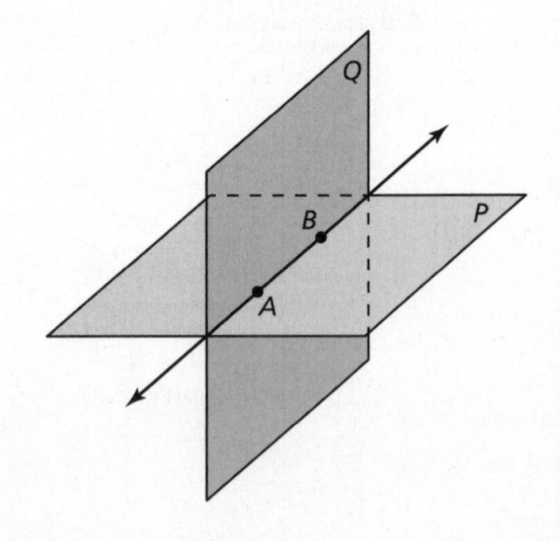

2 EXPLORATION: Intersections of Lines and Planes (continued)

b. Describe and sketch the ways in which a line and a plane can intersect or not intersect. Give examples of each using the walls, floor, and ceiling in your classroom.

c. Describe and sketch the ways in which two planes can intersect or not intersect. Give examples of each using the walls, floor, and ceiling in your classroom.

3 EXPLORATION: Exploring Dynamic Geometry Software

Go to *BigIdeasMath.com* for an interactive tool to investigate this exploration.

Work with a partner. Use dynamic geometry software to explore geometry. Use the software to find a term or concept that is unfamiliar to you. Then use the capabilities of the software to determine the meaning of the term or concept.

Communicate Your Answer

4. How can you use dynamic geometry software to visualize geometric concepts?

Name ____ Date __________

1.1 Notetaking with Vocabulary
For use after Lesson 1.1

In your own words, write the meaning of each vocabulary term.

undefined terms

point

line

plane

collinear points

coplanar points

defined terms

line segment, or segment

endpoints

ray

opposite rays

intersection

Core Concepts

Undefined Terms: Point, Line, and Plane

Point A **point** has no dimension.
A dot represents a point.

A
point A

Line A **line** has one dimension. It is represented by a line with two arrowheads, but it extends without end.

Through any two points, there is exactly one line. You can use any two points on a line to name it.

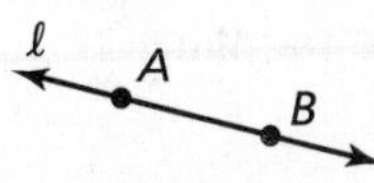

line ℓ, line AB ($\overleftrightarrow{AB}$), or line BA ($\overleftrightarrow{BA}$)

Plane A **plane** has two dimensions. It is represented by a shape that looks like a floor or a wall, but it extends without end.

Through any three points not on the same line, there is exactly one plane. You can use three points that are not all on the same line to name a plane.

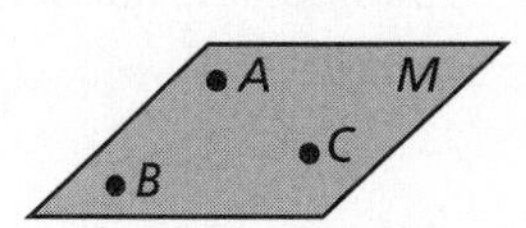

plane M, or plane ABC

Notes:

Defined Terms: Segment and Ray

The definitions below use line AB (written as $\overleftrightarrow{AB}$) and points A and B.

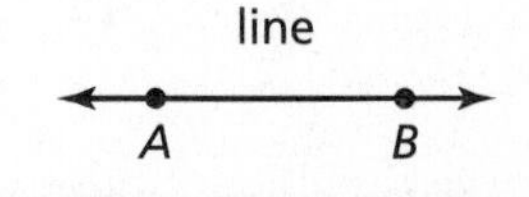

Segment The **line segment** AB, or **segment** AB (written as $\overline{AB}$) consists of the **endpoints** A and B and all points on $\overleftrightarrow{AB}$ that are between A and B. Note that $\overline{AB}$ can also be named $\overline{BA}$.

segment
endpoint endpoint
A B

Ray The **ray** AB (written as $\overrightarrow{AB}$) consists of the endpoint A and all points on $\overleftrightarrow{AB}$ that lie on the same side of A as B.

Note that $\overrightarrow{AB}$ and $\overrightarrow{BA}$ are different rays.

ray
endpoint
A B
endpoint
A B

Opposite Rays If point C lies on $\overleftrightarrow{AB}$ between A and B, then $\overrightarrow{CA}$ and $\overrightarrow{CB}$ are **opposite rays**.

A C B

Notes:

Name ______________________________ Date __________

Extra Practice

In Exercises 1–4, use the diagram.

1. Give two other names for $\overleftrightarrow{CD}$.

2. Give another name for plane M.

3. Name three points that are collinear. Then name a fourth point that is not collinear with these three points.

4. Name a point that is not coplanar with points A, C, E.

In Exercises 5–8, use the diagram.

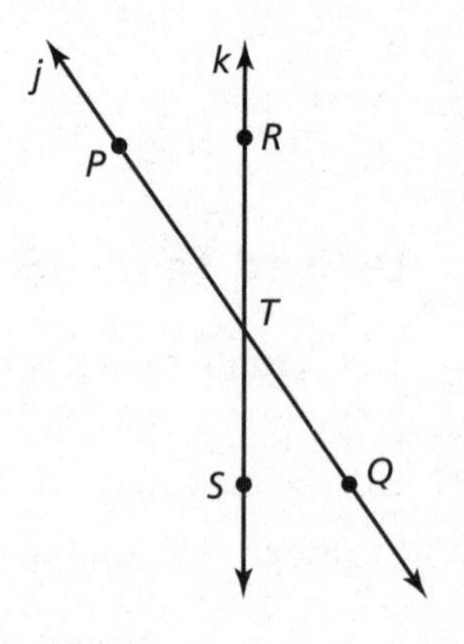

5. What is another name for $\overleftrightarrow{PQ}$?

6. What is another name for $\overleftrightarrow{RS}$?

7. Name all rays with endpoint T. Which of these rays are opposite rays?

8. On the diagram, draw planes M and N that intersect at line k.

In Exercises 9 and 10, sketch the figure described.

9. $\overline{AB}$ and $\overrightarrow{BC}$

10. line k in plane M

Name______________________________ Date__________

1.2 Measuring and Constructing Segments

For use with Exploration 1.2

Essential Question How can you measure and construct a line segment?

1 EXPLORATION: Measuring Line Segments Using Nonstandard Units

Work with a partner.

a. Draw a line segment that has a length of 6 inches.

b. Use a standard-sized paper clip to measure the length of the line segment. Explain how you measured the line segment in "paper clips."

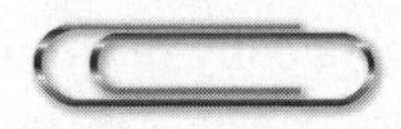

c. Write conversion factors from paper clips to inches and vice versa.

1 paper clip = ____ in.

1 in. = ____ paper clip

d. A *straightedge* is a tool that you can use to draw a straight line. An example of a straightedge is a ruler. Use only a pencil, straightedge, paper clip, and paper to draw another line segment that is 6 inches long. Explain your process.

2 EXPLORATION: Measuring Line Segments Using Nonstandard Units

Work with a partner.

a. Fold a 3-inch by 5-inch index card on one of its diagonals.

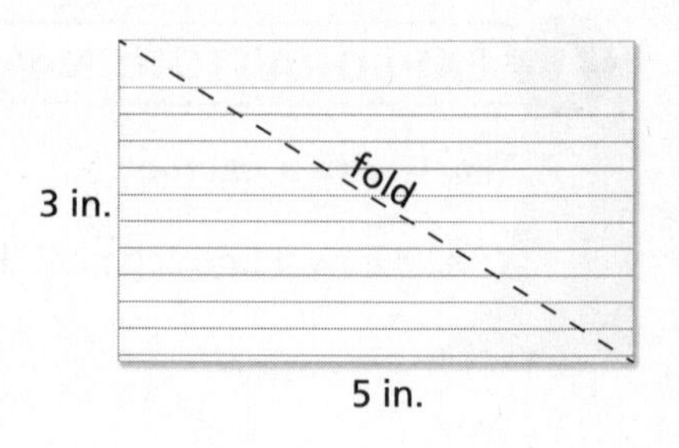

b. Use the Pythagorean Theorem to algebraically determine the length of the diagonal in inches. Use a ruler to check your answer.

c. Measure the length and width of the index card in paper clips.

d. Use the Pythagorean Theorem to algebraically determine the length of the diagonal in paper clips. Then check your answer by measuring the length of the diagonal in paper clips. Does the Pythagorean Theorem work for any unit of measure? Justify your answer.

3 EXPLORATION: Measuring Heights Using Nonstandard Units

Work with a partner. Consider a unit of length that is equal to the length of the diagonal you found in Exploration 2. Call this length "1 diag." How tall are you in diags? Explain how you obtained your answer.

Communicate Your Answer

4. How can you measure and construct a line segment?

Name____ Date__________

1.2 Notetaking with Vocabulary
For use after Lesson 1.2

In your own words, write the meaning of each vocabulary term.

postulate

axiom

coordinate

distance

construction

congruent segments

between

Postulate 1.1 Ruler Postulate

The points on a line can be matched one to one with the real numbers. The real number that corresponds to a point is the **coordinate** of the point.

The **distance** between points A and B, written as AB, is the absolute value of the difference of the coordinates of A and B.

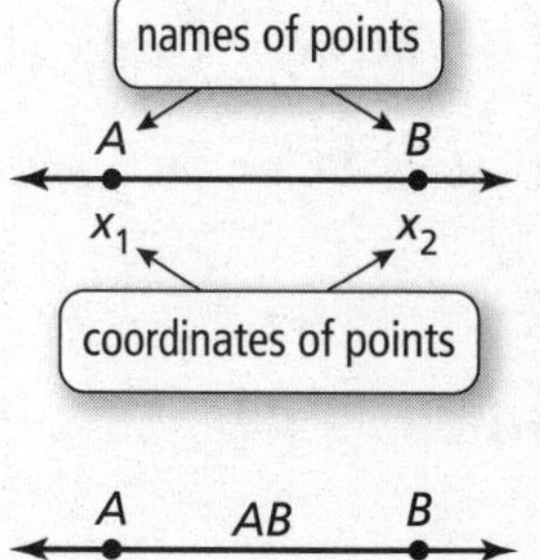

Notes:

A AB B

x_1 x_2

$AB = |x_2 - x_1|$

Name ______________________________ Date __________

1.2 Notetaking with Vocabulary (continued)

Core Concepts

Congruent Segments

Line segments that have the same length are called **congruent segments**. You can say "the length of $\overline{AB}$ is equal to the length of $\overline{CD}$," or you can say "$\overline{AB}$ *is congruent to* $\overline{CD}$." The symbol $\cong$ means "is congruent to."

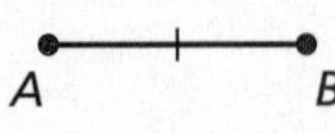

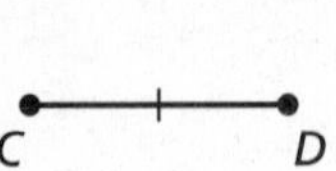

Lengths are equal.

$AB = CD$

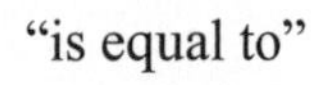

"is equal to"

Segments are congruent.

$\overline{AB} \cong \overline{CD}$

"is congruent to"

Notes:

Postulate 1.2 Segment Addition Postulate

If B is between A and C, then $AB + BC = AC$.

If $AB + BC = AC$, then B is between A and C.

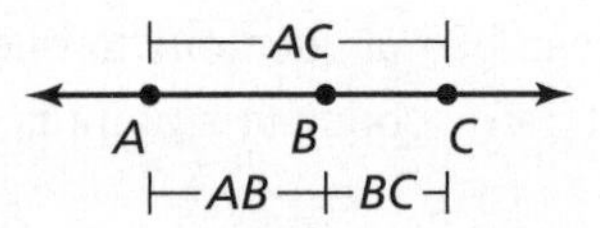

Notes:

Name____ Date__________

Extra Practice

In Exercises 1–3, plot the points in the coordinate plane. Then determine whether $\overline{AB}$ and $\overline{CD}$ are congruent.

1. $A(-5, 5), B(-2, 5)$
$C(2, -4), D(-1, -4)$

2. $A(4, 0), B(4, 3)$
$C(-4, -4), D(-4, 1)$

3. $A(-1, 5), B(5, 5)$
$C(1, 3), D(1, -3)$

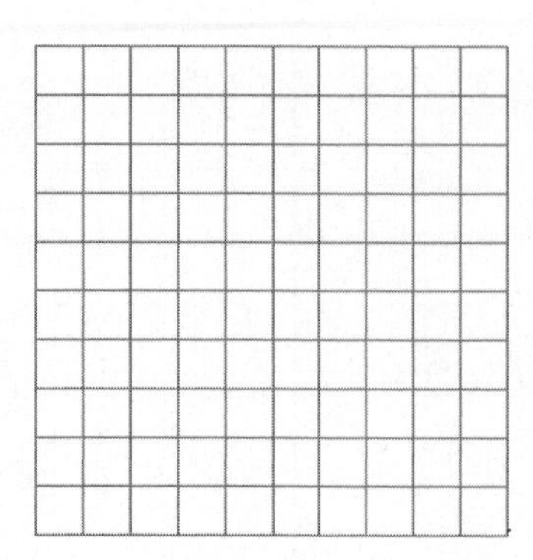

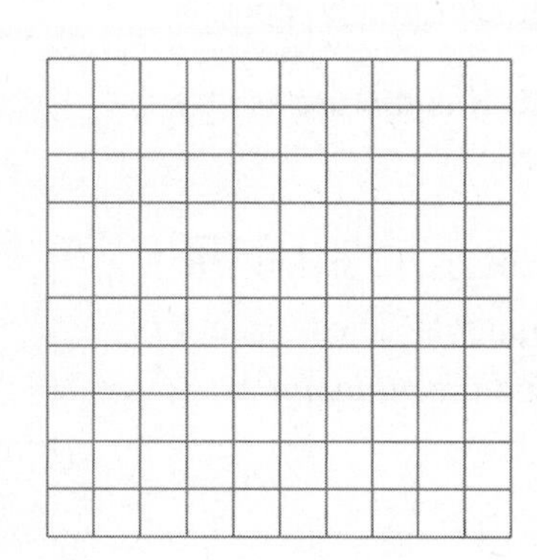

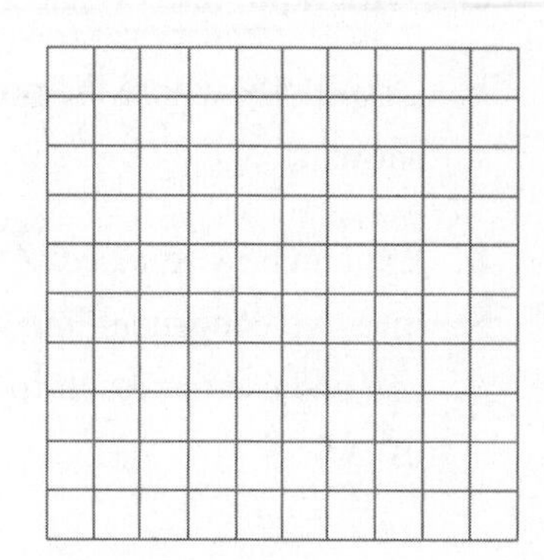

In Exercises 4–6, find *VW*.

4.

5.

6.

7. A bookstore and a movie theater are 6 kilometers apart along the same street. A florist is located between the bookstore and the theater on the same street. The florist is 2.5 kilometers from the theater. How far is the florist from the bookstore?

Name ____ Date __________

1.3 Using Midpoint and Distance Formulas

For use with Exploration 1.3

Essential Question How can you find the midpoint and length of a line segment in a coordinate plane?

1 EXPLORATION: Finding the Midpoint of a Line Segment

Work with a partner. Use centimeter graph paper.

a. Graph $\overline{AB}$, where the points A and B are as shown.

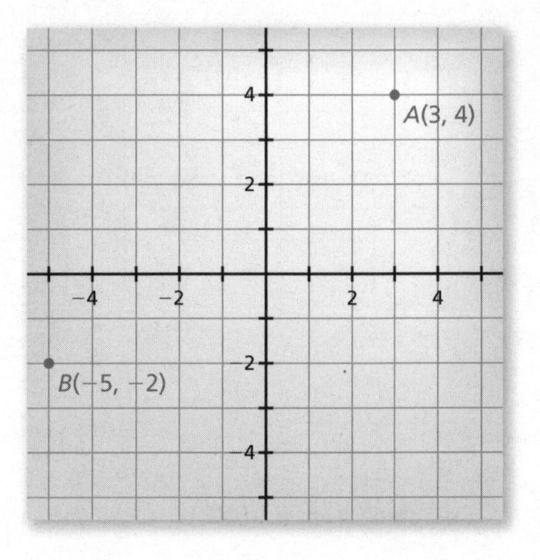

b. Explain how to *bisect* $\overline{AB}$, that is, to divide $\overline{AB}$ into two congruent line segments. Then bisect $\overline{AB}$ and use the result to find the *midpoint* M of $\overline{AB}$.

c. What are the coordinates of the midpoint M?

d. Compare the x-coordinates of A, B, and M. Compare the y-coordinates of A, B, and M. How are the coordinates of the midpoint M related to the coordinates of A and B?

2 EXPLORATION: Finding the Length of a Line Segment

Work with a partner. Use centimeter graph paper.

a. Add point C to your graph as shown.

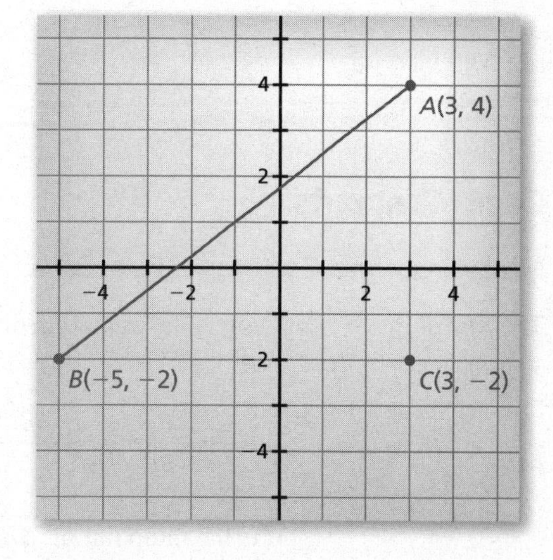

b. Use the Pythagorean Theorem to find the length of $\overline{AB}$.

c. Use a centimeter ruler to verify the length you found in part (b).

d. Use the Pythagorean Theorem and point M from Exploration 1 to find the lengths of $\overline{AM}$ and $\overline{MB}$. What can you conclude?

Communicate Your Answer

3. How can you find the midpoint and length of a line segment in a coordinate plane?

4. Find the coordinates of the midpoint M and the length of the line segment whose endpoints are given.

a. $D(-10, -4), E(14, 6)$

b. $F(-4, 8), G(9, 0)$

Name ______________________________ Date __________

1.3 Notetaking with Vocabulary

For use after Lesson 1.3

In your own words, write the meaning of each vocabulary term.

midpoint

segment bisector

Core Concepts

Midpoints and Segment Bisectors

The **midpoint** of a segment is the point that divides the segment into two congruent segments.

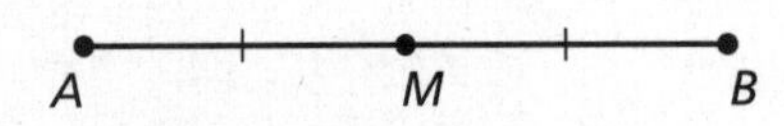

M is the midpoint of $\overline{AB}$.
So, $\overline{AM} \cong \overline{MB}$ and $AM = MB$.

A **segment bisector** is a point, ray, line, line segment, or plane that intersects the segment at its midpoint. A midpoint or a segment bisector *bisects* a segment.

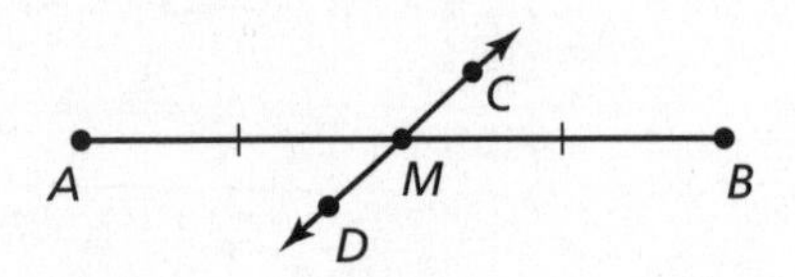

$\overleftrightarrow{CD}$ is a segment bisector of $\overline{AB}$.
So, $\overline{AM} \cong \overline{MB}$ and $AM = MB$.

Notes:

Name____ Date__________

The Midpoint Formula

The coordinates of the midpoint of a segment are the averages of the x-coordinates and of the y-coordinates of the endpoints.

If $A(x_1, y_1)$ and $B(x_2, y_2)$ are points in a coordinate plane, then the midpoint M of $\overline{AB}$ has coordinates

$$\left(\frac{x_1 + x_2}{2}, \frac{y_1 + y_2}{2}\right).$$

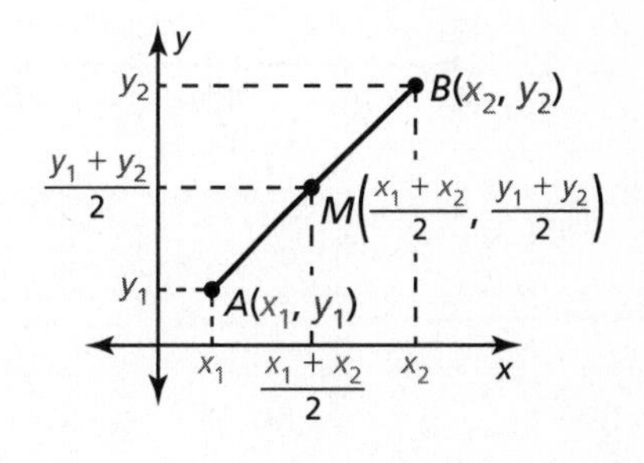

Notes:

The Distance Formula

If $A(x_1, y_1)$ and $B(x_2, y_2)$ are points in a coordinate plane, then the distance between A and B is

$$AB = \sqrt{(x_2 - x_1)^2 + (y_2 - y_1)^2}.$$

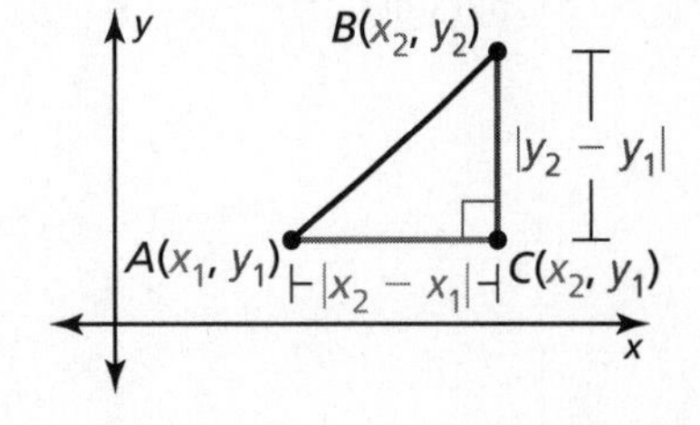

Notes:

Extra Practice

In Exercises 1–3, identify the segment bisector of $\overline{AB}$. Then find AB.

1. A 13 M B

2.

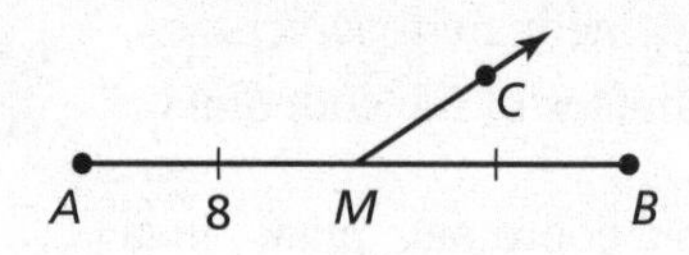

3.

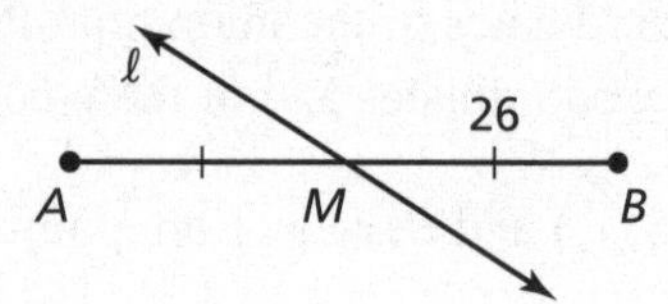

In Exercises 4-6, identify the segment bisector of $\overline{EF}$. Then find EF.

4.

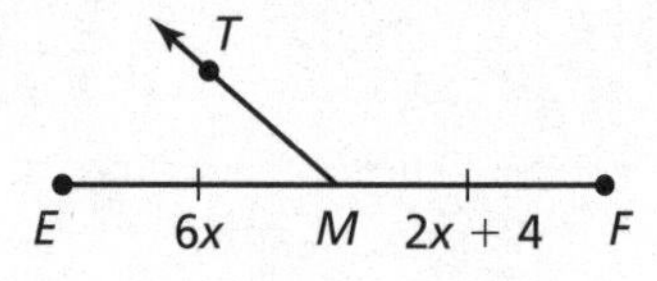

5.

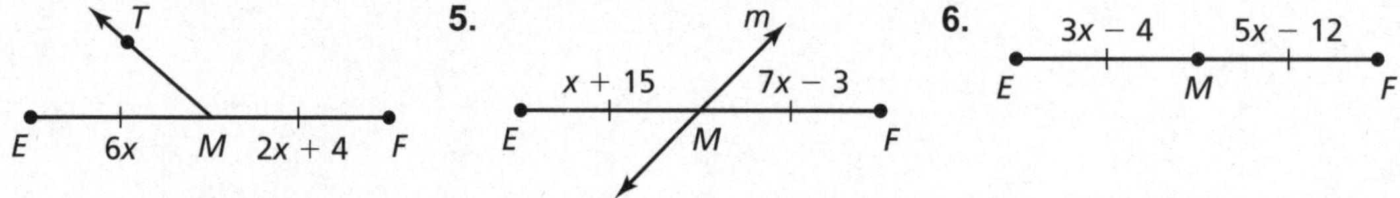

6.

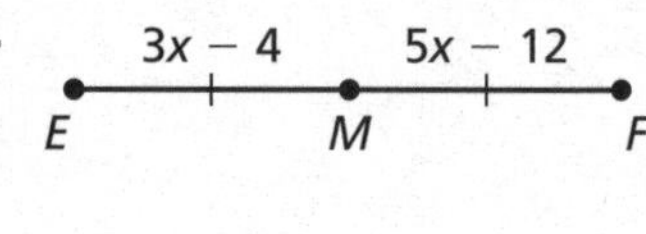

In Exercises 7–9, the endpoints of $\overline{PQ}$ are given. Find the coordinates of the midpoint M.

7. $P(-4, 3)$ and $Q(0, 5)$

8. $P(-2, 7)$ and $Q(10, -3)$

9. $P(3, -15)$ and $Q(9, -3)$

In Exercises 10–12, the midpoint M and one endpoint of $\overline{JK}$ are given. Find the coordinates of the other endpoint.

10. $J(7, 2)$ and $M(1, -2)$

11. $J(5, -2)$ and $M(0, -1)$

12. $J(2, 16)$ and $M\left(-\frac{9}{2}, 7\right)$

Name____ Date__________

1.4 Perimeter and Area in the Coordinate Plane

For use with Exploration 1.4

Essential Question How can you find the perimeter and area of a polygon in a coordinate plane?

1 EXPLORATION: Finding the Perimeter and Area of a Quadrilateral

Work with a partner.

a. On the centimeter graph paper, draw quadrilateral *ABCD* in a coordinate plane. Label the points $A(1, 4)$, $B(-3, 1)$, $C(0, -3)$, and $D(4, 0)$.

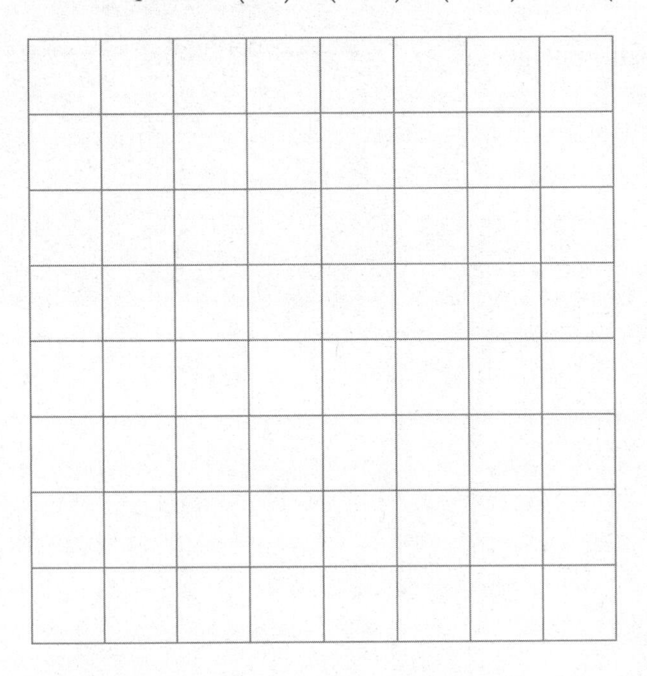

b. Find the perimeter of quadrilateral *ABCD*.

c. Are adjacent sides of quadrilateral *ABCD* perpendicular to each other? How can you tell?

d. What is the definition of a square? Is quadrilateral *ABCD* a square? Justify your answer. Find the area of quadrilateral *ABCD*.

Name ______________________________ Date __________

2 EXPLORATION: Finding the Area of a Polygon

Work with a partner.

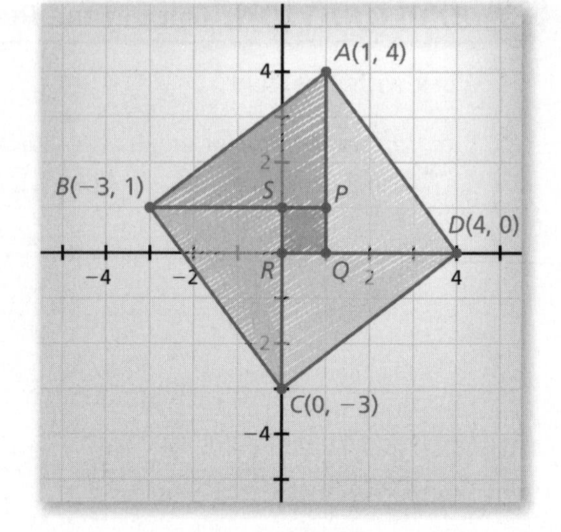

a. Quadrilateral *ABCD* is partitioned into four right triangles and one square, as shown. Find the coordinates of the vertices for the five smaller polygons.

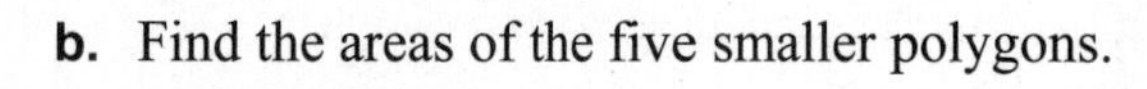

b. Find the areas of the five smaller polygons.

Area of Triangle *BPA*:

Area of Triangle *AQD*:

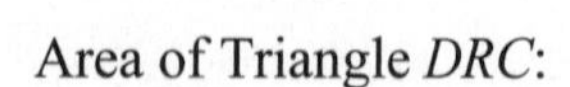

Area of Triangle *DRC*:

Area of Triangle *CSB*:

Area of Square *PQRS*:

c. Is the sum of the areas of the five smaller polygons equal to the area of quadrilateral *ABCD*? Justify your answer.

Communicate Your Answer

3. How can you find the perimeter and area of a polygon in a coordinate plane?

4. Repeat Exploration 1 for quadrilateral *EFGH*, where the coordinates of the vertices are $E(-3, 6)$, $F(-7, 3)$, $G(-1, -5)$, and $H(3, -2)$.

Name____ Date__________

1.4 Notetaking with Vocabulary

For use after Lesson 1.4

In your own words, write the meaning of each vocabulary term.

polygon

side

vertex

n-gon

convex

concave

Core Concepts

Polygons

In geometry, a figure that lies in a plane is called a plane figure. Recall that a *polygon* is a closed plane figure formed by three or more line segments called *sides*. Each side intersects exactly two sides, one at each *vertex*, so that no two sides with a common vertex are collinear. You can name a polygon by listing the vertices in consecutive order.

Notes:

Extra Practice

In Exercises 1–4, classify the polygon by the number of sides. Tell whether it is *convex* or *concave*.

1.

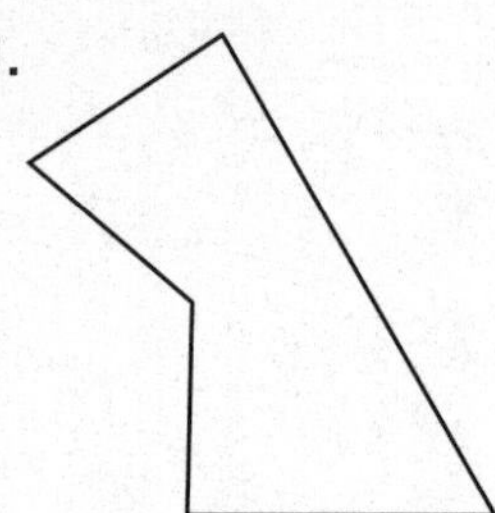

2.

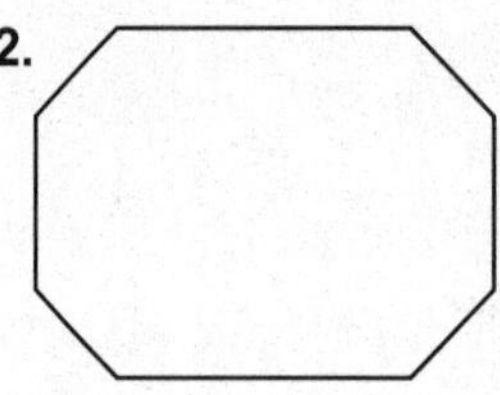

3.

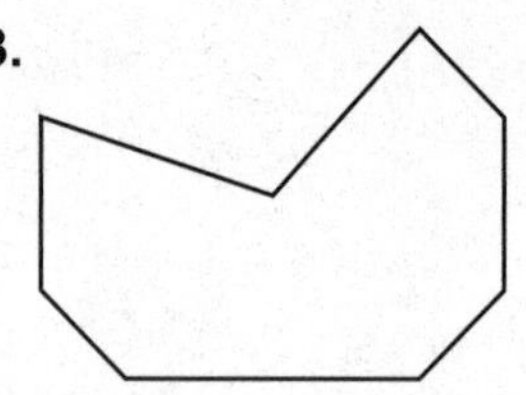

4.

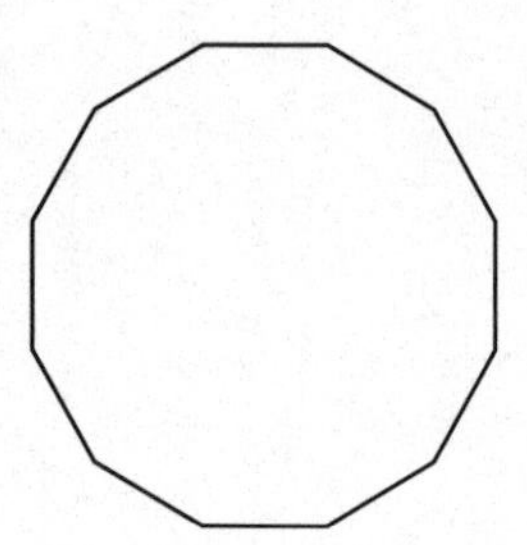

In Exercises 5–8, find the perimeter and area of the polygon with the given vertices.

5. $X(2, 4), Y(0, -2), Z(2, -2)$

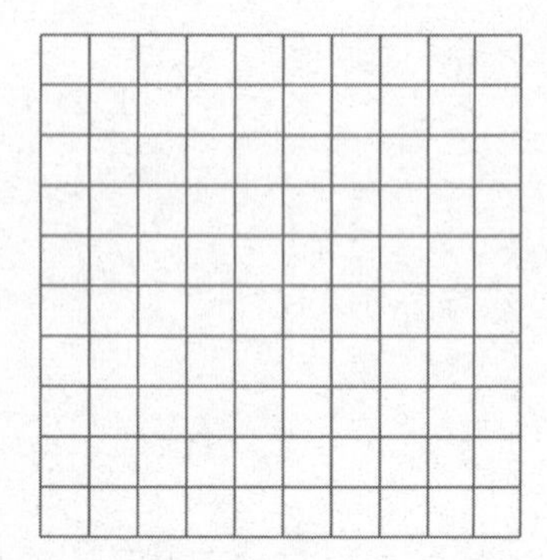

6. $P(1, 3), Q(1, 1), R(-4, 2)$

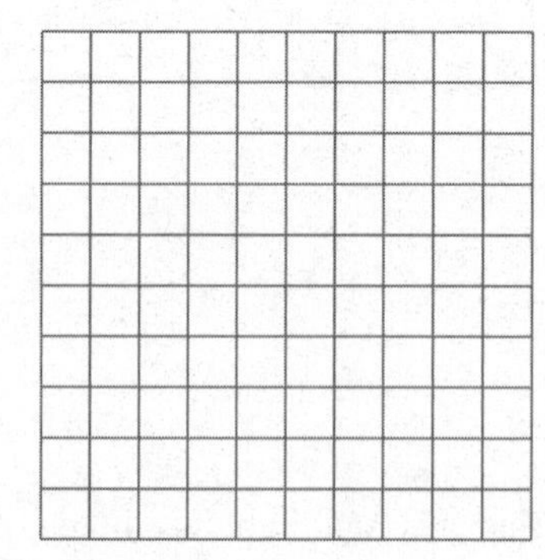

7. $J(-4, 1), K(-4, -2), L(6, -2), M(6, 1)$

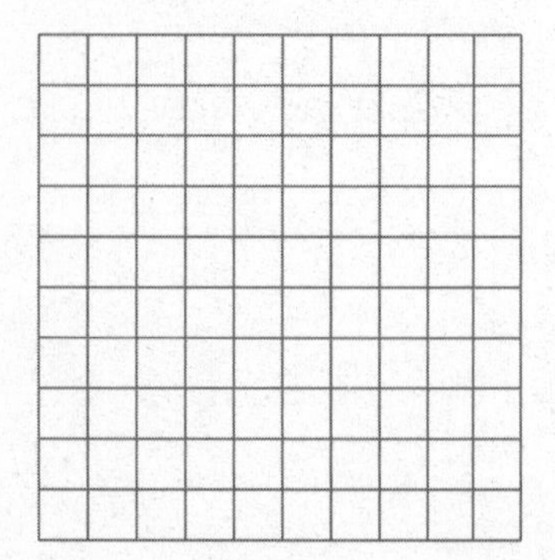

8. $D(5, -3), E(5, -6), F(2, -6), G(2, -3)$

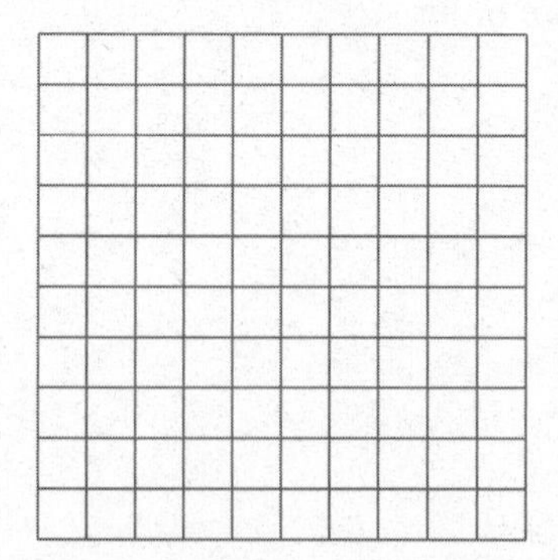

Name____ Date__________

In Exercises 9–14, use the diagram.

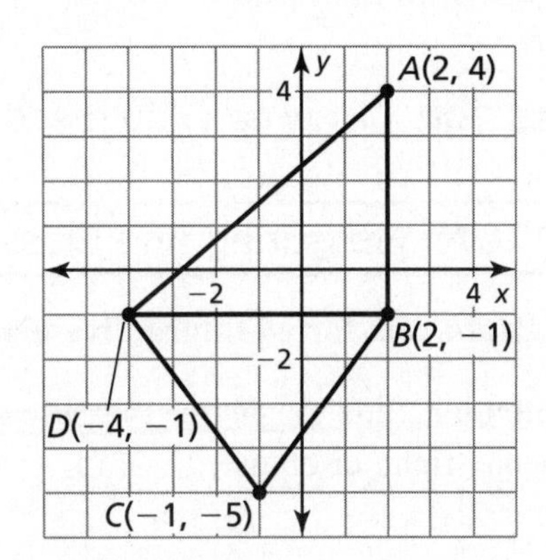

9. Find the perimeter of $\triangle ABD$.

10. Find the perimeter of $\triangle BCD$.

11. Find the perimeter of quadrilateral *ABCD*.

12. Find the area of $\triangle ABD$.

13. Find the area of $\triangle BCD$.

14. Find the area of quadrilateral *ABCD*.

Name ______________________________ Date __________

1.5 Measuring and Constructing Angles

For use with Exploration 1.5

Essential Question How can you measure and classify an angle?

1 EXPLORATION: Measuring and Classifying Angles

Go to *BigIdeasMath.com* for an interactive tool to investigate this exploration.

Work with a partner. Find the degree measure of each of the following angles. Classify each angle as acute, right, or obtuse.

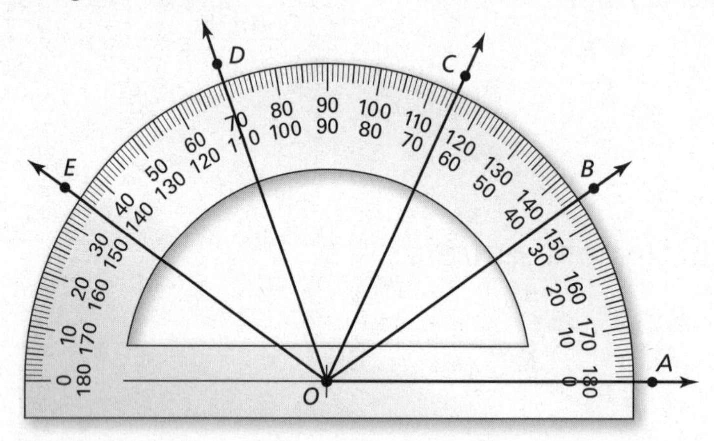

a. $\angle AOB$

b. $\angle AOC$

c. $\angle BOC$

d. $\angle BOE$

e. $\angle COE$

f. $\angle COD$

g. $\angle BOD$

h. $\angle AOE$

2 EXPLORATION: Drawing a Regular Polygon

Go to *BigIdeasMath.com* for an interactive tool to investigate this exploration.

Work with a partner.

a. On a separate sheet of paper or an index card, use a ruler and protractor to draw the triangular pattern shown at the right.

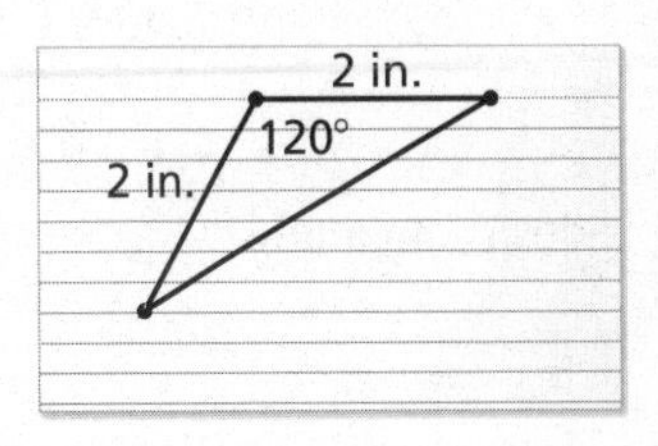

b. Cut out the pattern and use it to draw three regular hexagons, as shown in your book.

c. The sum of the angle measures of a polygon with n sides is equal to $180(n - 2)°$. Do the angle measures of your hexagons agree with this rule? Explain.

d. Partition your hexagons into smaller polygons, as shown in your book. For each hexagon, find the sum of the angle measures of the smaller polygons. Does each sum equal the sum of the angle measures of a hexagon? Explain.

Communicate Your Answer

3. How can you measure and classify an angle?

Name ______________________________ Date __________

1.5 Notetaking with Vocabulary
For use after Lesson 1.5

In your own words, write the meaning of each vocabulary term.

angle

vertex

sides of an angle

interior of an angle

exterior of an angle

measure of an angle

acute angle

right angle

obtuse angle

straight angle

congruent angles

angle bisector

Name____ Date__________

1.5 Notetaking with Vocabulary (continued)

Postulate 1.3 Protractor Postulate

Consider $\overleftrightarrow{OB}$ and a point A on one side of $\overleftrightarrow{OB}$. The rays of the form $\overrightarrow{OA}$ can be matched one to one with the real numbers from 0 to 180.

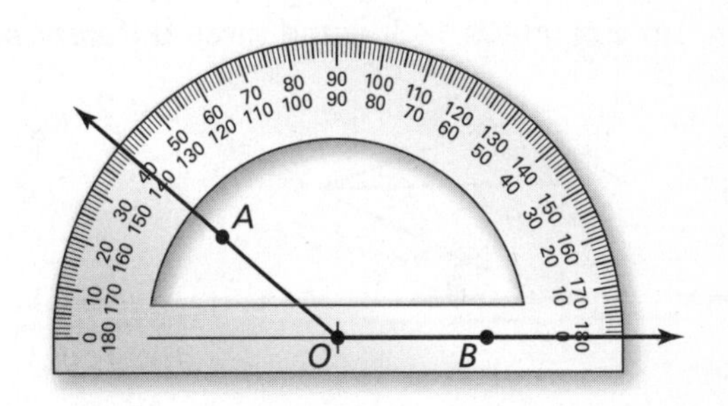

The **measure** of $\angle AOB$, which can be written as $m\angle AOB$, is equal to the absolute value of the difference between the real numbers matched with $\overrightarrow{OA}$ and $\overrightarrow{OB}$ on a protractor.

Notes:

Core Concepts

Types of Angles

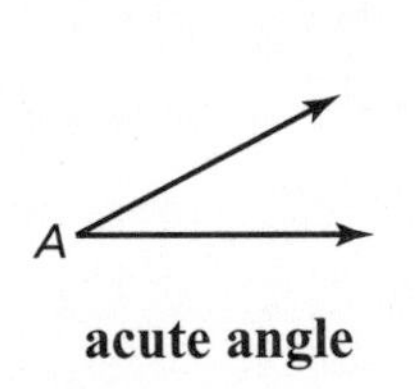

acute angle

Measures greater than 0° and less than 90°

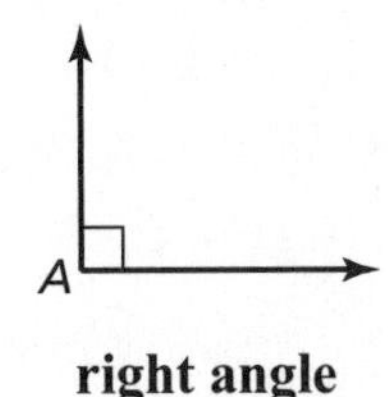

right angle

Measures 90°

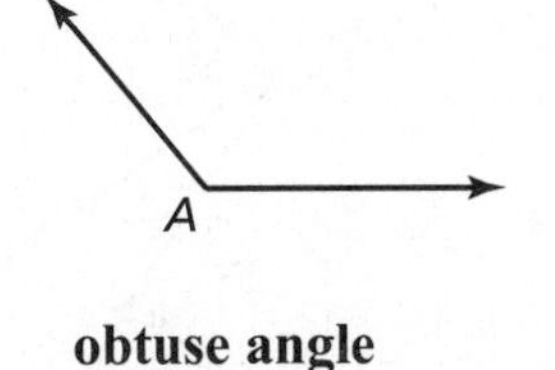

obtuse angle

Measures greater than 90° and less than 180°

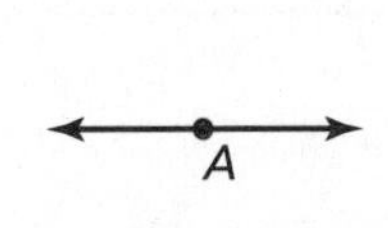

straight angle

Measures 180°

Notes:

Postulate 1.4 Angle Addition Postulate

Words If P is the interior of $\angle RST$, then the measure of $\angle RST$ is equal to the sum of the measures of $\angle RSP$ and $\angle PST$.

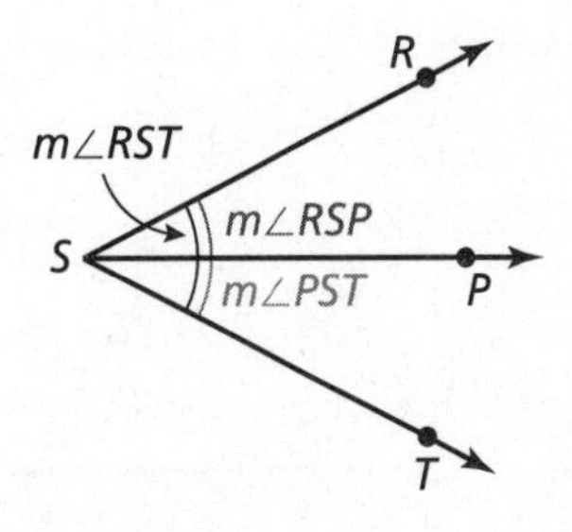

Symbols If P is in the interior of $\angle RST$, then $m\angle RST = m\angle RSP + m\angle PST$.

Notes:

Name ______________________________ Date __________

Extra Practice

In Exercises 1–3, name three different angles in the diagram.

1. 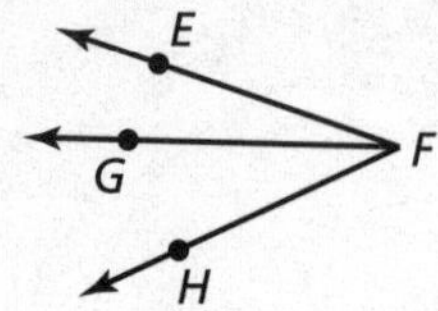

2. 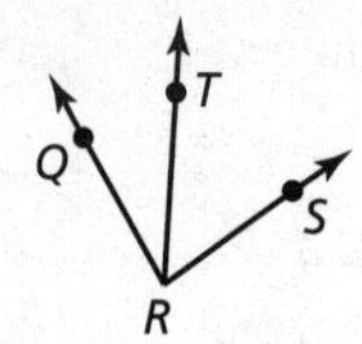

3. 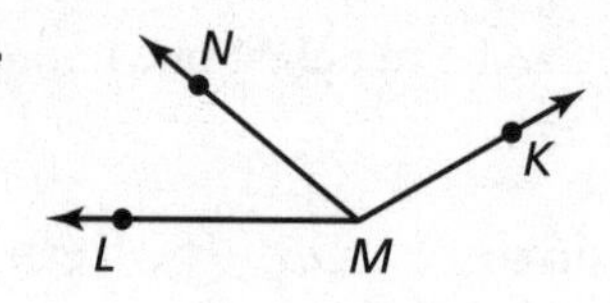

In Exercises 4–9, find the indicated angle measure(s).

4. Find $m\angle JKL$.

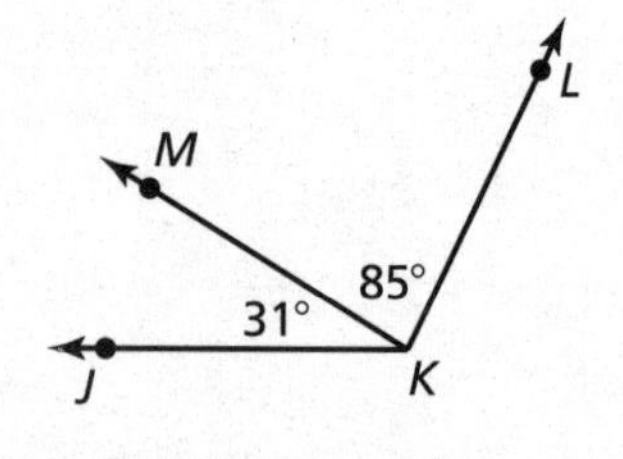

5. $m\angle RSU = 91°$. Find $m\angle RST$.

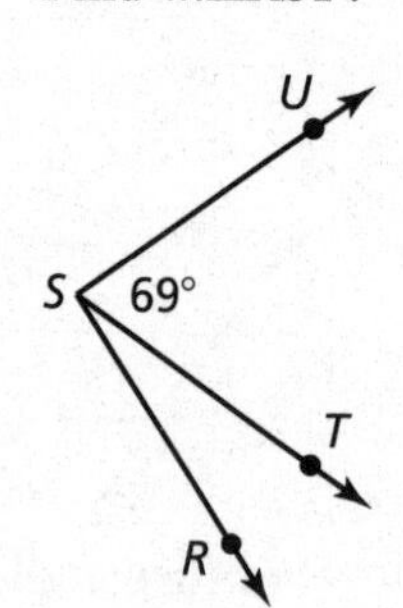

6. $\angle UWX$ is a straight angle. Find $m\angle UWV$ and $m\angle XWV$.

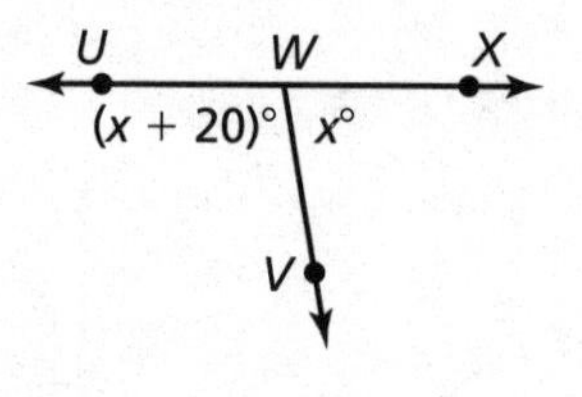

7. Find $m\angle CAD$ and $m\angle BAD$.

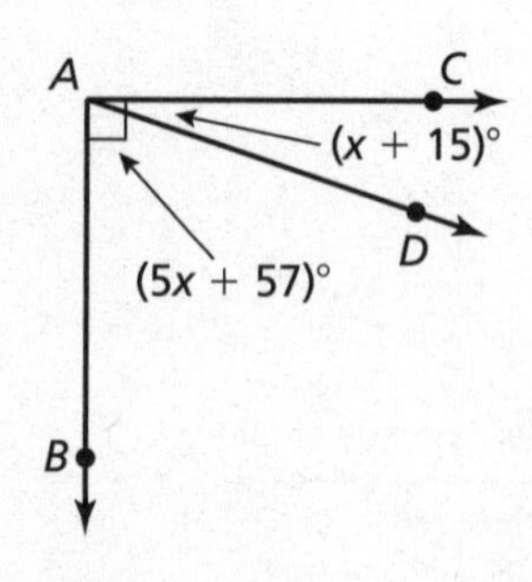

8. $\overrightarrow{EG}$ bisects $\angle DEF$. Find $m\angle DEG$ and $m\angle GEF$.

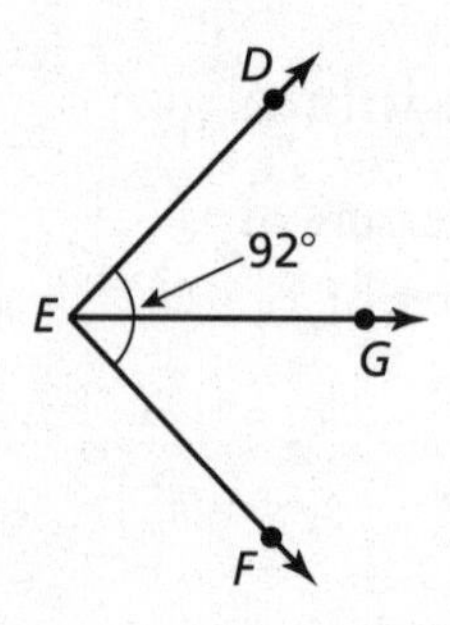

9. $\overrightarrow{QR}$ bisects $\angle PQS$. Find $m\angle PQR$ and $m\angle PQS$.

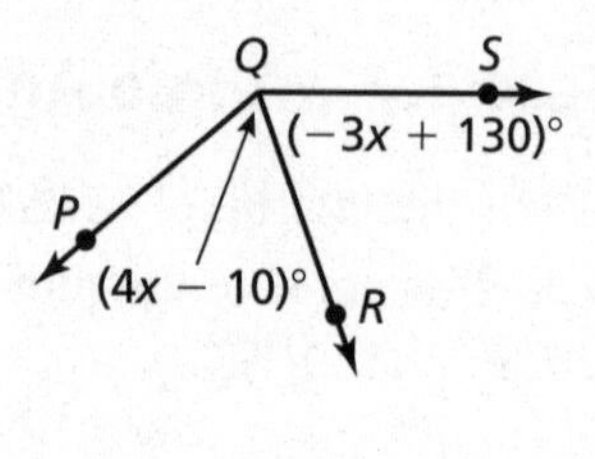

Name____ Date__________

1.6 Describing Pairs of Angles

For use with Exploration 1.6

Essential Question How can you describe angle pair relationships and use these descriptions to find angle measures?

1 EXPLORATION: Finding Angle Measures

Work with a partner. The five-pointed star has a regular pentagon at its center.

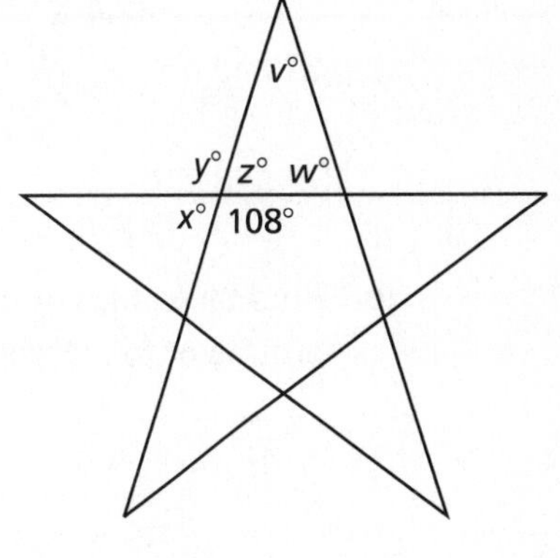

a. What do you notice about the following angle pairs?

$x°$ and $y°$

$y°$ and $z°$

$x°$ and $z°$

b. Find the values of the indicated variables. Do not use a protractor to measure the angles.

$x =$

$y =$

$z =$

$w =$

$v =$

Explain how you obtained each answer.

Name ______________________________ Date ________

2 EXPLORATION: Finding Angle Measures

Work with a partner. A square is divided by its diagonals into four triangles.

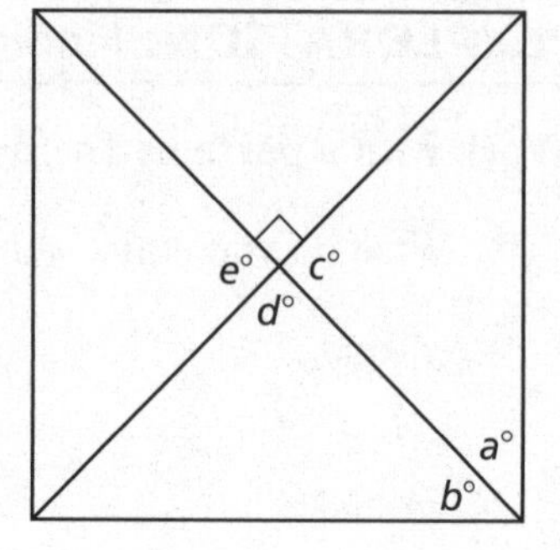

a. What do you notice about the following angle pairs?

$a°$ and $b°$

$c°$ and $d°$

$c°$ and $e°$

b. Find the values of the indicated variables. Do not use a protractor to measure the angles.

$c =$

$d =$

$e =$

Explain how you obtained each answer.

Communicate Your Answer

3. How can you describe angle pair relationships and use these descriptions to find angle measures?

4. What do you notice about the angle measures of complementary angles, supplementary angles, and vertical angles?

Name____ Date__________

1.6 Notetaking with Vocabulary
For use after Lesson 1.6

In your own words, write the meaning of each vocabulary term.

complementary angles

supplementary angles

adjacent angles

linear pair

vertical angles

Core Concepts

Complementary and Supplementary Angles

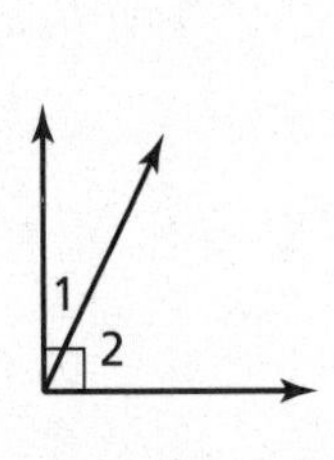

$\angle 1$ and $\angle 2$

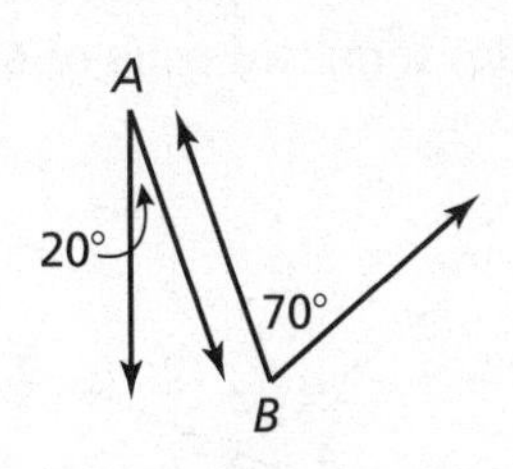

$\angle A$ and $\angle B$

complementary angles

Two positive angles whose measures have a sum of 90°. Each angle is the *complement* of the other.

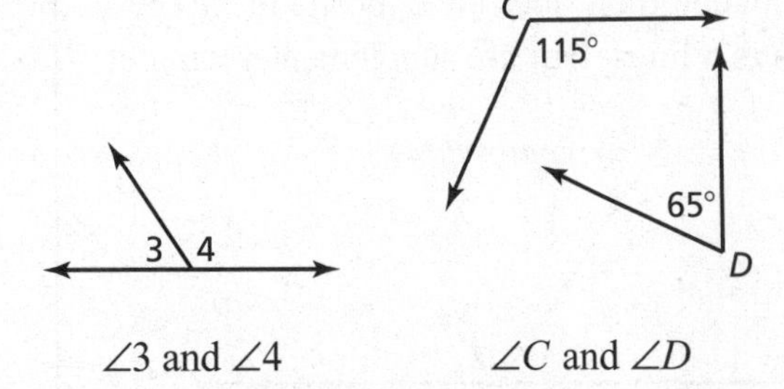

$\angle 3$ and $\angle 4$ $\angle C$ and $\angle D$

supplementary angles

Two positive angles whose measures have a sum of 180°. Each angle is the *supplement* of the other.

Notes:

Adjacent Angles

Complementary angles and supplementary angles can be *adjacent angles* or *nonadjacent angles*. **Adjacent angles** are two angles that share a common vertex and side, but have no common interior points.

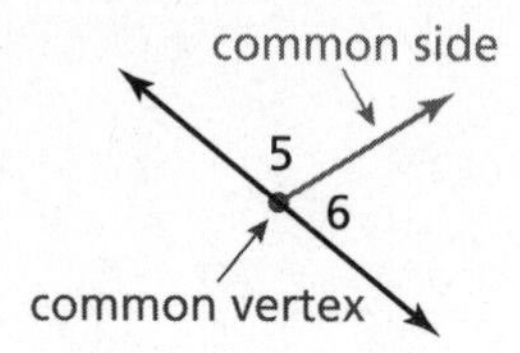

∠5 and ∠6 are adjacent angles

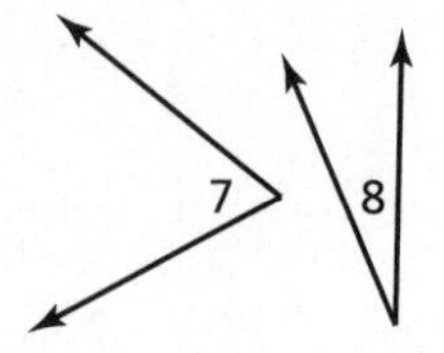

∠7 and ∠8 are nonadjacent angles.

Notes:

Linear Pairs and Vertical Angles

Two adjacent angles are a **linear pair** when their noncommon sides are opposite rays. The angles in a linear pair are supplementary angles.

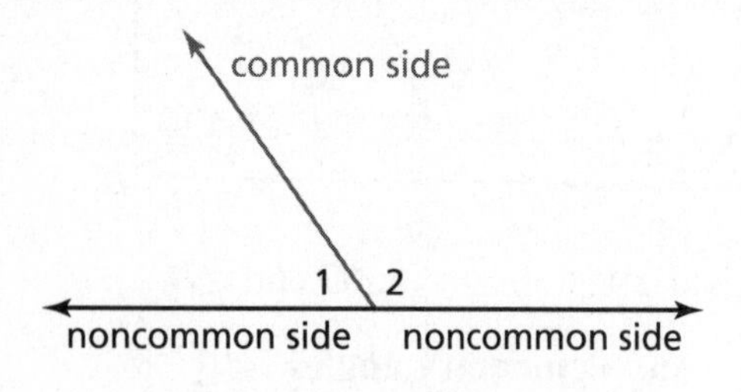

∠1 and ∠2 are a linear pair.

Two angles are **vertical angles** when their sides form two pairs of opposite rays.

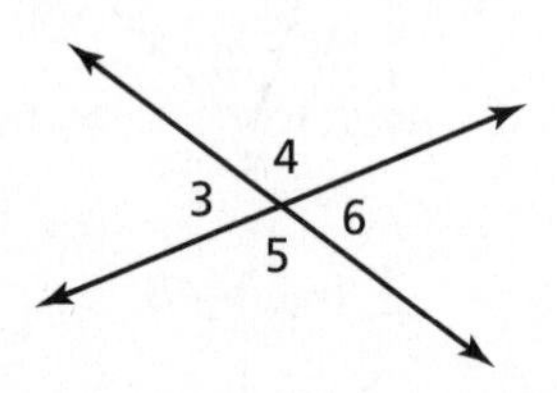

∠3 and ∠6 are vertical angles.

∠4 and ∠5 are vertical angles.

Notes:

Extra Practice

In Exercises 1 and 2, use the figure.

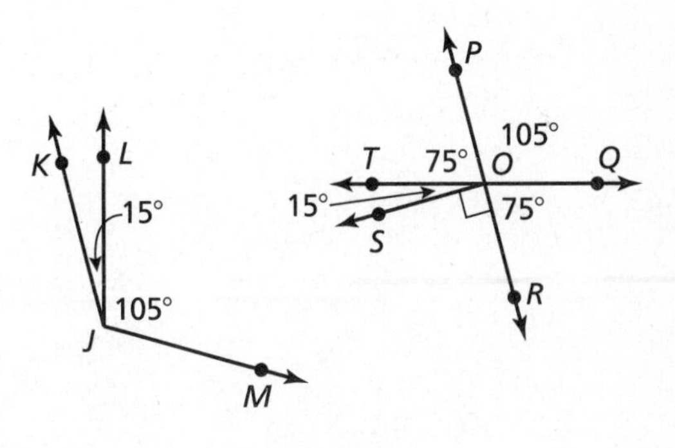

1. Name the pair(s) of adjacent complementary angles.

2. Name the pair(s) of nonadjacent supplementary angles.

In Exercises 3 and 4, find the angle measure.

3. $\angle A$ is a complement of $\angle B$ and $m\angle A = 36°$. Find $m\angle B$.

4. $\angle C$ is a supplement of $\angle D$ and $m\angle D = 117°$. Find $m\angle C$.

In Exercises 5 and 6, find the measure of each angle.

5.

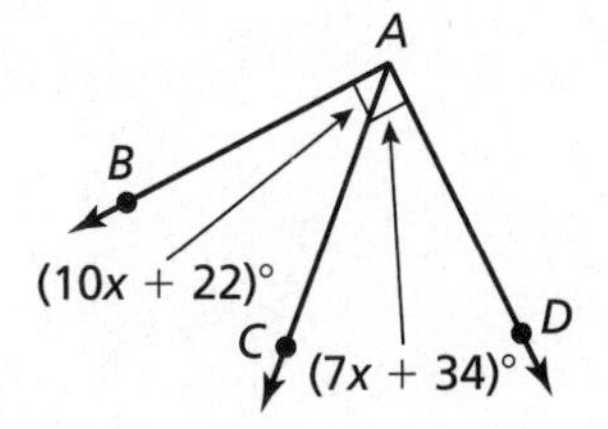

6.

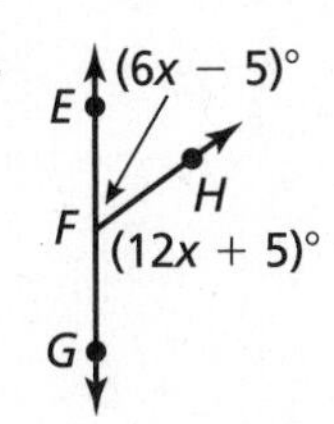

In Exercises 7–9, use the figure.

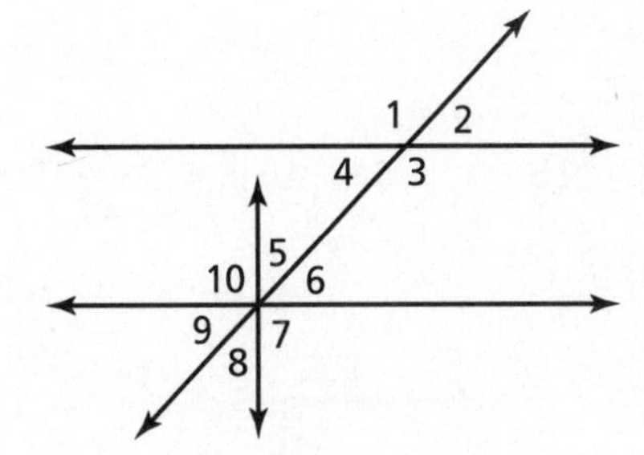

7. Identify the linear pair(s) that include $\angle 1$.

8. Identify the vertical angles.

9. Are $\angle 6$ and $\angle 7$ a linear pair? Explain.

Name ______________________________ Date __________

Chapter 2 Maintaining Mathematical Proficiency

Write an equation for the *n*th term of the arithmetic sequence. Then find a_{20}.

1. 5, 11, 17, 23, ...

2. 22, 34, 46, 58, ...

3. −13, 0, 13, 26, ...

4. −4.5, −4.0, −3.5, −3.0, ...

5. 40, 25, 10, −5, ...

6. $-\frac{1}{2}, \frac{1}{2}, \frac{3}{2}, \frac{5}{2}, \ldots$

Solve the literal equation for *x*.

7. $3x - 9y = 12$

8. $16y - 4x = 40$

9. $6x + 5 = 30y - 7$

10. $2x - y = 11x - 18$

11. $10y = 2x + 3zx + 1$

12. $14z = 2x + 4xy$

Name____________________ Date__________

2.1 Conditional Statements
For use with Exploration 2.1

Essential Question When is a conditional statement true or false?

A *conditional statement*, symbolized by $p \to q$, can be written as an "if-then statement" in which p is the *hypothesis* and q is the *conclusion*. Here is an example.

If a polygon is a triangle, then the sum of its angle measures is 180°.

hypothesis, p — conclusion, q

1 EXPLORATION: Determining Whether a Statement Is True or False

Work with a partner. A hypothesis can either be true or false. The same is true of a conclusion. For a conditional statement to be true, the hypothesis and conclusion do not necessarily both have to be true. Determine whether each conditional statement is true or false. Justify your answer.

a. If yesterday was Wednesday, then today is Thursday.

b. If an angle is acute, then it has a measure of 30°.

c. If a month has 30 days, then it is June.

d. If an even number is not divisible by 2, then 9 is a perfect cube.

2 EXPLORATION: Determining Whether a Statement Is True or False?

Work with a partner. Use the points in the coordinate plane to determine whether each statement is true or false. Justify your answer.

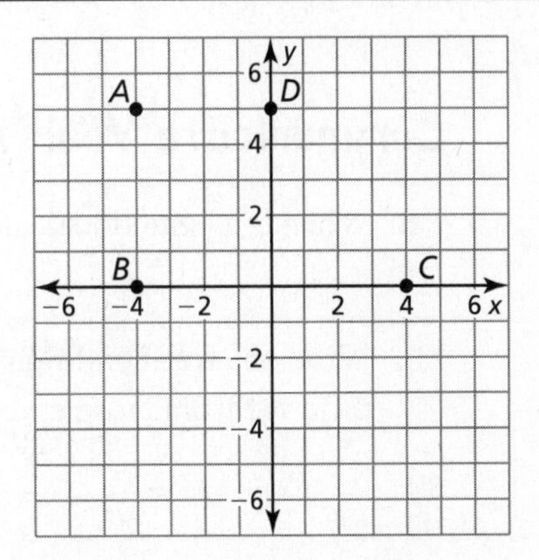

a. $\triangle ABC$ is a right triangle.

2 EXPLORATION: Determining Whether a Statement Is True or False (continued)

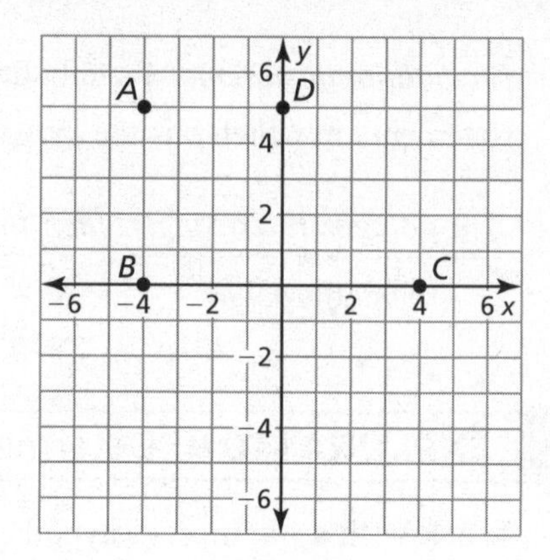

b. $\triangle BDC$ is an equilateral triangle.

c. $\triangle BDC$ is an isosceles triangle.

d. Quadrilateral *ABCD* is a trapezoid.

e. Quadrilateral *ABCD* is a parallelogram.

3 EXPLORATION: Determining Whether a Statement Is True or False

Work with a partner. Determine whether each conditional statement is true or false. Justify your answer.

a. If $\triangle ADC$ is a right triangle, then the Pythagorean Theorem is valid for $\triangle ADC$.

b. If $\angle A$ and $\angle B$ are complementary, then the sum of their measures is 180°.

c. If figure *ABCD* is a quadrilateral, then the sum of its angle measures is 180°.

d. If points *A*, *B*, and *C* are collinear, then they lie on the same line.

e. If $\overleftrightarrow{AB}$ and $\overleftrightarrow{BD}$ intersect at a point, then they form two pairs of vertical angles.

Communicate Your Answer

4. When is a conditional statement true or false?

5. Write one true conditional statement and one false conditional statement that are different from those given in Exploration 3. Justify your answer.

Name______________________________ Date__________

2.1 Notetaking with Vocabulary

For use after Lesson 2.1

In your own words, write the meaning of each vocabulary term.

conditional statement

if-then form

hypothesis

conclusion

negation

converse

inverse

contrapositive

equivalent statements

perpendicular lines

biconditional statement

truth value

truth table

Core Concepts

Conditional Statement

A **conditional statement** is a logical statement that has two parts, a *hypothesis* p and a *conclusion* q. When a conditional statement is written in **if-then form**, the "if" part contains the **hypothesis** and the "then" part contains the **conclusion**.

Words If p, then q. **Symbols** $p \rightarrow q$ (read as "p implies q")

Notes:

Negation

The **negation** of a statement is the *opposite* of the original statement. To write the negation of a statement p, you write the symbol for negation ($\sim$) before the letter. So, "not p" is written $\sim p$.

Words not p **Symbols** $\sim p$

Notes:

Related Conditionals

Consider the conditional statement below.

Words If p, then q. **Symbols** $p \rightarrow q$

Converse To write the **converse** of a conditional statement, exchange the hypothesis and the conclusion.

Words If q, then p. **Symbols** $q \rightarrow p$

Inverse To write the **inverse** of a conditional statement, negate both the hypothesis and the conclusion.

Words If not p, then not q. **Symbols** $\sim p \rightarrow \sim q$

Related Conditionals (continued)

Contrapositive To write the **contrapositive** of a conditional statement, first write the converse. Then negate both the hypothesis and the conclusion.

Words If not q, then not p. **Symbols** $\sim q \rightarrow \sim p$

A conditional statement and its contrapositive are either both true or both false. Similarly, the converse and inverse of a conditional statement are either both true or both false. In general, when two statements are both true or both false, they are called **equivalent statements**.

Notes:

Biconditional Statement

When a conditional statement and its converse are both true, you can write them as a single *biconditional statement.* A **biconditional statement** is a statement that contains the phrase "if and only if."

Words p if and only if q **Symbols** $p \leftrightarrow q$

Any definition can be written as a biconditional statement.

Notes:

Name ______________________________ Date __________

Extra Practice

In Exercises 1 and 2, rewrite the conditional statement in if-then form.

1. $13x - 5 = -18$, because $x = -1$.

2. The sum of the measures of interior angles of a triangle is 180°.

3. Let p be "Quadrilateral $ABCD$ is a rectangle" and let q be "the sum of the angle measures is 360°." Write the conditional statement $p \rightarrow q$, the converse $q \rightarrow p$, the inverse $\sim p \rightarrow \sim q$, and the contrapositive $\sim q \rightarrow \sim p$ in words. Then decide whether each statement is true or false.

In Exercises 4–6, decide whether the statement about the diagram is true. Explain your answer using the definitions you have learned.

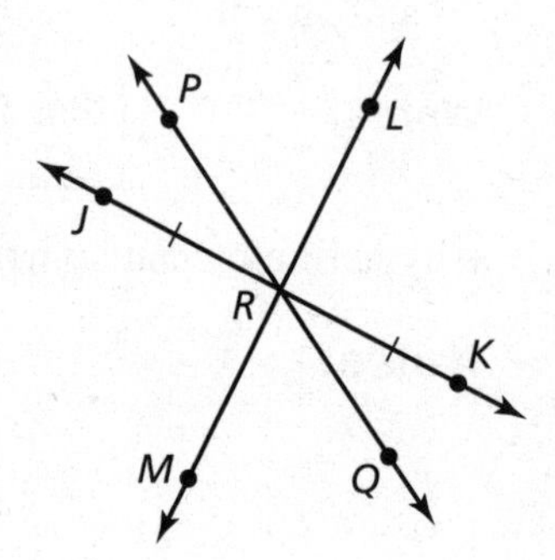

4. $\overline{LM}$ bisects $\overline{JK}$

5. $\angle JRP$ and $\angle PRL$ are complementary.

6. $\angle MRQ \cong \angle PRL$

Name______________________________ Date__________

2.2 Inductive and Deductive Reasoning

For use with Exploration 2.2

Essential Question How can you use reasoning to solve problems?

A **conjecture** is an unproven statement based on observations.

1 EXPLORATION: Writing a Conjecture

Work with a partner. Write a conjecture about the pattern. Then use your conjecture to draw the 10th object in the pattern.

a. 1 2 3 4 5 6 7

b.

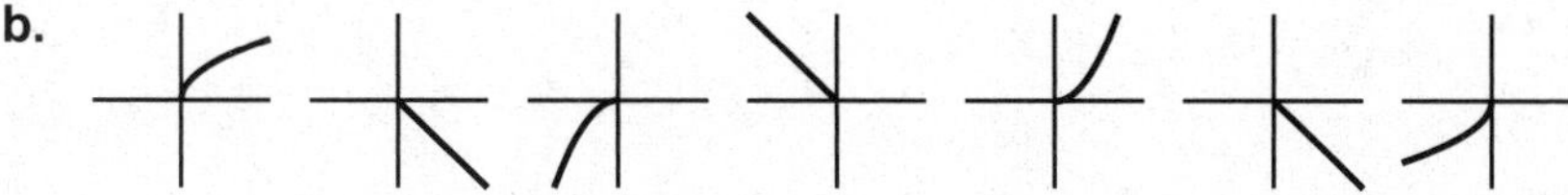

c. ○△ ○△ ○□ △○ △○ □○ ○△
□△ ○□ △□ △□ □○ □△ □△

2 EXPLORATION: Using a Venn Diagram

Work with a partner. Use the Venn diagram to determine whether the statement is true or false. Justify your answer. Assume that no region of the Venn diagram is empty.

a. If an item has Property B, then it has Property A.

b. If an item has Property A, then it has Property B.

2 EXPLORATION: Using a Venn Diagram (continued)

c. If an item has Property A, then it has Property C.

d. Some items that have Property A do not have Property B.

e. If an item has Property C, then it does not have Property B.

f. Some items have both Properties A and C.

g. Some items have both Properties B and C.

3 EXPLORATION: Reasoning and Venn Diagrams

Work with a partner. Draw a Venn diagram that shows the relationship between different types of quadrilaterals: squares, rectangles, parallelograms, trapezoids, rhombuses, and kites. Then write several conditional statements that are shown in your diagram, such as "If a quadrilateral is a square, then it is a rectangle."

Communicate Your Answer

4. How can you use reasoning to solve problems?

5. Give an example of how you used reasoning to solve a real-life problem.

Name____ Date__________

2.2 Notetaking with Vocabulary
For use after Lesson 2.2

In your own words, write the meaning of each vocabulary term.

conjecture

inductive reasoning

counterexample

deductive reasoning

Core Concepts

Inductive Reasoning

A **conjecture** is an unproven statement that is based on observations. You use **inductive reasoning** when you find a pattern in specific cases and then write a conjecture for the general case.

Notes:

Counterexample

To show that a conjecture is true, you must show that it is true for all cases. You can show that a conjecture is false, however, by finding just one *counterexample*. A **counterexample** is a specific case for which the conjecture is false.

Notes:

Name ______________________ Date __________

Deductive Reasoning

Deductive reasoning uses facts, definitions, accepted properties, and the laws of logic to form a logical argument. This is different from *inductive reasoning*, which uses specific examples and patterns to form a conjecture.

Laws of Logic

Law of Detachment

If the hypothesis of a true conditional statement is true, then the conclusion is also true.

Law of Syllogism

If hypothesis p, then conclusion q. ← If these statements are true,

If hypothesis q, then conclusion r. ←

If hypothesis p, then conclusion r. ← then this statement is true.

Notes:

Extra Practice

In Exercises 1–4, describe the pattern. Then write or draw the next two numbers, letters, or figures.

1. 20, 19, 17, 14, 10, …

2. 2, −3, 5, −7, 11, …

3. C, E, G, I, K, …

4.

Name______________________________ Date__________

2.2 Notetaking with Vocabulary (continued)

In Exercises 5 and 6, make and test a conjecture about the given quantity.

5. the sum of two negative integers

6. the product of three consecutive nonzero integers

In Exercises 7 and 8, find a counterexample to show that the conjecture is false.

7. If n is a rational number, then n^2 is always less than n.

8. Line k intersects plane P at point Q on the plane. Plane P is perpendicular to line k.

In Exercises 9 and 10, use the Law of Detachment to determine what you can conclude from the given information, if possible.

9. If a triangle has equal side lengths, then each interior angle measure is $60°$. $\triangle ABC$ has equal side lengths.

10. If a quadrilateral is a rhombus, then it has two pairs of opposite sides that are parallel. Quadrilateral $PQRS$ has two pairs of opposite sides that are parallel.

In Exercises 11 and 12, use the Law of Syllogism to write a new conditional statement that follows from the pair of true statements, if possible.

11. If it does not rain, then I will walk to school.

If I walk to school, then I will wear my walking shoes.

12. If $x > 1$, then $3x > 3$.

If $3x > 3$, then $(3x)^2 > 9$.

Name _______________________________________ Date __________

2.3 Postulates and Diagrams
For use with Exploration 2.3

Essential Question In a diagram, what can be assumed and what needs to be labeled?

1 EXPLORATION: Looking at a Diagram

Work with a partner. On a piece of paper, draw two perpendicular lines. Label them $\overleftrightarrow{AB}$ and $\overleftrightarrow{CD}$. Look at the diagram from different angles. Do the lines appear perpendicular regardless of the angle at which you look at them? Describe *all* the angles at which you can look at the lines and have them appear perpendicular.

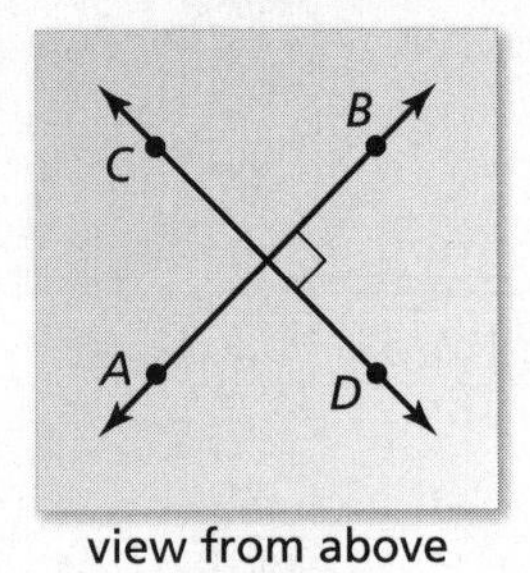

view from above

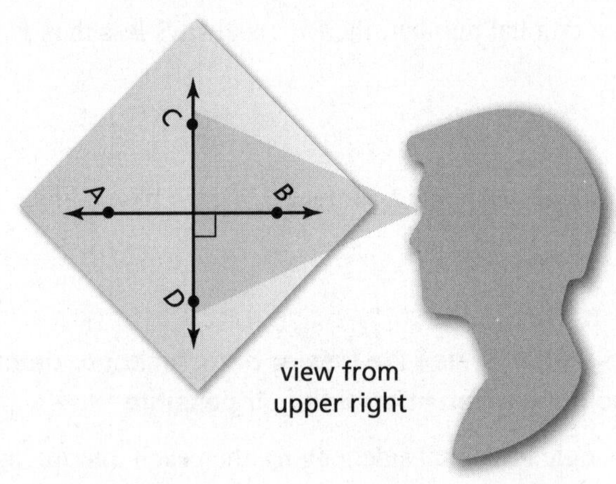

view from upper right

2 EXPLORATION: Interpreting a Diagram

Work with a partner. When you draw a diagram, you are communicating with others. It is important that you include sufficient information in the diagram. Use the diagram to determine which of the following statements you can assume to be true. Explain your reasoning.

a. All the points shown are coplanar.

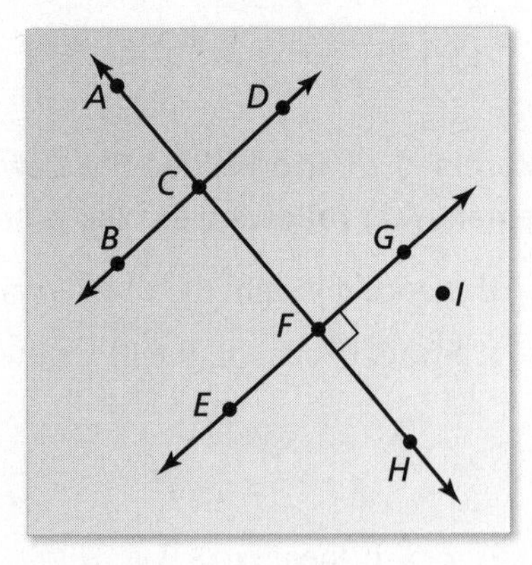

b. Points D, G, and I are collinear.

c. Points A, C, and H are collinear.

d. $\overleftrightarrow{EG}$ and $\overleftrightarrow{AH}$ are perpendicular.

Name ____________________ Date __________

2 EXPLORATION: Interpreting a Diagram (continued)

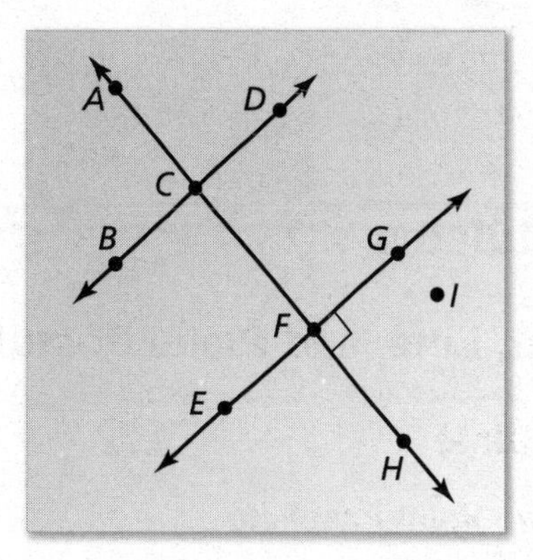

e. $\angle BCA$ and $\angle ACD$ are a linear pair.

f. $\overleftrightarrow{AF}$ and $\overleftrightarrow{BD}$ are perpendicular.

g. $\overleftrightarrow{EG}$ and $\overleftrightarrow{BD}$ are parallel.

h. $\overleftrightarrow{AF}$ and $\overleftrightarrow{BD}$ are coplanar.

i. $\overleftrightarrow{EG}$ and $\overleftrightarrow{BD}$ do not intersect.

j. $\overleftrightarrow{AF}$ and $\overleftrightarrow{BD}$ intersect.

k. $\overleftrightarrow{EG}$ and $\overleftrightarrow{BD}$ are perpendicular.

l. $\angle ACD$ and $\angle BCF$ are vertical angles.

m. $\overleftrightarrow{AC}$ and $\overleftrightarrow{FH}$ are the same line.

Communicate Your Answer

3. In a diagram, what can be assumed and what needs to be labeled?

4. Use the diagram in Exploration 2 to write two statements you can assume to be true and two statements you cannot assume to be true. Your statements should be different from those given in Exploration 2. Explain your reasoning.

Name ______________________________ Date __________

2.3 Notetaking with Vocabulary
For use after Lesson 2.3

In your own words, write the meaning of each vocabulary term.

line perpendicular to a plane

Postulates

Point, Line, and Plane Postulates

Postulate	Example	
2.1 Two Point Postulate Through any two points, there exists exactly one line.		Through points A and B, there is exactly one line ℓ. Line ℓ contains at least two points.
2.2 Line-Point Postulate A line contains at least two points.		
2.3 Line Intersection Postulate If two lines intersect, then their intersection is exactly one point.	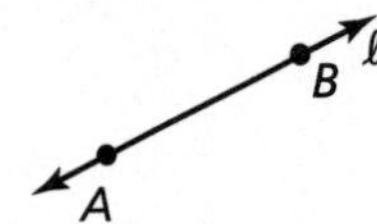	The intersection of line m and line n is point C.
2.4 Three Point Postulate Through any three noncollinear points, there exists exactly one plane.	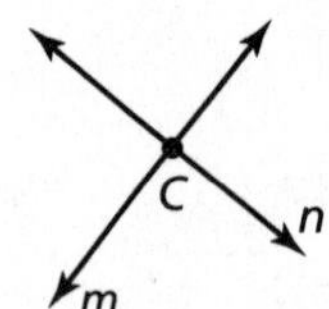	Through points D, E, and F, there is exactly one plane, plane R. Plane R contains at least three noncollinear points.
2.5 Plane-Point Postulate A plane contains at least three noncollinear points		

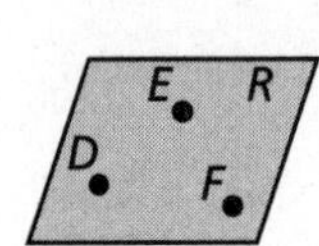

Notes:

Point, Line, and Plane Postulates (continued)

Postulate	Example	
2.6 Plane-Line Postulate If two points lie in a plane, then the line containing them lies in the plane.	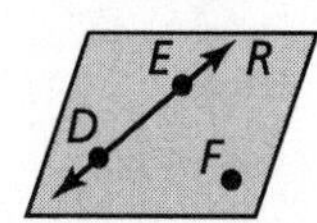	Points D and E lie in plane R, so $\overleftrightarrow{DE}$ lies in plane R.
2.7 Plane Intersection Postulate If two planes intersect, then their intersection is a line.	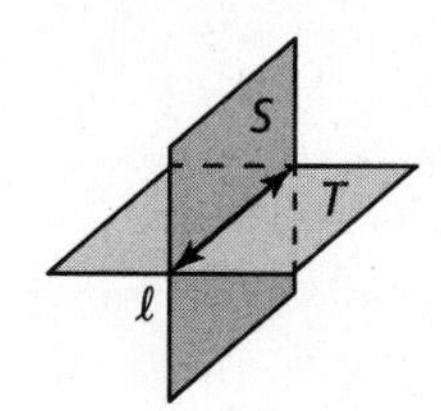	The intersection of plane S and plane T is line ℓ.

Notes:

Extra Practice

In Exercises 1 and 2, state the postulate illustrated by the diagram.

1.

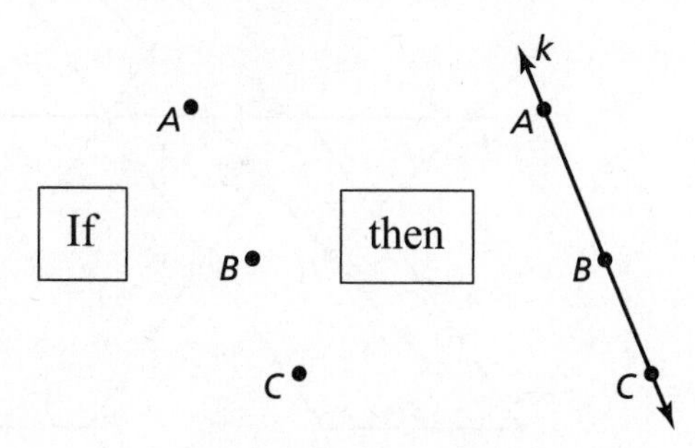

2.

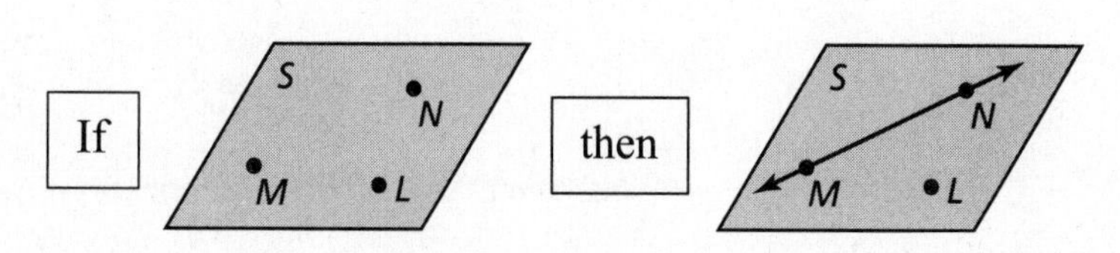

Name ______________________________ Date __________

2.3 Notetaking with Vocabulary (continued)

In Exercises 3–6, use the diagram to write an example of the postulate.

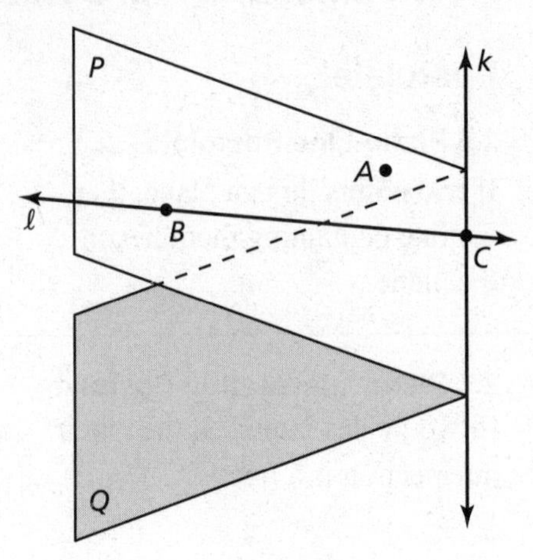

3. Two Point Postulate (Postulate 2.1)

4. Line-Point Postulate (Postulate 2.2)

5. Plane-Point Postulate (Postulate 2.5)

6. Plane Intersection Postulate (Postulate 2.7)

In Exercises 7 and 8, sketch a diagram of the description.

7. $\overrightarrow{RS}$ bisecting $\overline{KL}$ at point R

8. $\overleftrightarrow{AB}$ in plane U intersecting $\overleftrightarrow{CD}$ at point E, and point C not on plane U

In Exercises 9–14, use the diagram to determine whether you can assume the statement.

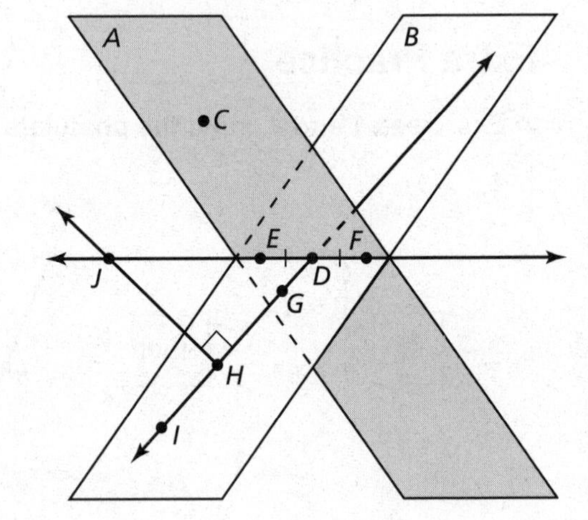

9. Planes A and B intersect at $\overleftrightarrow{EF}$.

10. Points C and D are collinear.

11. $\overrightarrow{HJ}$ and $\overrightarrow{ID}$ are perpendicular.

12. $\overleftrightarrow{GD}$ is a bisector of $\overleftrightarrow{EF}$ at point D.

13. 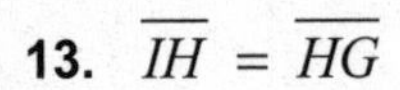 $\overline{IH} = \overline{HG}$

14. $\angle HJD$ and $\angle HDJ$ are complementary angles.

Name____ Date__________

2.4 Algebraic Reasoning

For use with Exploration 2.4

Essential Question How can algebraic properties help you solve an equation?

1 EXPLORATION: Justifying Steps in a Solution

Work with a partner. In previous courses, you studied different properties, such as the properties of equality and the Distributive, Commutative, and Associative Properties. Write the property that justifies each of the following solution steps.

Algebraic Step	Justification
$2(x + 3) - 5 = 5x + 4$	Write given equation.
$2x + 6 - 5 = 5x + 4$	______________________
$2x + 1 = 5x + 4$	______________________
$2x - \mathbf{2x} + 1 = 5x - \mathbf{2x} + 4$	______________________
$1 = 3x + 4$	______________________
$1 - \mathbf{4} = 3x + 4 - \mathbf{4}$	______________________
$-3 = 3x$	______________________
$\frac{-3}{\mathbf{3}} = \frac{3x}{\mathbf{3}}$	______________________
$-1 = x$	______________________
$x = -1$	______________________

Name ______________________________ Date __________

2 EXPLORATION: Stating Algebraic Properties

Work with a partner. The symbols ◆ and ● represent addition and multiplication (not necessarily in that order). Determine which symbol represents which operation. Justify your answer. Then state each algebraic property being illustrated.

Example of Property	Name of Property
5 ◆ 6 = 6 ◆ 5	______________
5 ● 6 = 6 ● 5	______________
4 ◆ (5 ◆ 6) = (4 ◆ 5) ◆ 6	______________
4 ● (5 ● 6) = (4 ● 5) ● 6	______________
0 ◆ 5 = 0	______________
0 ● 5 = 5	______________
1 ◆ 5 = 5	______________
4 ◆ (5 ● 6) = 4 ◆ 5 ● 4 ◆ 6	______________

Communicate Your Answer

3. How can algebraic properties help you solve an equation?

4. Solve $3(x + 1) - 1 = -13$. Justify each step.

Name____ Date__________

2.4 Notetaking with Vocabulary
For use after Lesson 2.4

In your own words, write the meaning of each vocabulary term.

equation

solve an equation

formula

Core Concepts

Algebraic Properties of Equality

Let a, b, and c be real numbers.

Addition Property of Equality If $a = b$, then $a + c = b + c$.

Subtraction Property of Equality If $a = b$, then $a - c = b - c$.

Multiplication Property of Equality If $a = b$, then $a \bullet c = b \bullet c, c \neq 0$.

Division Property of Equality If $a = b$, then $\frac{a}{c} = \frac{b}{c}, c \neq 0$.

Substitution Property of Equality If $a = b$, then a can be substituted for b (or b for a) in any equation or expression.

Notes:

2.4 Notetaking with Vocabulary (continued)

Distributive Property

Let a, b, and c be real numbers.

Sum $a(b + c) = ab + ac$ **Difference** $a(b - c) = ab - ac$

Notes:

Reflexive, Symmetric, and Transitive Properties of Equality

	Real Numbers	Segment Lengths	Angle Measures
Reflexive Property	$a = a$	$AB = AB$	$m\angle A = m\angle A$
Symmetric Property	If $a = b$, then $b = a$.	If $AB = CD$, then $CD = AB$.	If $m\angle A = m\angle B$, then $m\angle B = m\angle A$.
Transitive Property	If $a = b$ and $b = c$, then $a = c$.	If $AB = CD$ and $CD = EF$, then $AB = EF$.	If $m\angle A = m\angle B$ and $m\angle B = m\angle C$, then $m\angle A = m\angle C$.

Notes:

Extra Practice

In Exercises 1–3, solve the equation. Justify each step.

1. $3x - 7 = 5x - 19$

2. $20 - 2(3x - 1) = x - 6$

3. $-5(2u + 10) = 2(u - 7)$

In Exercises 4 and 5, solve the equation for the given variable. Justify each step.

4. $9x + 2y = 5; y$

5. $\frac{1}{15}s - \frac{2}{3}t = -2; s$

6. The formula for the surface area S of a cone is $S = \pi r^2 + \pi rs$, where r is the radius and s is the slant height. Solve the formula for s. Justify each step. Then find the slant height of the cone when the surface area is 220 square feet and the radius is 7 feet. Approximate to the nearest tenth.

Name ______________________________ Date __________

2.5 Proving Statements about Segments and Angles
For use with Exploration 2.5

Essential Question How can you prove a mathematical statement?

A **proof** is a logical argument that uses deductive reasoning to show that a statement is true.

1 EXPLORATION: Writing Reasons in a Proof

Work with a partner. Four steps of a proof are shown. Write the reasons for each statement.

Given $AC = AB + AB$

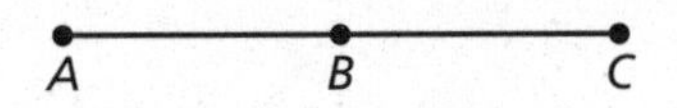

Prove $AB = BC$

STATEMENTS	REASONS
1. $AC = AB + AB$	1. Given
2. $AB + BC = AC$	2. ________________
3. $AB + AB = AB + BC$	3. ________________
4. $AB = BC$	4. ________________

2 EXPLORATION: Writing Steps in a Proof

Work with a partner. Six steps of a proof are shown. Complete the statements that correspond to each reason.

Given $m\angle 1 = m\angle 3$

Prove $m\angle EBA = m\angle CBD$

2 EXPLORATION: Writing Steps in a Proof (continued)

STATEMENTS	REASONS
1. ______________	1. Given
2. $m\angle EBA = m\angle 2 + m\angle 3$	2. Angle Addition Postulate (Post. 1.4)
3. $m\angle EBA = m\angle 2 + m\angle 1$	3. Substitution Property of Equality
4. $m\angle EBA =$ ______________	4. Commutative Property of Addition
5. $m\angle 1 + m\angle 2 =$ ______________	5. Angle Addition Postulate (Post. 1.4)
6. ______________	6. Transitive Property of Equality

Communicate Your Answer

3. How can you prove a mathematical statement?

4. Use the given information and the figure to write a proof for the statement.

Given B is the midpoint of $\overline{AC}$.

C is the midpoint of $\overline{BD}$.

Prove $AB = CD$

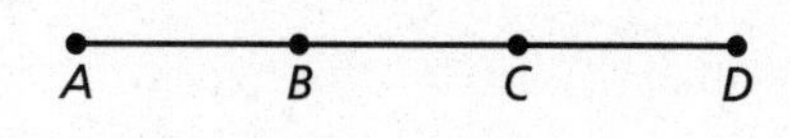

Name ______________________________ Date __________

2.5 Notetaking with Vocabulary

For use after Lesson 2.5

In your own words, write the meaning of each vocabulary term.

proof

two-column proof

theorem

Theorems

Theorem 2.1 Properties of Segment Congruence

Segment congruence is reflexive, symmetric, and transitive.

Reflexive For any segment AB, $\overline{AB} \cong \overline{AB}$.

Symmetric If $\overline{AB} \cong \overline{CD}$, then $\overline{CD} \cong \overline{AB}$.

Transitive If $\overline{AB} \cong \overline{CD}$ and $\overline{CD} \cong \overline{EF}$, then $\overline{AB} \cong \overline{EF}$.

Theorem 2.2 Properties of Angle Congruence

Angle congruence is reflexive, symmetric, and transitive.

Reflexive For any angle A, $\angle A \cong \angle A$.

Symmetric If $\angle A \cong \angle B$, then $\angle B \cong \angle A$.

Transitive If $\angle A \cong \angle B$ and $\angle B \cong \angle C$, then $\angle A \cong \angle C$.

Notes:

Name____ Date__________

Core Concepts

Writing a Two-Column Proof

In a proof, you make one statement at a time until you reach the conclusion. Because you make statements based on facts, you are using deductive reasoning. Usually the first statement-and-reason pair you write is given information.

Copy or draw diagrams and label given information to help develop proofs. Do not mark or label the information in the Prove statement on the diagram.

Proof of the Symmetric Property of Angle Congruence

Given $\angle 1 \cong \angle 2$ **Prove** $\angle 2 \cong \angle 1$

STATEMENTS	REASONS
1. $\angle 1 \cong \angle 2$	1. Given
2. $m\angle 1 = m\angle 2$	2. Definition of congruent angles
3. $m\angle 2 = m\angle 1$	3. Symmetric Property of Equality
4. $\angle 2 \cong \angle 1$	4. Definition of congruent angles

Statements: statements based on facts that you know or on conclusions from deductive reasoning

Reasons: definitions, postulates, or proven theorems that allow you to state the corresponding statement

The number of statements will vary.

Remember to give a reason for the last statement.

Notes:

Extra Practice

In Exercises 1 and 2, complete the proof.

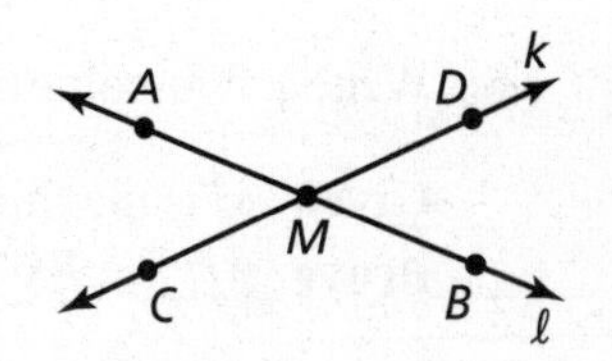

1. **Given** $\overline{AB}$ and $\overline{CD}$ bisect each other at point M and $\overline{BM} \cong \overline{CM}$.
   Prove $AB = AM + DM$

STATEMENTS	REASONS
1. $\overline{BM} \cong \overline{CM}$	1. Given
2. $\overline{CM} \cong \overline{DM}$	2. ________________
3. $\overline{BM} \cong \overline{DM}$	3. ________________
4. $BM = DM$	4. ________________
5. ________________	5. Segment Addition Postulate (Post. 1.2)
6. $AB = AM + DM$	6. ________________

Name ______________________________ Date __________

2.5 Notetaking with Vocabulary (continued)

2. Given $\angle AEB$ is a complement of $\angle BEC$.
Prove $m\angle AED = 90°$

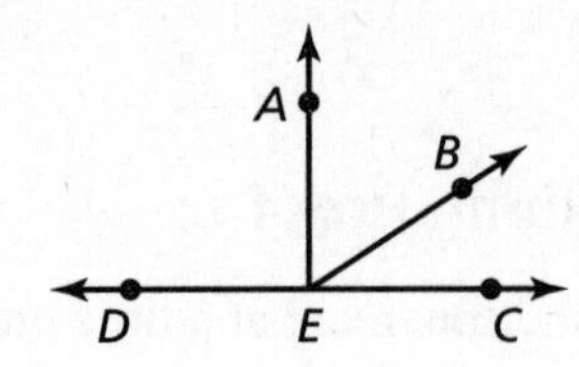

STATEMENTS	REASONS
1. $\angle AEB$ is a complement of $\angle BEC$.	**1.** Given
2. ______________	**2.** Definition of complementary angles
3. $m\angle AEC = m\angle AEB + m\angle BEC$	**3.** ______________
4. $m\angle AEC = 90°$	**4.** ______________
5. $m\angle AED + m\angle AEC = 180°$	**5.** Definition of supplementary angles
6. ______________	**6.** Substitution Property of Equality
7. $m\angle AED = 90°$	**7.** ______________

In Exercises 3 and 4, name the property that the statement illustrates.

3. If $\angle RST \cong \angle TSU$ and $\angle TSU \cong \angle VWX$, then $\angle RST \cong \angle VWX$.

4. If $\overline{GH} \cong \overline{JK}$, then $\overline{JK} \cong \overline{GH}$.

5. Write a two-column proof.

Given M is the midpoint of $\overline{RT}$.
Prove $MT = RS + SM$

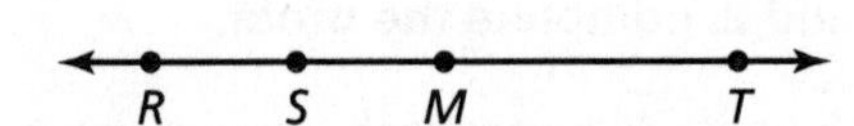

STATEMENTS	REASONS

Name_______________________________________ Date__________

2.6 Proving Geometric Relationships

For use with Exploration 2.6

Essential Question How can you use a flowchart to prove a mathematical statement?

1 EXPLORATION: Matching Reasons in a Flowchart Proof

Work with a partner. Match each reason with the correct step in the flowchart.

Given $AC = AB + AB$

Prove $AB = BC$

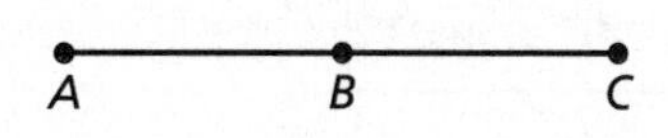

$AC = AB + AB$
____

$AB + BC = AC$ → $AB + AB = AB + BC$ → $AB = BC$

______________ ______________ ______________

A. Segment Addition Postulate (Post. 1.2)

B. Given

C. Transitive Property of Equality

D. Subtraction Property of Equality

Name ________________________ Date ________

2 EXPLORATION: Matching Reasons in a Flowchart Proof

Work with a partner. Match each reason with the correct step in the flowchart.

Given $m\angle 1 = m\angle 3$

Prove $m\angle EBA = m\angle CBD$

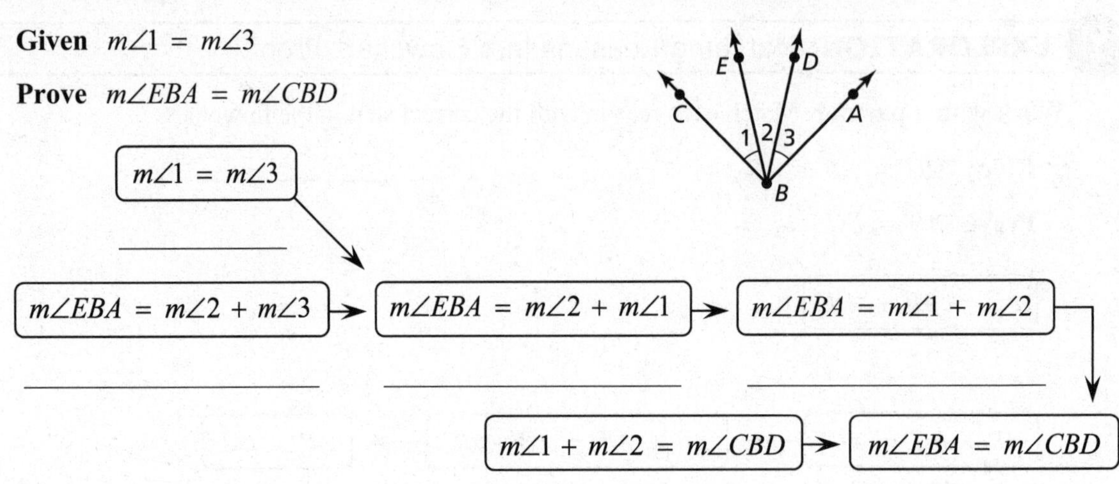

A. Angle Addition Postulate (Post. 1.4)

B. Transitive Property of Equality

C. Substitution Property of Equality

D. Angle Addition Postulate (Post. 1.4)

E. Given

F. Commutative Property of Addition

Communicate Your Answer

3. How can you use a flowchart to prove a mathematical statement?

4. Compare the flowchart proofs above with the two-column proofs in the Section 2.5 Explorations. Explain the advantages and disadvantages of each.

Name____ Date__________

2.6 Notetaking with Vocabulary

For use after Lesson 2.6

In your own words, write the meaning of each vocabulary term.

flowchart proof, or flow proof

paragraph proof

Theorems and Postulates

Theorem 2.3 Right Angles Congruence Theorem

All right angles are congruent.

Notes:

Theorem 2.4 Congruent Supplements Theorem

If two angles are supplementary to the same angle (or to congruent angles), then they are congruent.

If $\angle 1$ and $\angle 2$ are supplementary and $\angle 3$ and $\angle 2$ are supplementary, then $\angle 1 \cong \angle 3$.

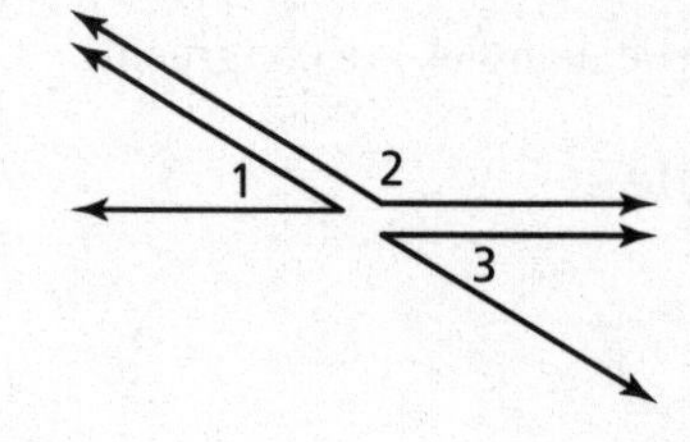

Notes:

2.6 Notetaking with Vocabulary (continued)

Theorem 2.5 Congruent Complements Theorem

If two angles are complementary to the same angle (or to congruent angles), then they are congruent.

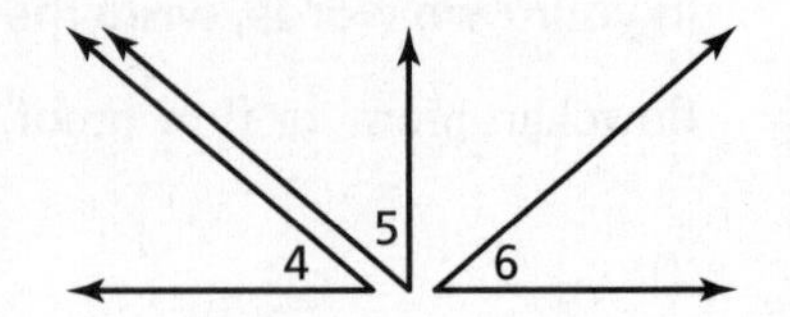

If $\angle 4$ and $\angle 5$ are complementary and $\angle 6$ and $\angle 5$ are complementary, then $\angle 4 \cong \angle 6$.

Notes:

Postulate 2.8 Linear Pair Postulate

If two angles form a linear pair, then they are supplementary.

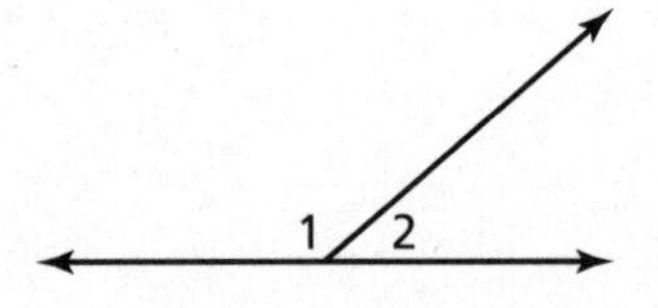

$\angle 1$ and $\angle 2$ form a linear pair, so $\angle 1$ and $\angle 2$ are supplementary and $m\angle 1 + m\angle 2 = 180°$.

Notes:

Theorem 2.6 Vertical Angles Congruence Theorem

Vertical angles are congruent.

Notes:

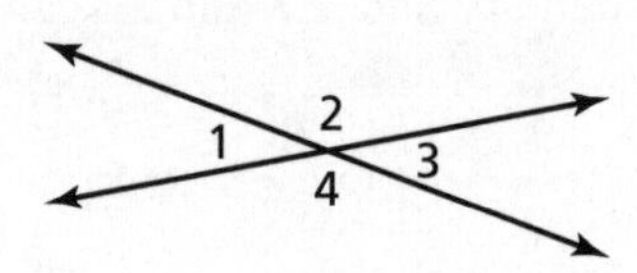

$\angle 1 \cong \angle 3, \angle 2 \cong \angle 4$

Name________________________________ Date__________

Extra Practice

1. Complete the flowchart proof. Then write a two-column proof.

   Given $\angle ACB$ and $\angle ACD$ are supplementary.
   $\angle EGF$ and $\angle ACD$ are supplementary.

   Prove $\angle ACB \cong \angle EGF$

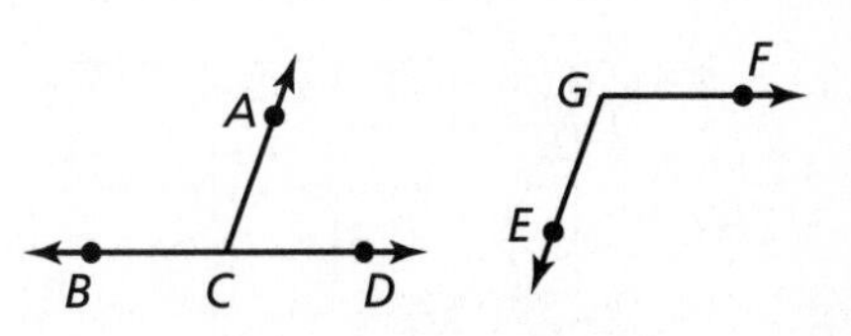

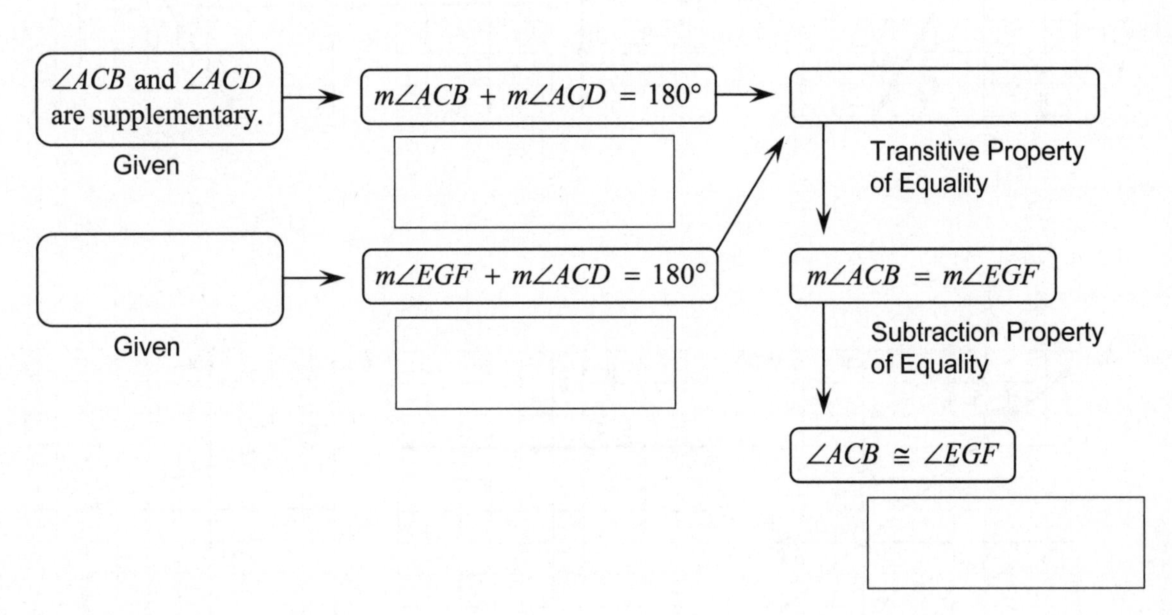

Two-Column Proof

STATEMENTS	REASONS

Name ______________________________ Date __________

Chapter 3 Maintaining Mathematical Proficiency

Find the slope of the line.

1.

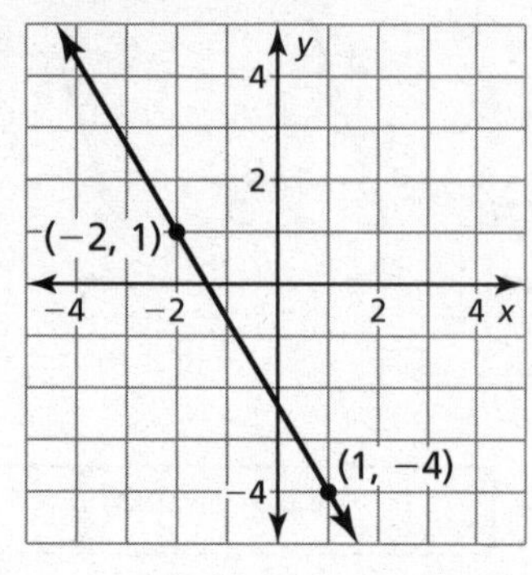

2.

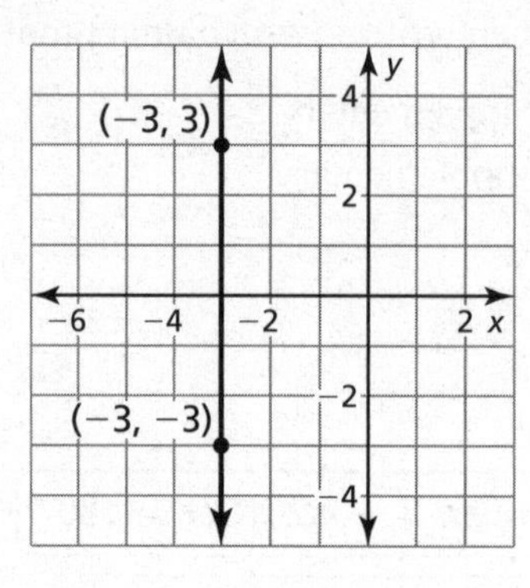

3.

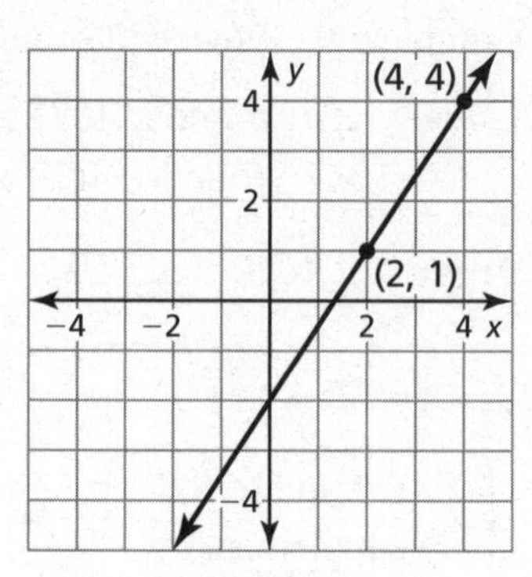

4.

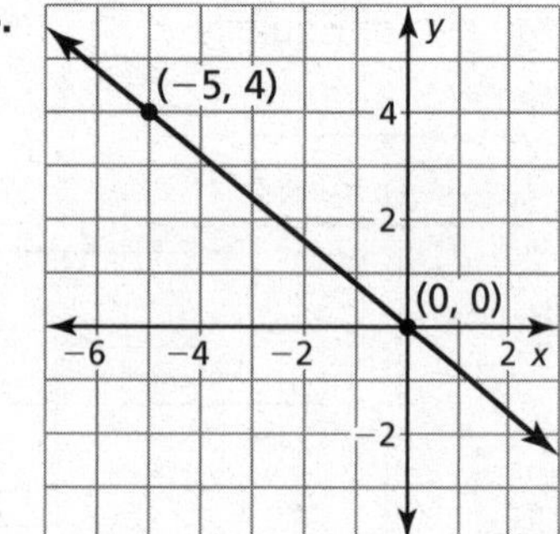

5.

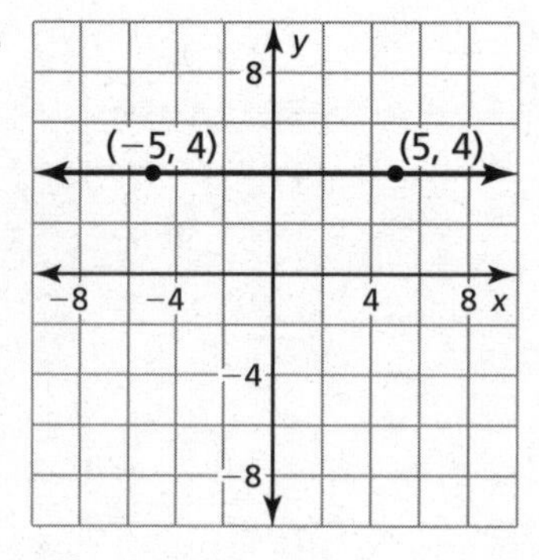

6.

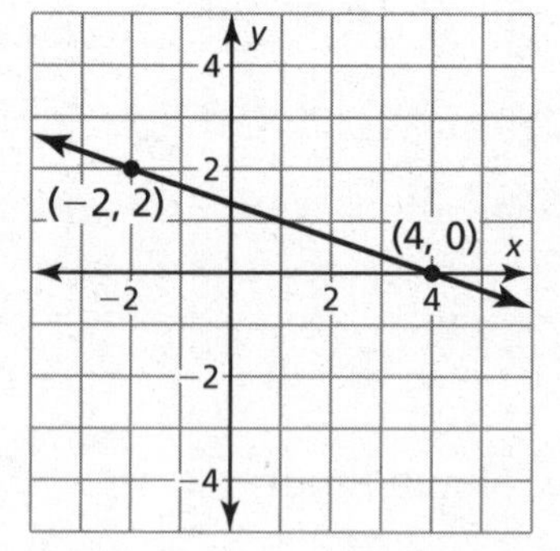

Write an equation of the line that passes through the given point and has the given slope.

7. $(0, -8);\ m = \frac{3}{5}$

8. $(-4, 3);\ m = \frac{1}{3}$

9. $(2, -1);\ m = 5$

Name ______________________________ Date __________

3.1 Pairs of Lines and Angles
For use with Exploration 3.1

Essential Question What does it mean when two lines are parallel, intersecting, coincident, or skew?

1 EXPLORATION: Points of Intersection

Work with a partner. Write the number of points of intersection of each pair of coplanar lines.

a. parallel lines ______________

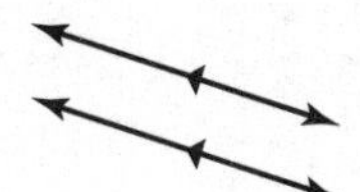

b. intersecting lines ______________

c. coincident lines ______________

2 EXPLORATION: Classifying Pairs of Lines

Work with a partner. The figure shows a *right rectangular prism.* All its angles are right angles. Classify each of the following pairs of lines as *parallel*, *intersecting*, *coincident*, or *skew*. Justify your answers. (Two lines are **skew lines** when they do not intersect and are not coplanar.)

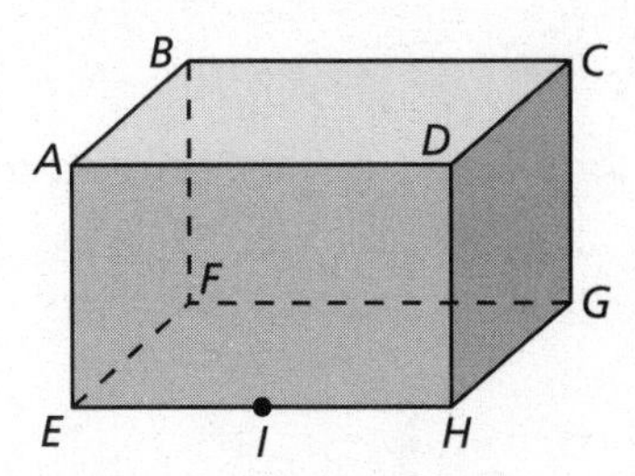

Pair of Lines	Classification	Reason
a. $\overleftrightarrow{AB}$ and $\overleftrightarrow{BC}$		
b. $\overleftrightarrow{AD}$ and $\overleftrightarrow{BC}$		
c. $\overleftrightarrow{EI}$ and $\overleftrightarrow{IH}$		
d. $\overleftrightarrow{BF}$ and $\overleftrightarrow{EH}$		
e. $\overleftrightarrow{EF}$ and $\overleftrightarrow{CG}$		
f. $\overleftrightarrow{AB}$ and $\overleftrightarrow{GH}$		

3 EXPLORATION: Identifying Pairs of Angles

Work with a partner. In the figure, two parallel lines are intersected by a third line called a *transversal*.

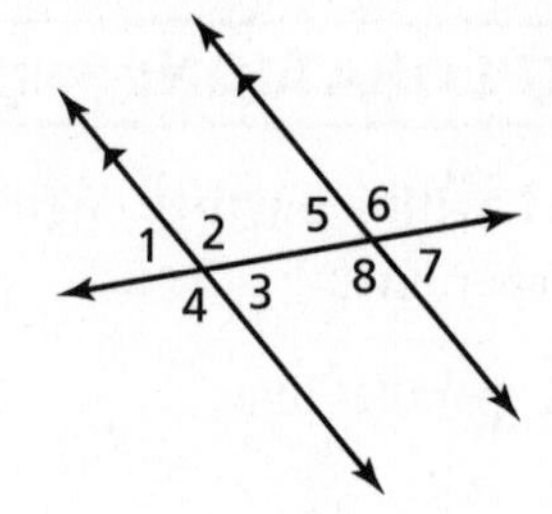

a. Identify all the pairs of vertical angles. Explain your reasoning.

b. Identify all the linear pairs of angles. Explain your reasoning.

Communicate Your Answer

4. What does it mean when two lines are parallel, intersecting, coincident, or skew?

5. In Exploration 2, find three more pairs of lines that are different from those given. Classify the pairs of lines as *parallel*, *intersecting*, *coincident*, or *skew*. Justify your answers.

Name_______________________________ Date__________

3.1 Notetaking with Vocabulary

For use after Lesson 3.1

In your own words, write the meaning of each vocabulary term.

parallel lines

skew lines

parallel planes

transversal

corresponding angles

alternate interior angles

alternate exterior angles

consecutive interior angles

Notes:

Core Concepts

Parallel Lines, Skew Lines, and Parallel Planes

Two lines that do not intersect are either *parallel lines* or *skew lines*. Two lines are **parallel lines** when they do not intersect and are coplanar. Two lines are **skew lines** when they do not intersect and are not coplanar. Also, two planes that do not intersect are **parallel planes**.

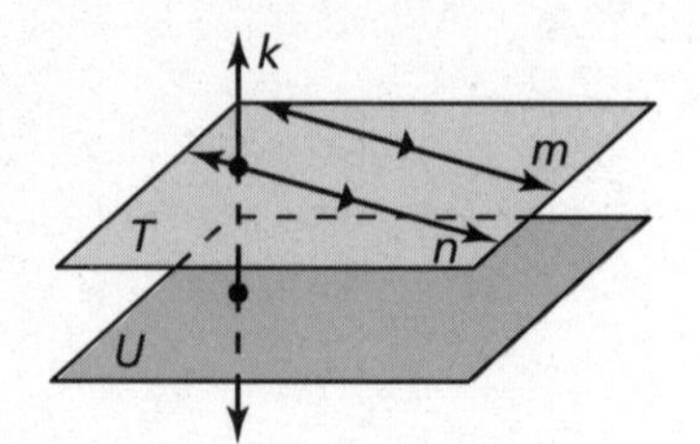

Lines m and n are parallel lines $(m \parallel n)$.

Lines m and k are skew lines.

Planes T and U are parallel planes $(T \parallel U)$.

Lines k and n are intersecting lines, and there is a plane (not shown) containing them.

Small directed arrows, as shown on lines m and n above, are used to show that lines are parallel. The symbol $\parallel$ means "is parallel to," as in $m \parallel n$.

Segments and rays are parallel when they lie in parallel lines. A line is parallel to a plane when the line is in a plane parallel to the given plane. In the diagram above, line n is parallel to plane U.

Notes:

Postulate 3.1 Parallel Postulate

If there is a line and a point not on the line, then there is exactly one line through the point parallel to the given line.

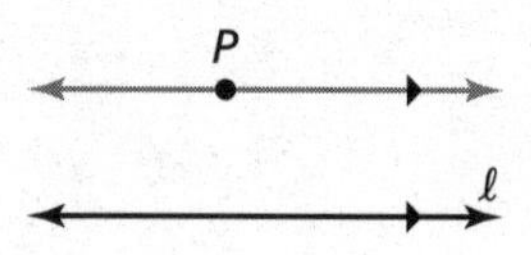

There is exactly one line through P parallel to ℓ.

Notes:

Postulate 3.2 Perpendicular Postulate

If there is a line and a point not on the line, then there is exactly one line through the point perpendicular to the given line.

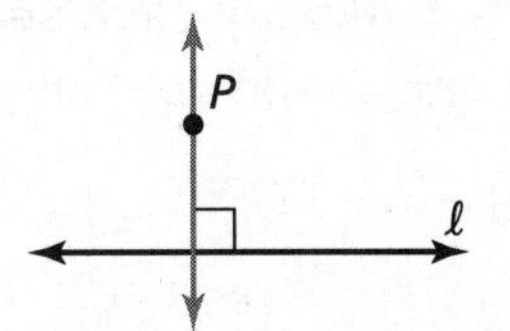

There is exactly one line through P perpendicular to ℓ.

Notes:

Angles Formed by Transversals

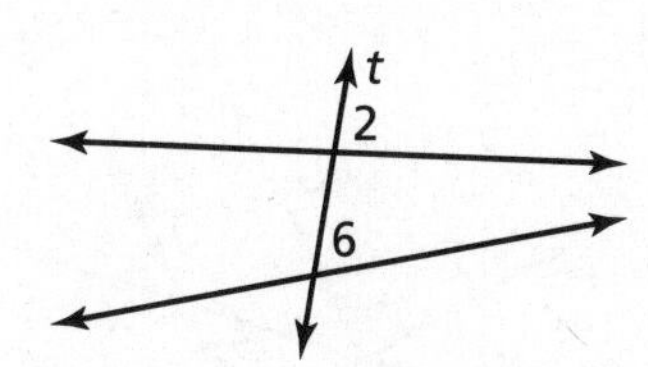

Two angles are **corresponding angles** when they have corresponding positions. For example, $\angle 2$ and $\angle 6$ are above the lines and to the right of the transversal t.

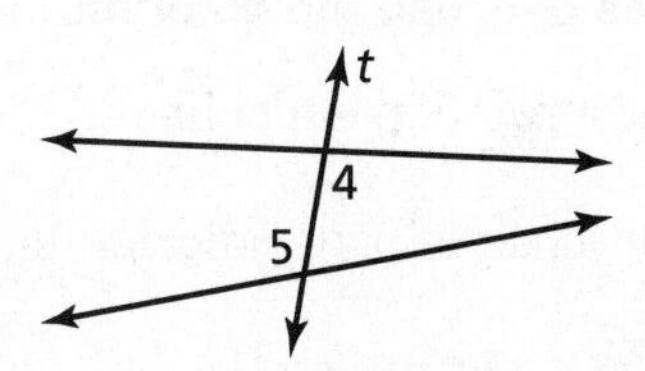

Two angles are **alternate interior angles** when they lie between the two lines and on opposite sides of the transversal t.

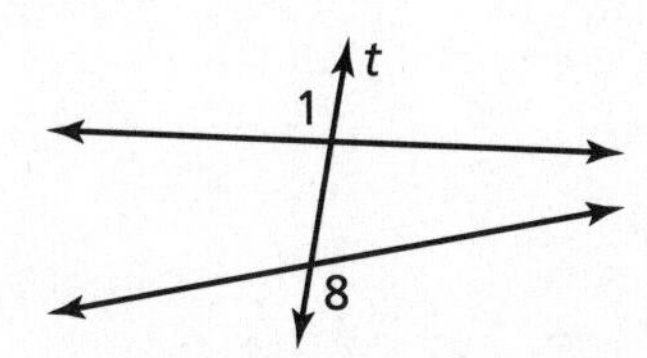

Two angles are **alternate exterior angles** when they lie outside the two lines and on opposite sides of the transversal t.

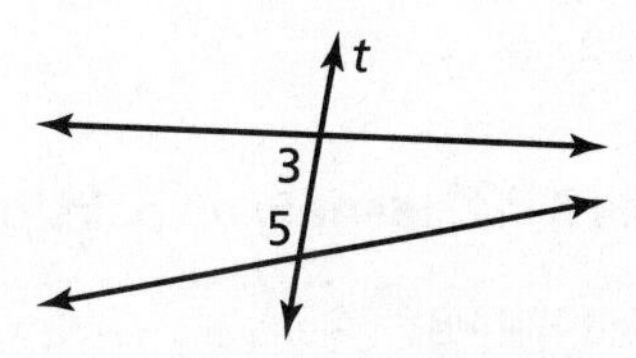

Two angles are **consecutive interior angles** when they lie between the two lines and on the same side of the transversal t.

Notes:

Name ______________________________ Date __________

Extra Practice

In Exercises 1–4, think of each segment in the diagram as part of a line. Which line(s) or plane(s) contain point *B* and appear to fit the description?

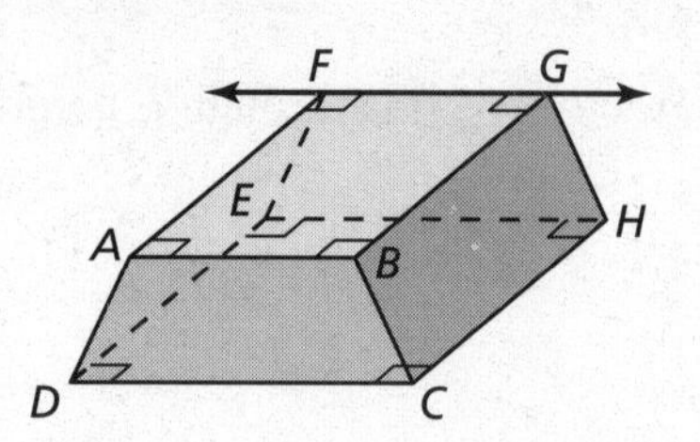

1. line(s) skew to $\overleftrightarrow{FG}$.

2. line(s) perpendicular to $\overleftrightarrow{FG}$.

3. line(s) parallel to $\overleftrightarrow{FG}$.

4. plane(s) parallel to plane *FGH*.

In Exercises 5–8, use the diagram.

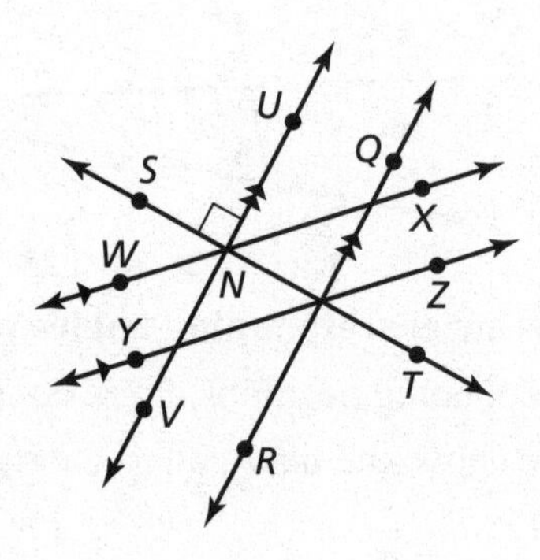

5. Name a pair of parallel lines.

6. Name a pair of perpendicular lines.

7. Is $\overleftrightarrow{WX} \parallel \overleftrightarrow{QR}$? Explain.

8. Is $\overleftrightarrow{ST} \perp \overleftrightarrow{NV}$? Explain.

In Exercises 9–12, identify all pairs of angles of the given type.

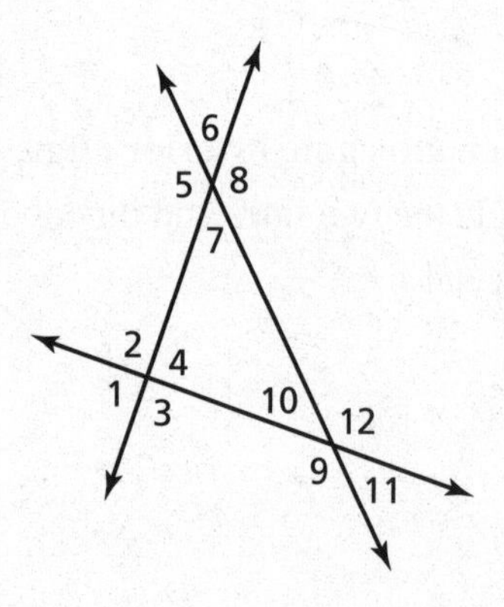

9. corresponding

10. alternate interior

11. alternate exterior

12. consecutive interior

Name____ Date__________

3.2 Parallel Lines and Transversals

For use with Exploration 3.2

Essential Question When two parallel lines are cut by a transversal, which of the resulting pairs of angles are congruent?

1 EXPLORATION: Exploring Parallel Lines

Go to *BigIdeasMath.com* for an interactive tool to investigate this exploration.

Work with a partner.
Use dynamic geometry software to draw two parallel lines. Draw a third line that intersects both parallel lines. Find the measures of the eight angles that are formed. What can you conclude?

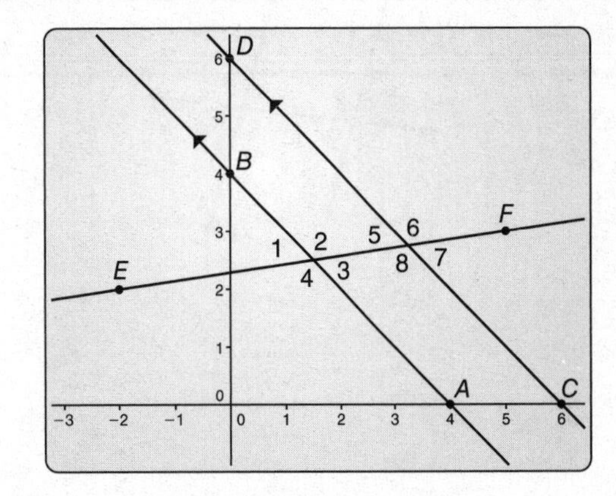

2 EXPLORATION: Writing Conjectures

Work with a partner. Use the results of Exploration 1 to write conjectures about the following pairs of angles formed by two parallel lines and a transversal.

a. corresponding angles

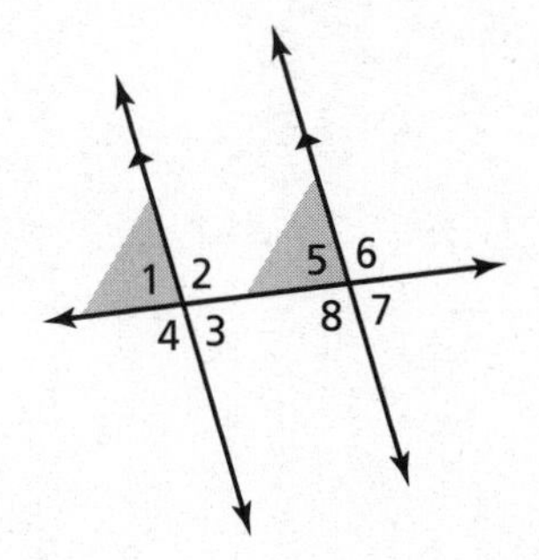

b. alternate interior angles

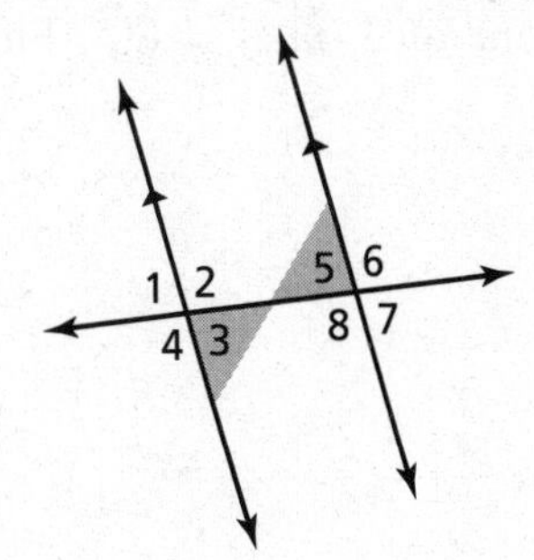

Name ________________________________ Date __________

2 EXPLORATION: Writing Conjectures (continued)

c. alternate exterior angles

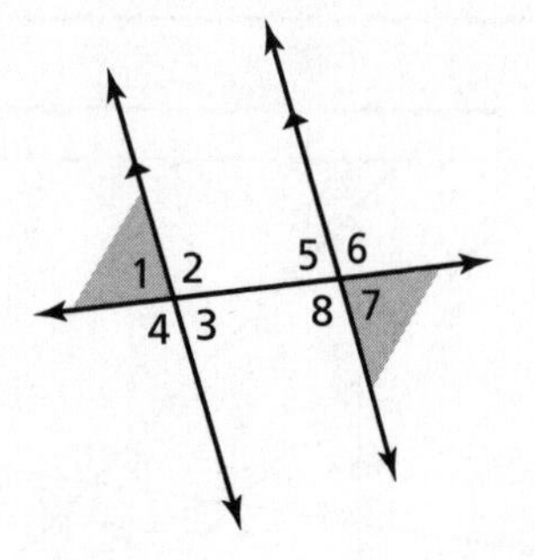

d. consecutive interior angles

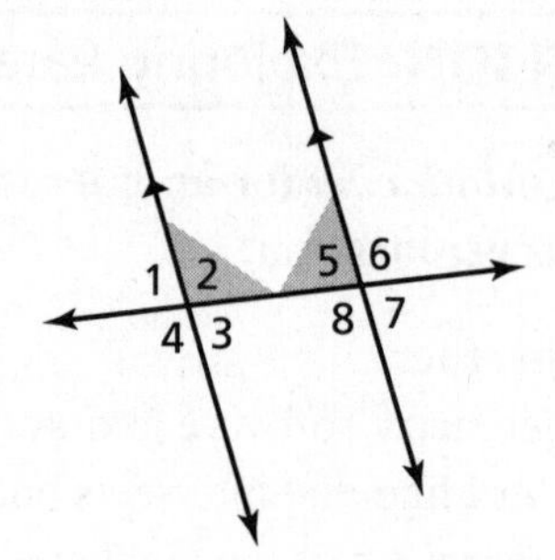

Communicate Your Answer

3. When two parallel lines are cut by a transversal, which of the resulting pairs of angles are congruent?

4. In Exploration 2, $m\angle 1 = 80°$. Find the other angle measures.

Name______________________________ Date__________

3.2 Notetaking with Vocabulary
For use after Lesson 3.2

In your own words, write the meaning of each vocabulary term.

corresponding angles

parallel lines

supplementary angles

vertical angles

Theorems

Theorem 3.1 Corresponding Angles Theorem

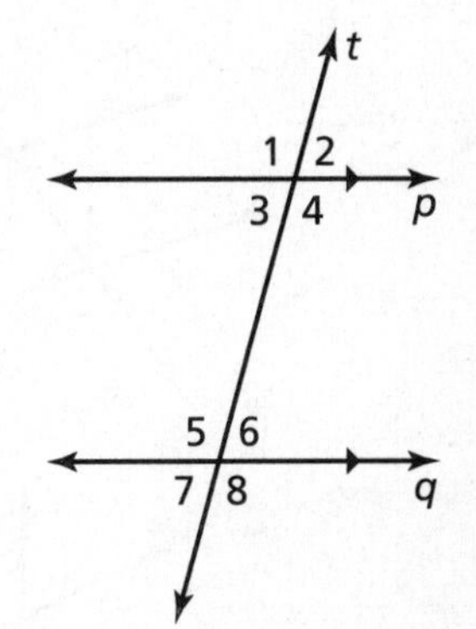

If two parallel lines are cut by a transversal, then the pairs of corresponding angles are congruent.

Examples In the diagram, $\angle 2 \cong \angle 6$ and $\angle 3 \cong \angle 7$.

Theorem 3.2 Alternate Interior Angles Theorem

If two parallel lines are cut by a transversal, then the pairs of alternate interior angles are congruent.

Examples In the diagram, $\angle 3 \cong \angle 6$ and $\angle 4 \cong \angle 5$.

Theorem 3.3 Alternate Exterior Angles Theorem

If two parallel lines are cut by a transversal, then the pairs of alternate exterior angles are congruent.

Examples In the diagram, $\angle 1 \cong \angle 8$ and $\angle 2 \cong \angle 7$.

Name ______________________ Date ________

3.2 Notetaking with Vocabulary (continued)

Theorem 3.4 Consecutive Interior Angles Theorem

If two parallel lines are cut by a transversal, then the pairs of consecutive interior angles are supplementary.

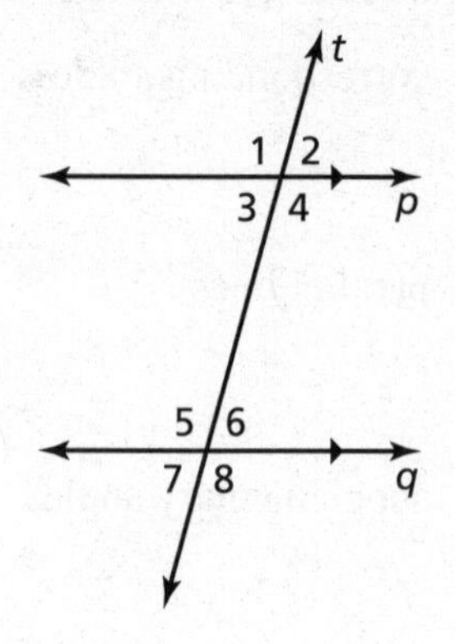

Examples In the diagram, $\angle 3$ and $\angle 5$ are supplementary, and $\angle 4$ and $\angle 6$ are supplementary.

Notes:

Extra Practice

In Exercises 1–4, find $m\angle 1$ and $m\angle 2$. Tell which theorem you use in each case.

1.

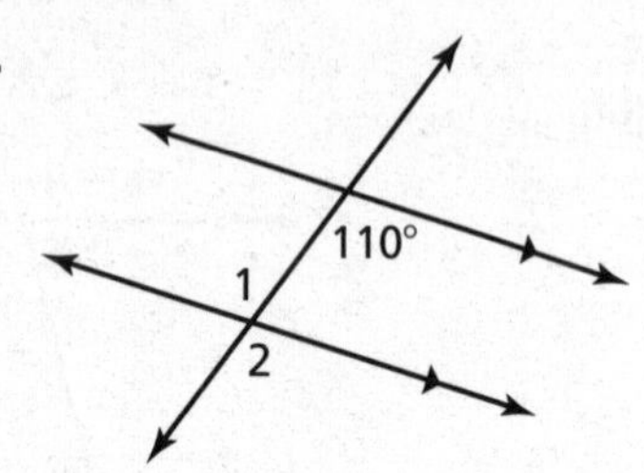

2.

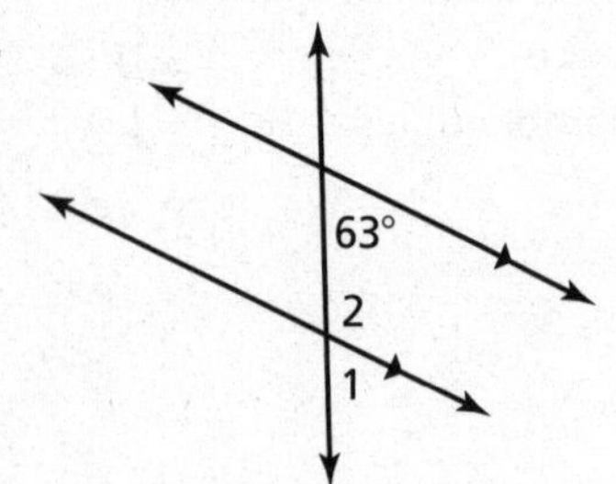

3.

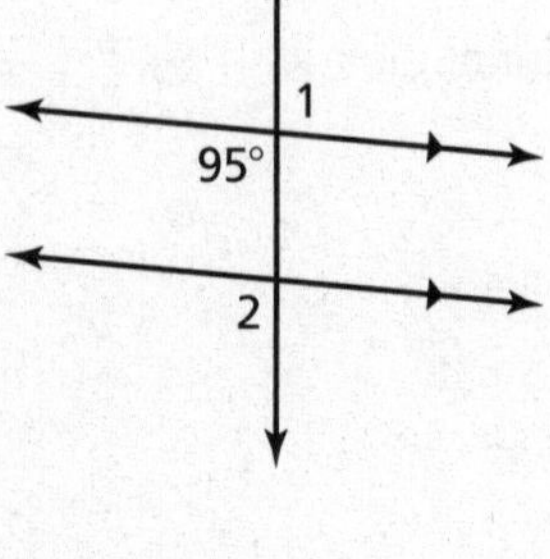

Name____ Date__________

3.2 Notetaking with Vocabulary (continued)

4.

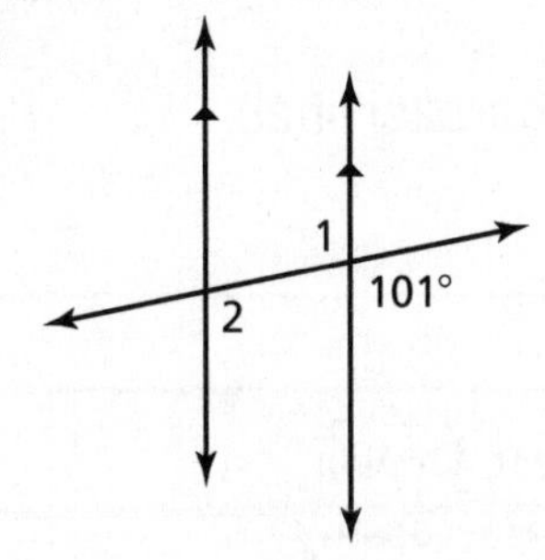

In Exercises 5–8, find the value of *x*. Show your steps.

5.

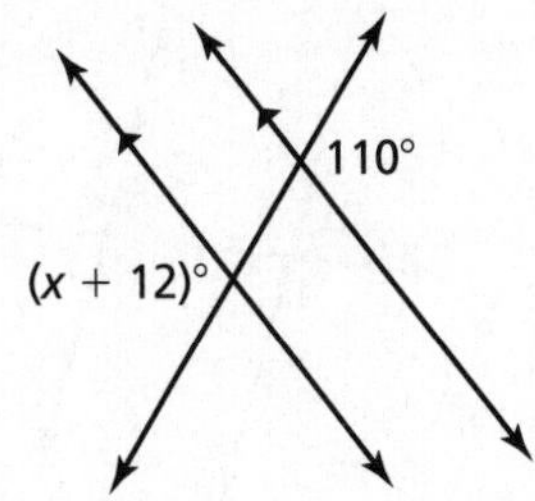

6.

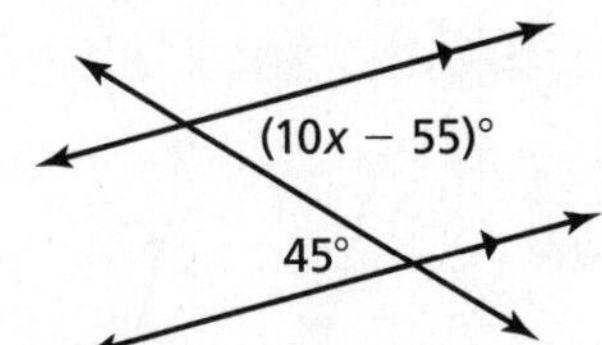

7.

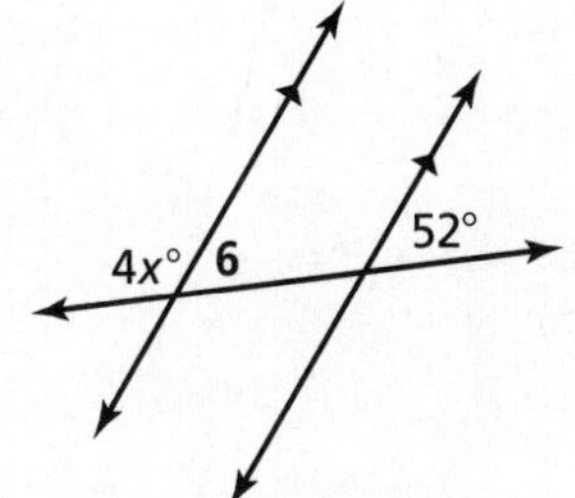

8.

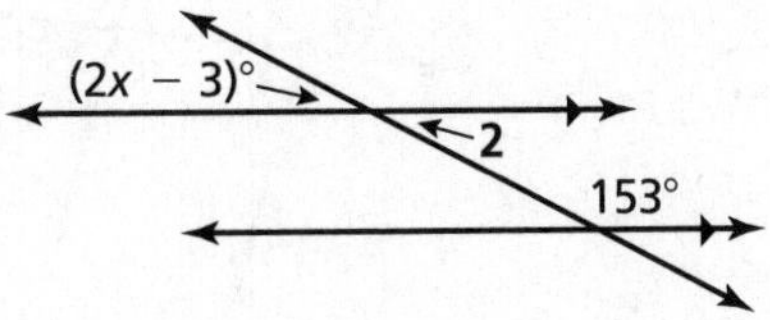

Name ______________________________ Date __________

3.3 Proofs with Parallel Lines

For use with Exploration 3.3

Essential Question For which of the theorems involving parallel lines and transversals is the converse true?

1 EXPLORATION: Exploring Converses

Work with a partner. Write the converse of each conditional statement. Draw a diagram to represent the converse. Determine whether the converse is true. Justify your conclusion.

a. Corresponding Angles Theorem (Theorem 3.1)

If two parallel lines are cut by a transversal, then the pairs of corresponding angles are congruent.

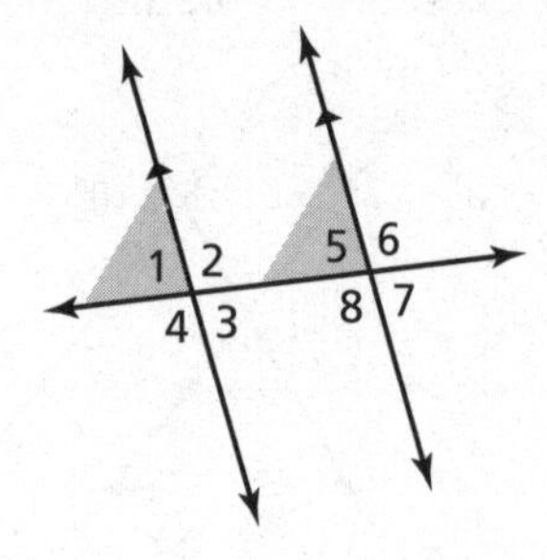

Converse

b. Alternate Interior Angles Theorem (Theorem 3.2)

If two parallel lines are cut by a transversal, then the pairs of alternate interior angles are congruent.

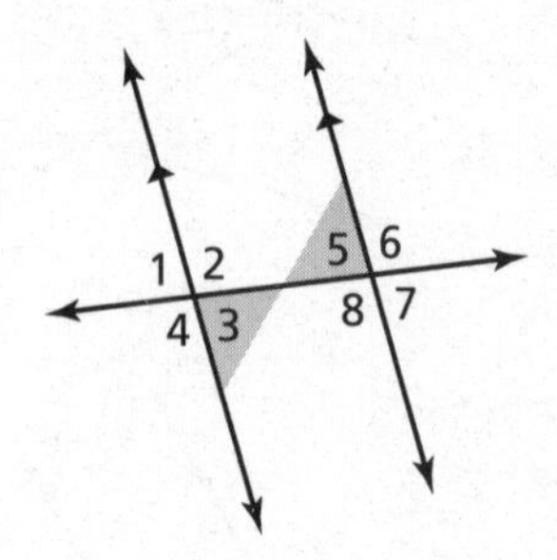

Converse

c. Alternate Exterior Angles Theorem (Theorem 3.3)

If two parallel lines are cut by a transversal, then the pairs of alternate exterior angles are congruent.

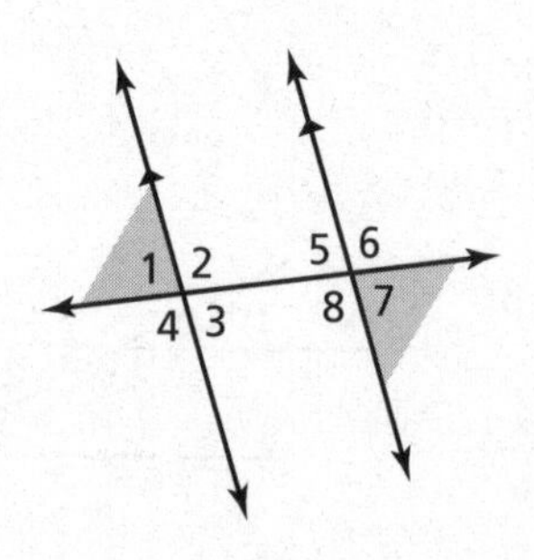

Converse

1 EXPLORATION: Exploring Converses (continued)

d. Consecutive Interior Angles Theorem (Theorem 3.4)

If two parallel lines are cut by a transversal, then the pairs of consecutive interior angles are supplementary.

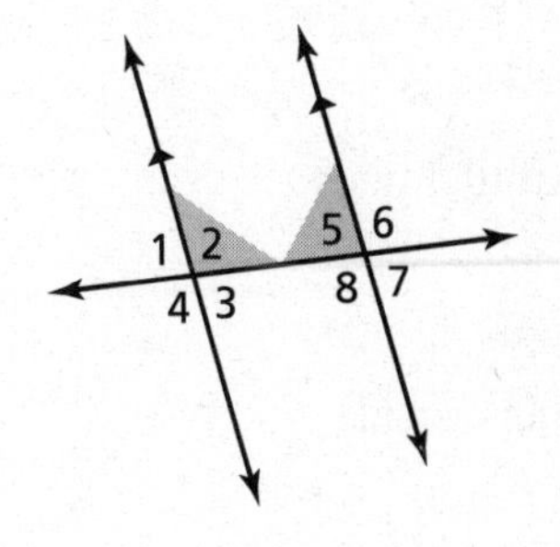

Converse

Communicate Your Answer

2. For which of the theorems involving parallel lines and transversals is the converse true?

3. In Exploration 1, explain how you would prove any of the theorems that you found to be true.

Name ______________________________ Date __________

3.3 Notetaking with Vocabulary
For use after Lesson 3.3

In your own words, write the meaning of each vocabulary term.

converse

parallel lines

transversal

corresponding angles

congruent

alternate interior angles

alternate exterior angles

consecutive interior angles

Theorems

Theorem 3.5 Corresponding Angles Converse

If two lines are cut by a transversal so the corresponding angles are congruent, then the lines are parallel.

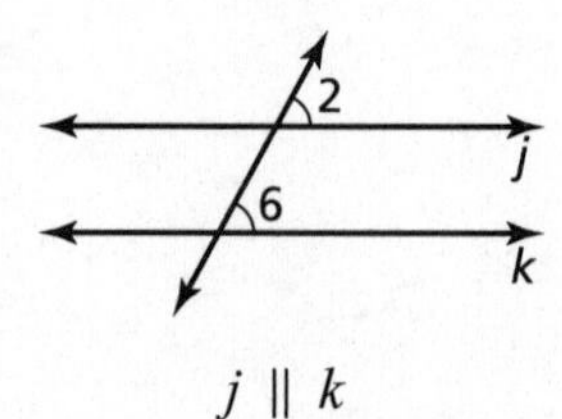

$j \parallel k$

Notes:

Theorem 3.6 Alternate Interior Angles Converse

If two lines are cut by a transversal so the alternate interior angles are congruent, then the lines are parallel.

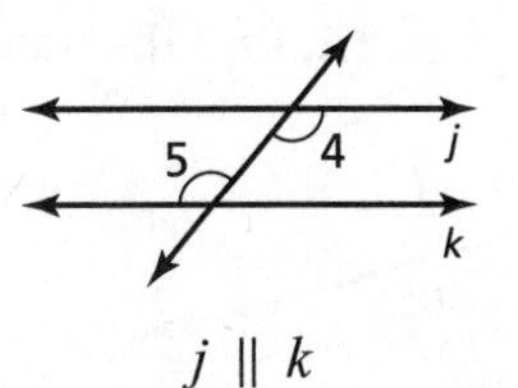

$j \parallel k$

Notes:

Theorem 3.7 Alternate Exterior Angles Converse

If two lines are cut by a transversal so the alternate exterior angles are congruent, then the lines are parallel.

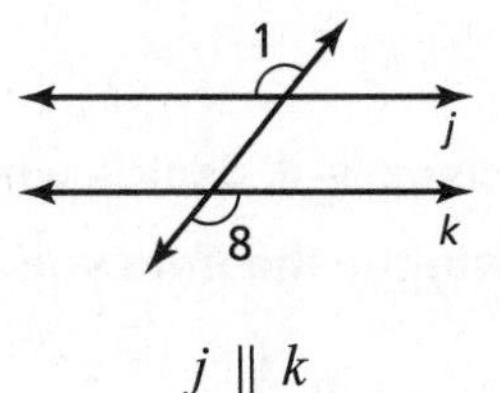

$j \parallel k$

Notes:

Theorem 3.8 Consecutive Interior Angles Converse

If two lines are cut by a transversal so the consecutive interior angles are supplementary, then the lines are parallel.

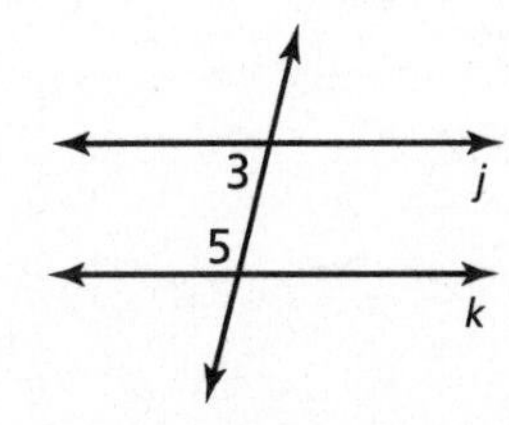

If $\angle 3$ and $\angle 5$ are supplementary, then $j \parallel k$.

Notes:

Theorem 3.9 Transitive Property of Parallel Lines

If two lines are parallel to the same line, then they are parallel to each other.

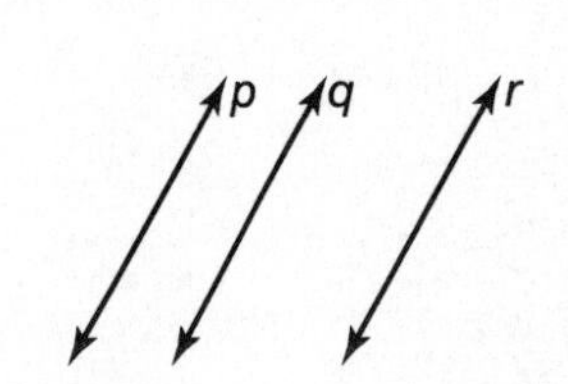

If $p \parallel q$ and $q \parallel r$, then $p \parallel r$.

Notes:

Extra Practice

In Exercises 1 and 2, find the value of x that makes $m \parallel n$. Explain your reasoning.

1.

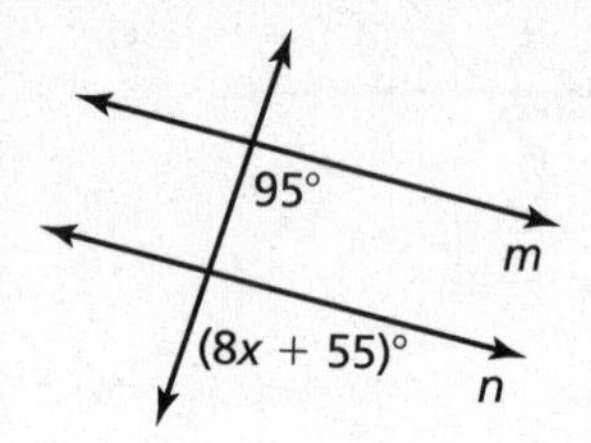

2.

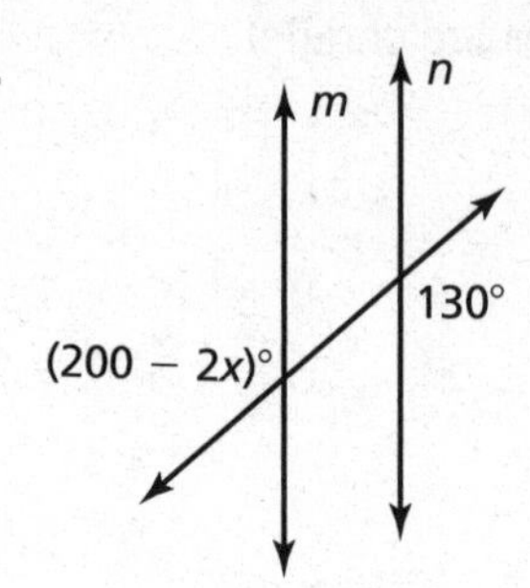

In Exercises 3–6, decide whether there is enough information to prove that $m \parallel n$. If so, state the theorem you would use.

3.

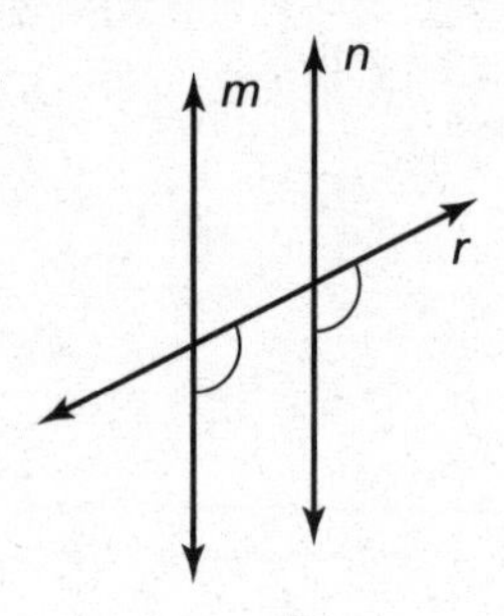

4.

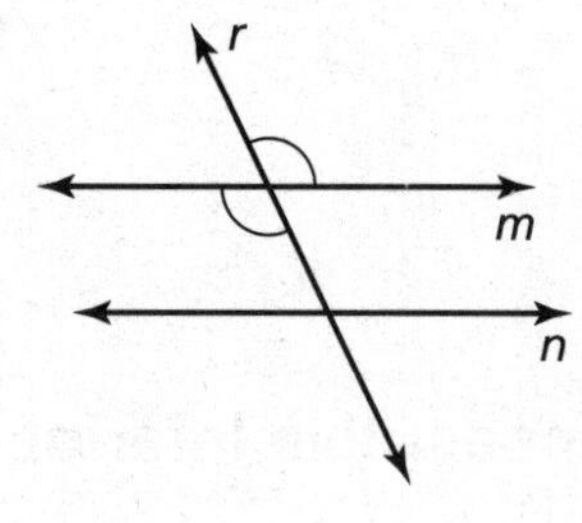

5.

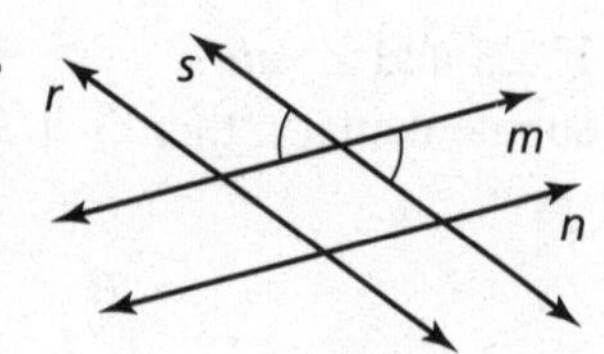

6.

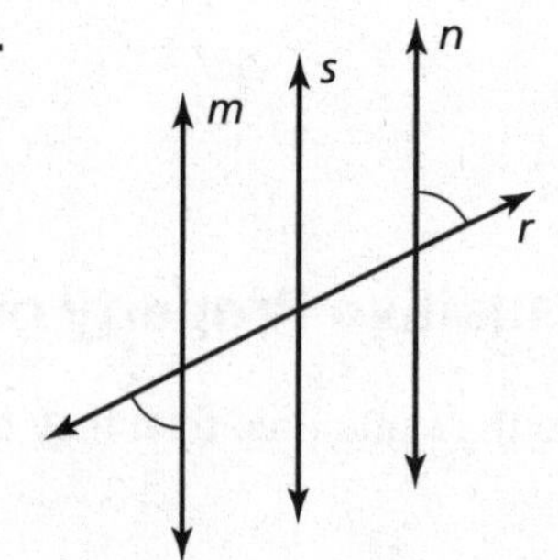

Name________________________________ Date__________

3.4 Proofs with Perpendicular Lines
For use with Exploration 3.4

Essential Question What conjectures can you make about perpendicular lines?

1 EXPLORATION: Writing Conjectures

Work with a partner. Fold a piece of paper in half twice. Label points on the two creases, as shown.

a. Write a conjecture about $\overline{AB}$ and $\overline{CD}$. Justify your conjecture.

b. Write a conjecture about $\overline{AO}$ and $\overline{OB}$. Justify your conjecture.

2 EXPLORATION: Exploring a Segment Bisector

Work with a partner. Fold and crease a piece of paper, as shown. Label the ends of the crease as A and B.

a. Fold the paper again so that point A coincides with point B. Crease the paper on that fold.

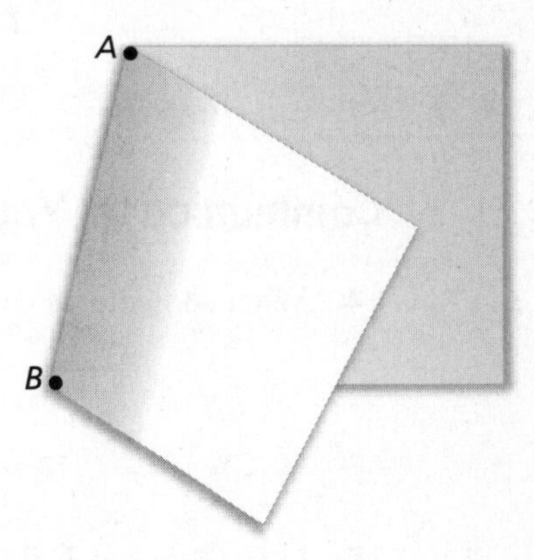

b. Unfold the paper and examine the four angles formed by the two creases. What can you conclude about the four angles?

Name ____________________ Date ________

3 EXPLORATION: Writing a Conjecture

Go to *BigIdeasMath.com* for an interactive tool to investigate this exploration.

Work with a partner.

a. Draw $\overline{AB}$, as shown.

b. Draw an arc with center A on each side of $\overline{AB}$. Using the same compass setting, draw an arc with center B on each side of $\overline{AB}$. Label the intersections of the arcs C and D.

c. Draw $\overline{CD}$. Label its intersection with $\overline{AB}$ as O. Write a conjecture about the resulting diagram. Justify your conjecture.

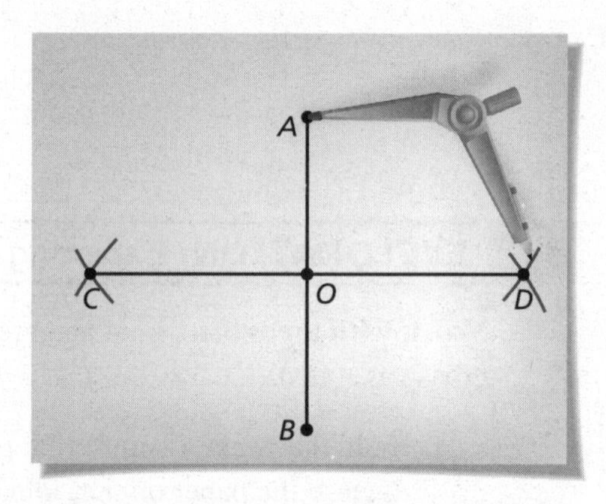

Communicate Your Answer

4. What conjectures can you make about perpendicular lines?

5. In Exploration 3, find AO and OB when $AB = 4$ units.

Name____ Date__________

3.4 Notetaking with Vocabulary

For use after Lesson 3.4

In your own words, write the meaning of each vocabulary term.

distance from a point to a line

perpendicular bisector

Theorems

Theorem 3.10 Linear Pair Perpendicular Theorem

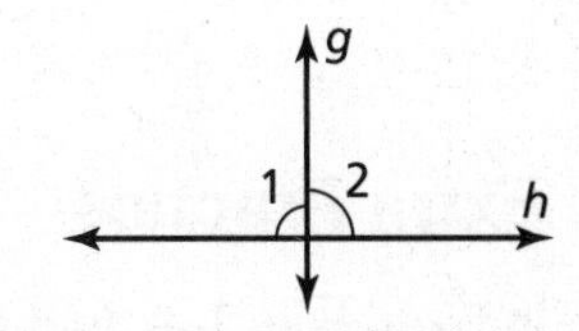

If two lines intersect to form a linear pair of congruent angles, then the lines are perpendicular.

If $\angle 1 \cong \angle 2$, then $g \perp h$.

Notes:

Theorem 3.11 Perpendicular Transversal Theorem

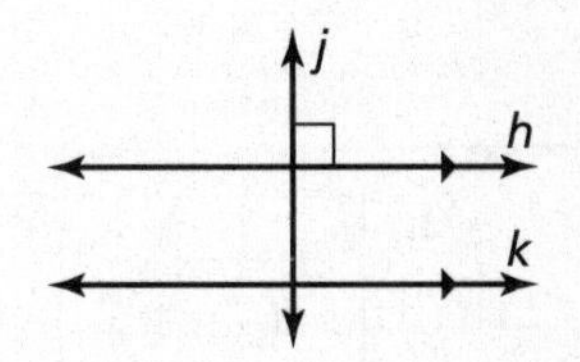

In a plane, if a transversal is perpendicular to one of two parallel lines, then it is perpendicular to the other line.

If $h \parallel k$ and $j \perp h$, then $j \perp k$.

Notes:

Name ______________________________ Date ________

3.4 Notetaking with Vocabulary (continued)

Theorem 3.12 Lines Perpendicular to a Transversal Theorem

In a plane, if two lines are perpendicular to the same line, then they are parallel to each other.

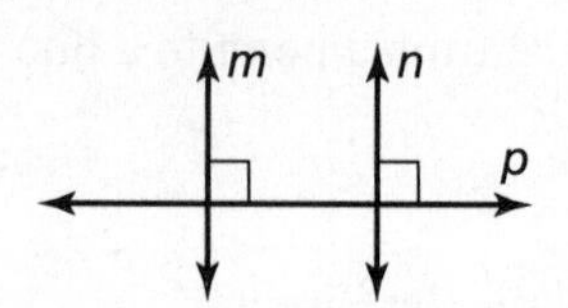

If $m \perp p$ and $n \perp p$, then $m \parallel n$.

Notes:

Extra Practice

In Exercises 1–4, find the distance from point A to $\overleftrightarrow{BC}$.

1.

2. 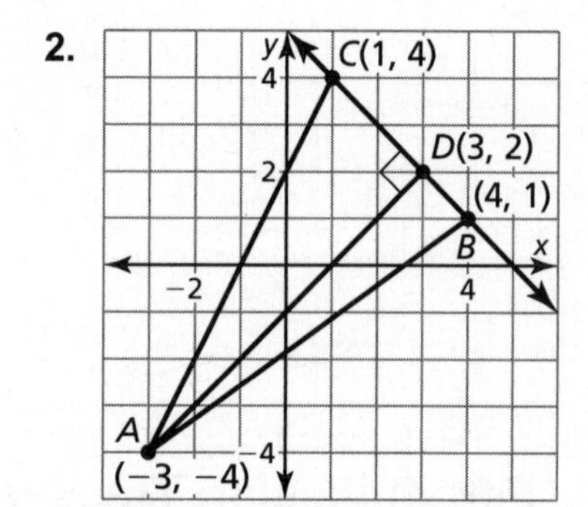

3. 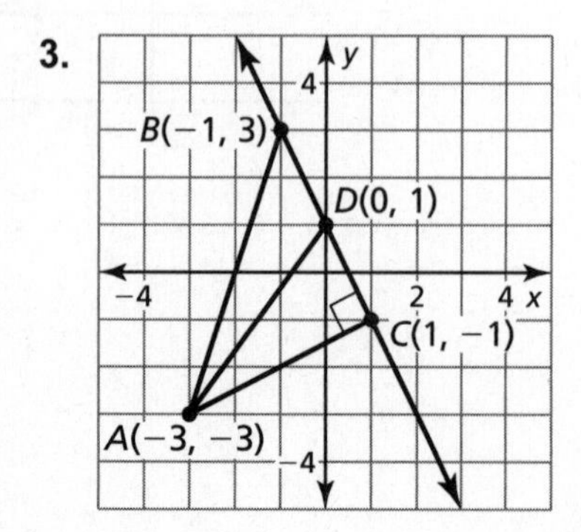

4. 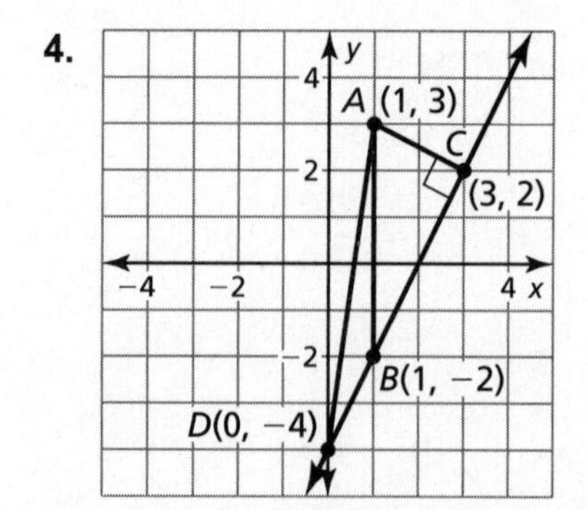

Name____ Date__________

In Exercises 5–8, determine which lines, if any, must be parallel. Explain your reasoning.

5.

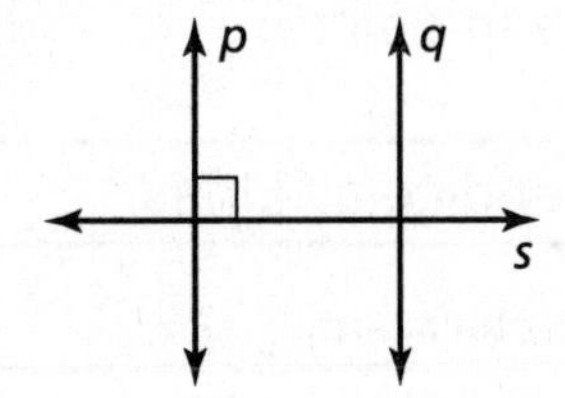

6.

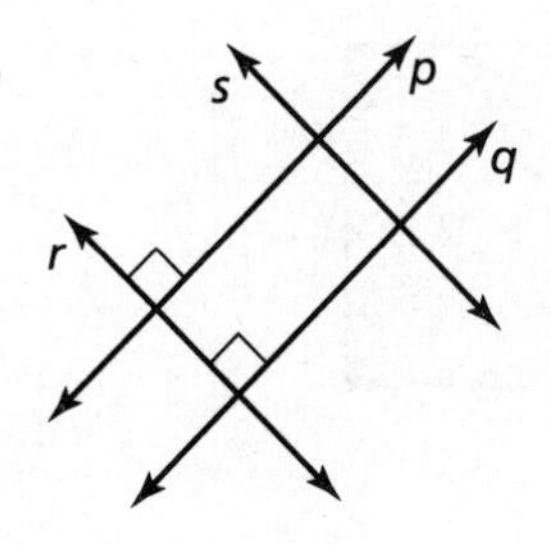

7.

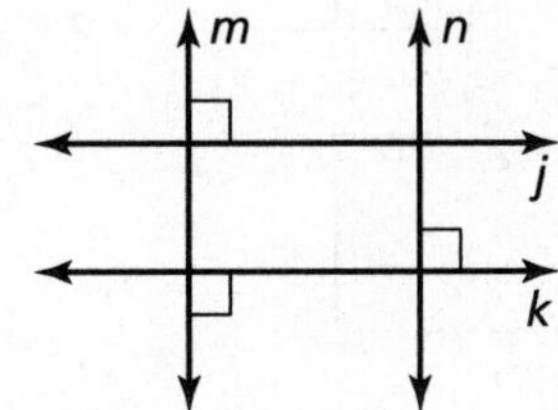

8.

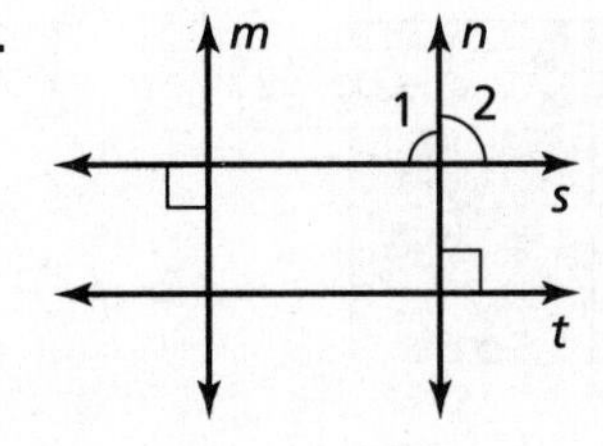

Name ______________________________ Date __________

3.5 Equations of Parallel and Perpendicular Lines

For use with Exploration 3.5

Essential Question How can you write an equation of a line that is parallel or perpendicular to a given line and passes through a given point?

1 EXPLORATION: Writing Equations of Parallel and Perpendicular Lines

Go to *BigIdeasMath.com* for an interactive tool to investigate this exploration.

Work with a partner. Write an equation of the line that is parallel or perpendicular to the given line and passes through the given point. Use a graphing calculator to verify your answer. What is the relationship between the slopes?

a.

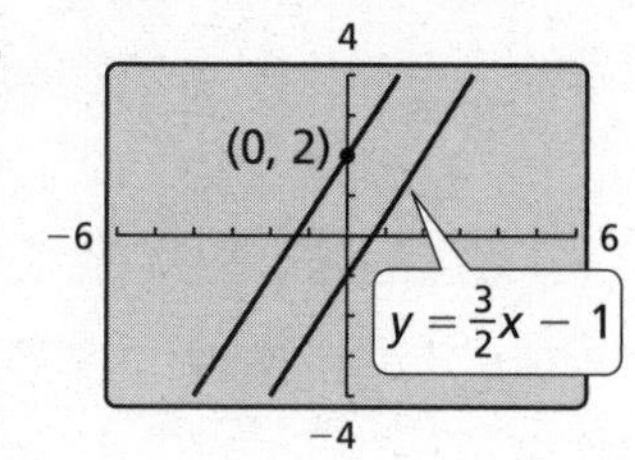

b.

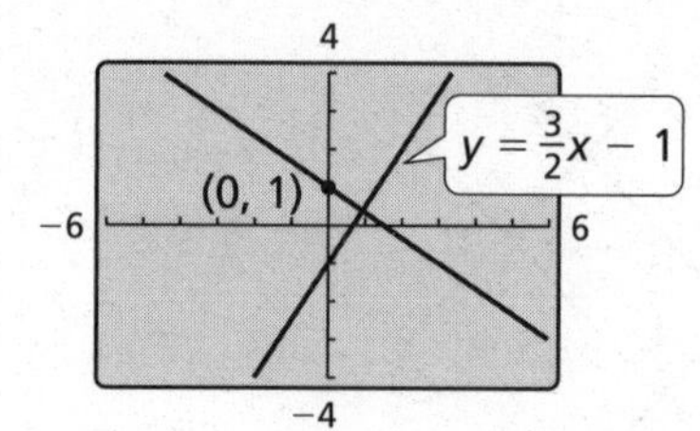

c.

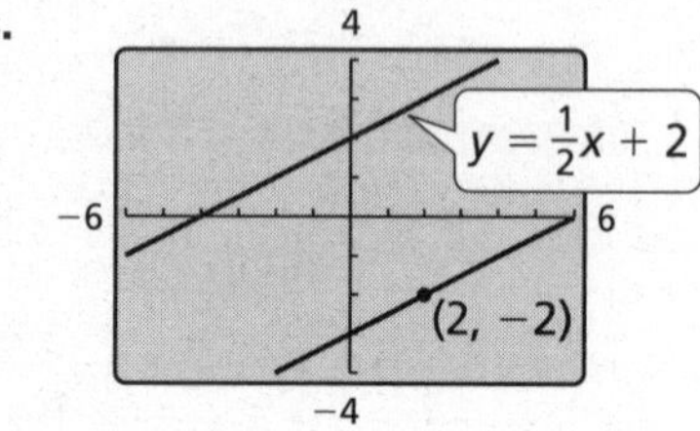

d.

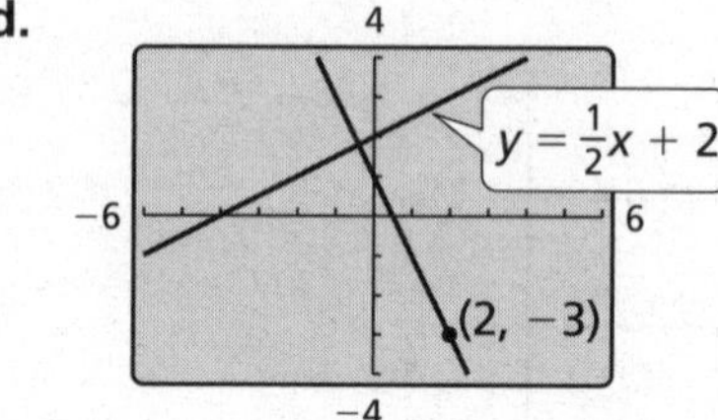

e.

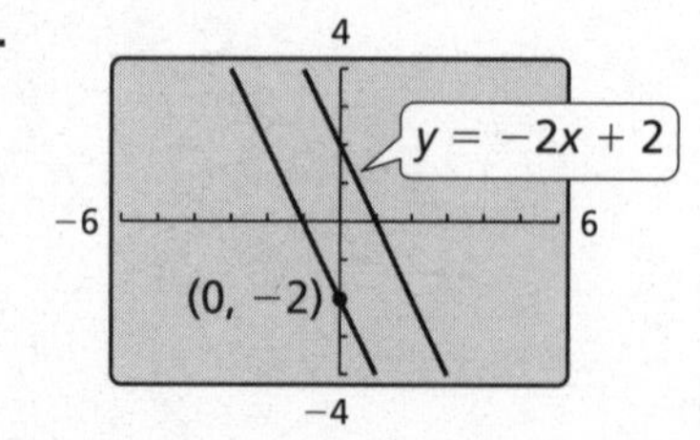

f.

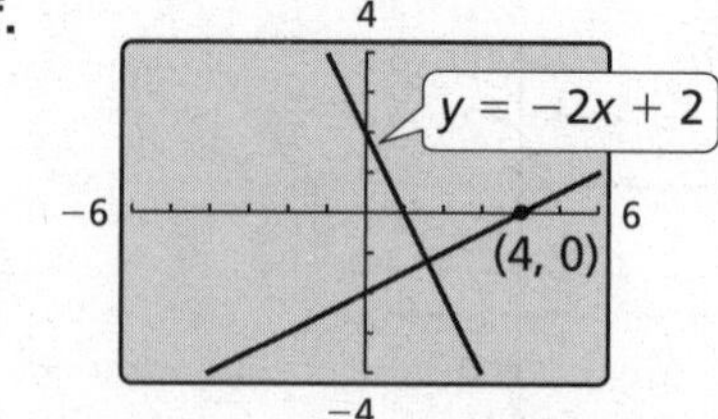

3.5 Equations of Parallel and Perpendicular Lines (continued)

2 EXPLORATION: Writing Equations of Parallel and Perpendicular Lines

Go to *BigIdeasMath.com* for an interactive tool to investigate this exploration.

Work with a partner. Write the equations of the parallel or perpendicular lines. Use a graphing calculator to verify your answers.

a.

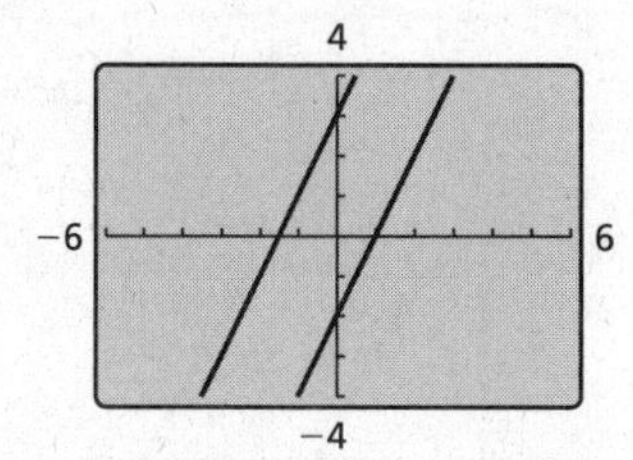

b.

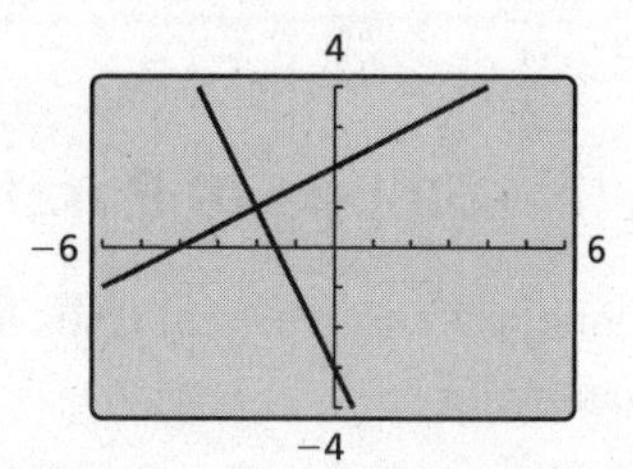

Communicate Your Answer

3. How can you write an equation of a line that is parallel or perpendicular to a given line and passes through a given point?

4. Write an equation of the line that is (a) parallel and (b) perpendicular to the line $y = 3x + 2$ and passes through the point $(1, -2)$.

Name ______________________________ Date __________

3.5 Notetaking with Vocabulary
For use after Lesson 3.5

In your own words, write the meaning of each vocabulary term.

directed line segment

Theorems

Theorem 3.13 Slopes of Parallel Lines

In a coordinate plane, two nonvertical lines are parallel if and only if they have the same slope.

Any two vertical lines are parallel.

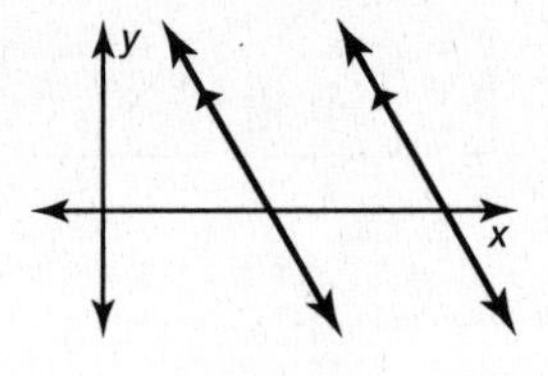

$m_1 = m_2$

Notes:

Theorem 3.14 Slopes of Perpendicular Lines

In a coordinate plane, two nonvertical lines are perpendicular if and only if the product of their slopes is -1.

Horizontal lines are perpendicular to vertical lines.

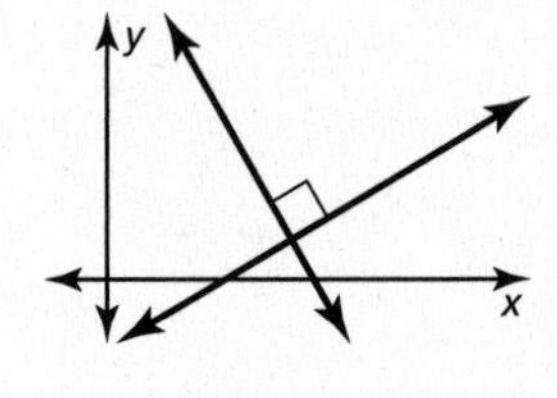

$m_1 \bullet m_2 = -1$

Notes:

Name____ Date__________

Extra Practice

In Exercises 1 and 2, find the coordinates of point *P* along the directed line segment *AB* so that *AP* to *PB* is the given ratio.

1. $A(-2, 7), B(-4, 1)$; 3 to 1

2. $A(3, 1), B(8, -2)$; 2 to 3

3. Determine which of the lines are parallel and which of the lines are perpendicular.

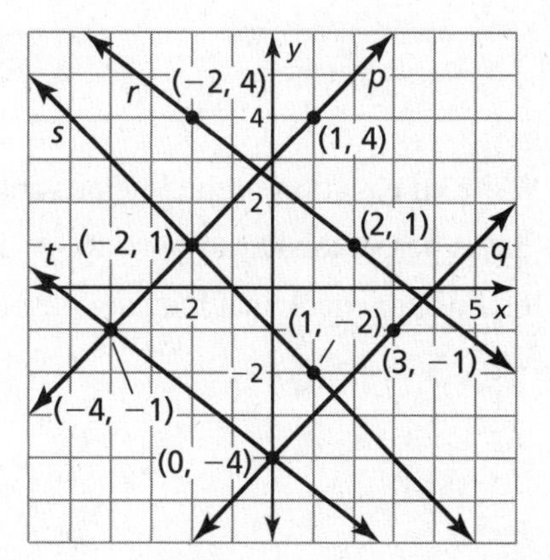

4. Tell whether the lines through the given points are *parallel*, *perpendicular*, or *neither*. Justify your answer.

Line 1: $(2, 0), (-2, 2)$

Line 2: $(1, -2), (4, 4)$

5. Write an equation of the line passing through point $P(3, -2)$ that is parallel to $y = \frac{2}{3}x - 1$. Graph the equations of the lines to check that they are parallel.

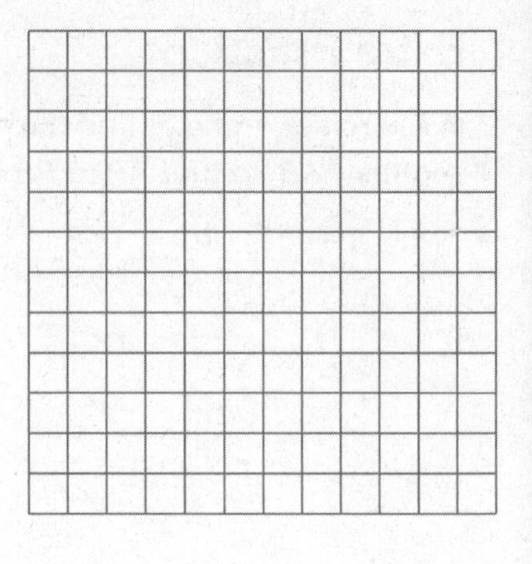

6. Write an equation of the line passing through point $P(-2, 2)$ that is perpendicular to $y = 2x + 3$. Graph the equations of the lines to check that they are perpendicular.

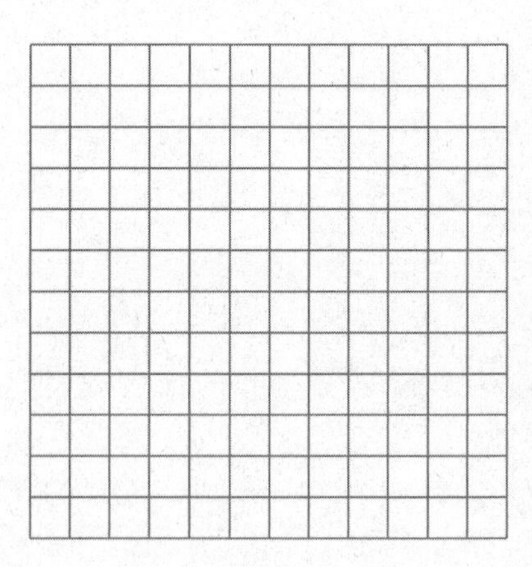

7. Find the distance from point $A(0, 5)$ to $y = -3x - 5$.

Name______________________________ Date__________

Maintaining Mathematical Proficiency

Tell whether the shaded figure is a translation, reflection, rotation, or dilation of the nonshaded figure.

1.

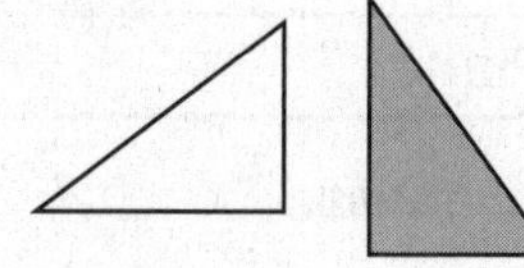

2.

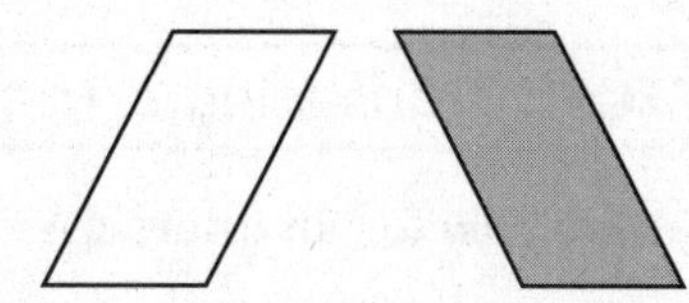

3.

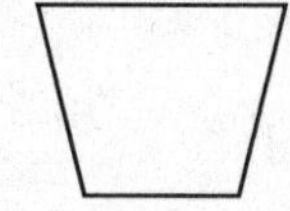

4.

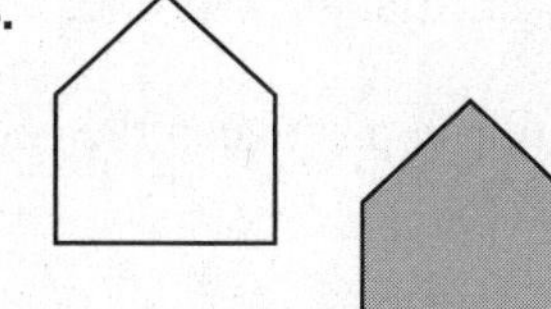

Tell whether the two figures are similar. Explain your reasoning.

5.

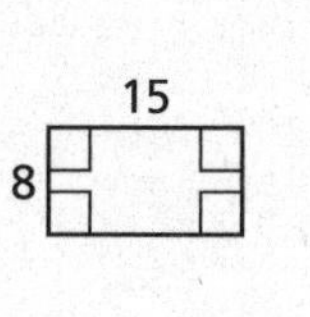

30
16

6.

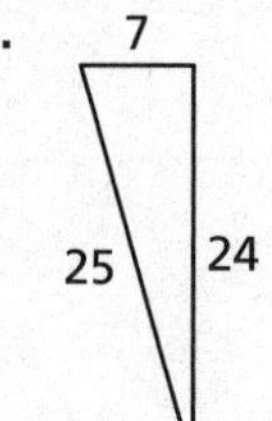

12
13
5

Name ______________________________ Date __________

4.1 Translations

For use with Exploration 4.1

Essential Question How can you translate a figure in a coordinate plane?

1 EXPLORATION: Translating a Triangle in a Coordinate Plane

Go to *BigIdeasMath.com* for an interactive tool to investigate this exploration.

Work with a partner.

a. Use dynamic geometry software to draw any triangle and label it $\triangle ABC$.

b. Copy the triangle and *translate* (or slide) it to form a new figure, called an *image*, $\triangle A'B'C'$. (read as "triangle A prime, B prime, C prime").

c. What is the relationship between the coordinates of the vertices of $\triangle ABC$ and those of $\triangle A'B'C'$?

d. What do you observe about the side lengths and angle measures of the two triangles?

Sample

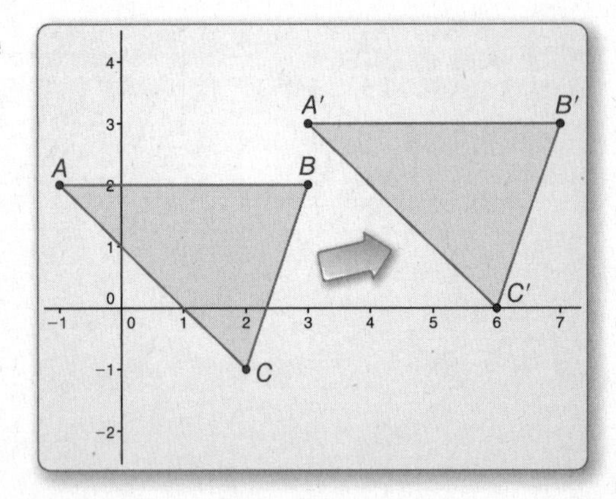

2 EXPLORATION: Translating a Triangle in a Coordinate Plane

Go to *BigIdeasMath.com* for an interactive tool to investigate this exploration.

Work with a partner.

a. The point (x, y) is translated a units horizontally and b units vertically. Write a rule to determine the coordinates of the image of (x, y).

$$(x, y) \rightarrow (\_\_\_\_, \_\_\_\_)$$

b. Use the rule you wrote in part (a) to translate ΔABC 4 units left and 3 units down. What are the coordinates of the vertices of the image, $\Delta A'B'C'$?

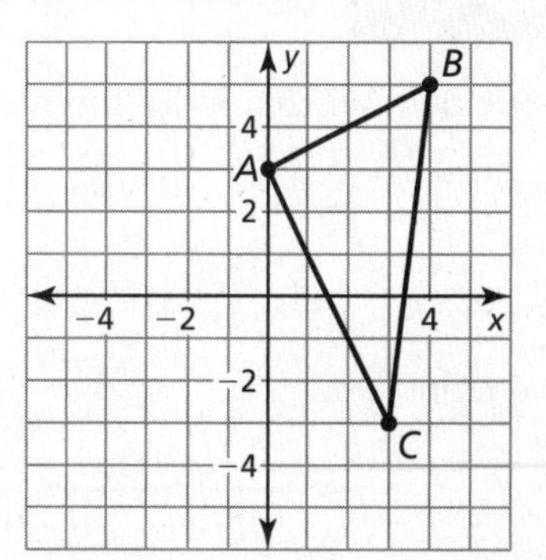

c. Draw $\Delta A'B'C'$. Are its side lengths the same as those of ΔABC? Justify your answer.

3 EXPLORATION: Comparing Angles of Translations

Work with a partner.

a. In Exploration 2, is ΔABC a right triangle? Justify your answer.

b. In Exploration 2, is $\Delta A'B'C'$ a right triangle? Justify your answer.

c. Do you think translations always preserve angle measures? Explain your reasoning.

Communicate Your Answer

4. How can you translate a figure in a coordinate plane?

5. In Exploration 2, translate $\Delta A'B'C'$ 3 units right and 4 units up. What are the coordinates of the vertices of the image, $\Delta A''B''C''$? How are these coordinates related to the coordinates of the vertices of the original triangle, ΔABC?

Name ______________________________ Date __________

4.1 Notetaking with Vocabulary
For use after Lesson 4.1

In your own words, write the meaning of each vocabulary term.

vector

initial point

terminal point

horizontal component

vertical component

component form

transformation

image

preimage

translation

rigid motion

composition of transformations

Name____ Date__________

4.1 Notetaking with Vocabulary (continued)

Core Concepts

Vectors

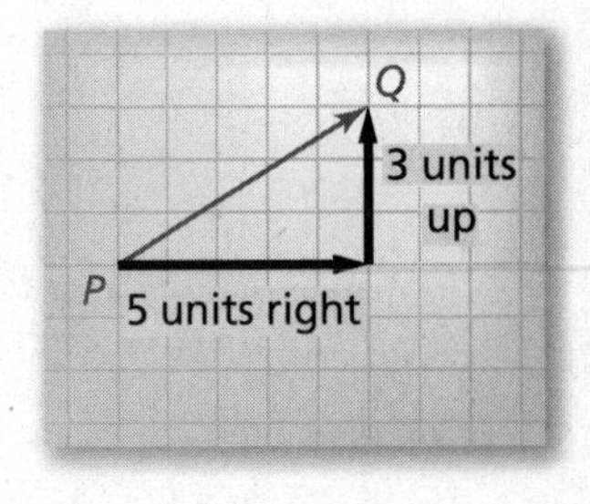

The diagram shows a vector. The **initial point**, or starting point, of the vector is P, and the **terminal point**, or ending point, is Q. The vector is named $\overrightarrow{PQ}$, which is read as "vector PQ." The **horizontal component** of $\overrightarrow{PQ}$ is 5, and the **vertical component** is 3. The **component form** of a vector combines the horizontal and vertical components. So, the component form of $\overrightarrow{PQ}$ is $\langle 5, 3 \rangle$.

Notes:

Translations

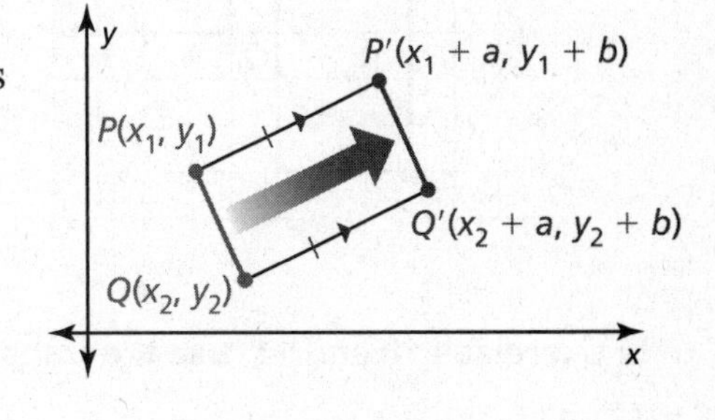

A translation moves every point of a figure the same distance in the same direction. More specifically, a translation *maps*, or moves the points P and Q of a plane figure along a vector $\langle a, b \rangle$ to the points P' and Q', so that one of the following statements is true.

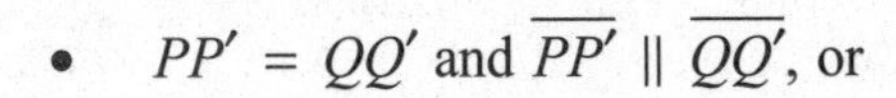

- $PP' = QQ'$ and $\overline{PP'} \parallel \overline{QQ'}$, or
- $PP' = QQ'$ and $\overline{PP'}$ and $\overline{QQ'}$ are collinear.

Notes:

Extra Practice

In Exercises 1–3, name the vector and write its component form.

1. 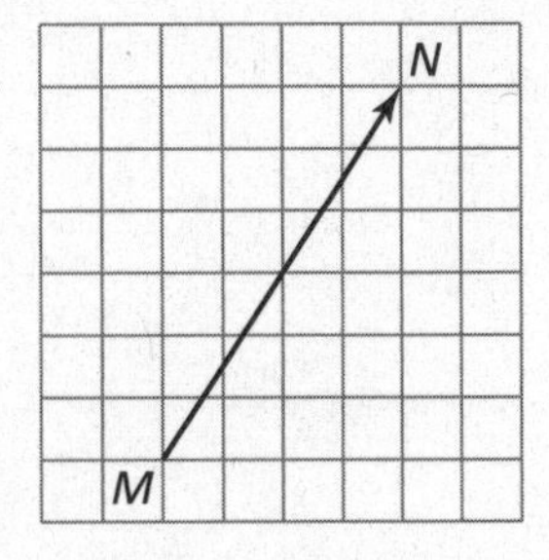

2. 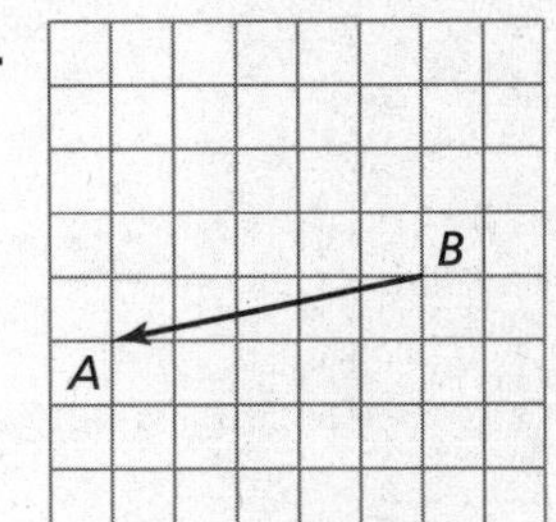

3. 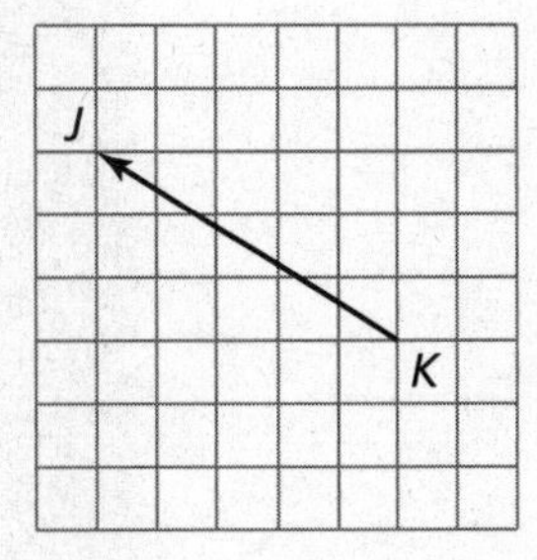

Name ______________________________ Date __________

4.1 Notetaking with Vocabulary (continued)

In Exercises 4–7, the vertices of $\triangle ABC$ are $A(1, 2)$, $B(5, 1)$, $C(5, 4)$. Translate $\triangle ABC$ using the given vector. Graph $\triangle ABC$ and its image.

4. $\langle -4, 0 \rangle$

5. $\langle -2, -4 \rangle$

6. $\langle 0, -5 \rangle$

7. $\langle 1, -3 \rangle$

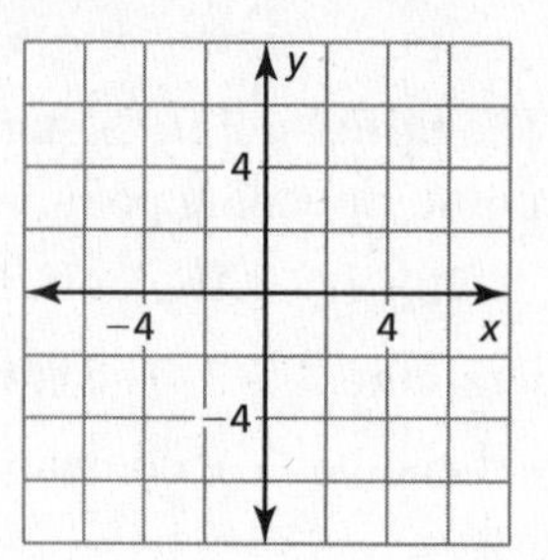

In Exercises 8 and 9, write a rule for the translation of quadrilateral *PQRS* to quadrilateral *P′Q′R′S′*.

8.

y
P′
S′
4
P
S
2
2
Q′
6
R′ x
−2
Q
R

9.

P
Q
y
4
P′
Q′
S
R
−8
−4
x
S′
R′
−2

In Exercises 10 and 11, use the translation.

$(x, y) \rightarrow (x + 6, y - 3)$

10. What is the image of $J(4, 5)$?

11. What is the image of $R'(0, -5)$?

12. In a video game, you move a spaceship 1 unit left and 4 units up. Then, you move the spaceship 2 units left. Rewrite the composition as a single transformation.

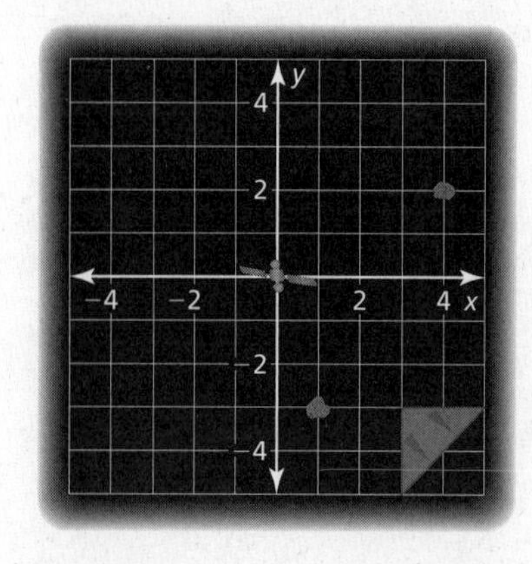

Name______________________________ Date__________

4.2 Reflections

For use with Exploration 4.2

Essential Question How can you reflect a figure in a coordinate plane?

1 EXPLORATION: Reflecting a Triangle Using a Reflective Device

Work with a partner. Use a straightedge to draw any triangle on paper. Label it $\triangle ABC$.

a. Use the straightedge to draw a line that does not pass through the triangle. Label it *m*.

b. Place a reflective device on line *m*.

c. Use the reflective device to plot the images of the vertices of $\triangle ABC$. Label the images of vertices A, B, and C as A', B', and C', respectively.

d. Use a straightedge to draw $\triangle A'B'C'$ by connecting the vertices.

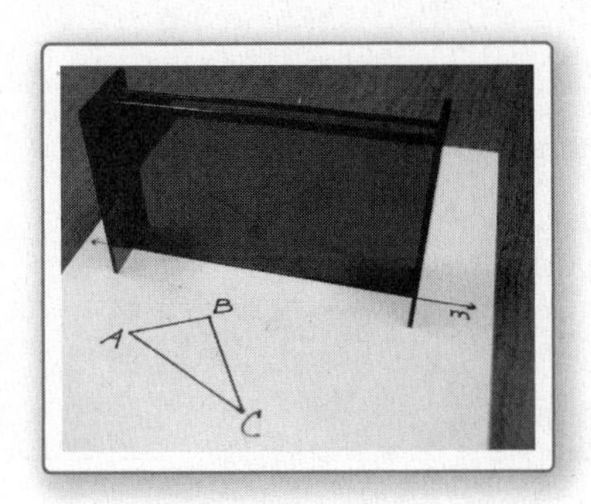

Name ____ Date __________

2 EXPLORATION: Reflecting a Triangle in a Coordinate Plane

Go to *BigIdeasMath.com* for an interactive tool to investigate this exploration.

Work with a partner. Use dynamic geometry software to draw any triangle and label it $\triangle ABC$.

a. *Reflect* $\triangle ABC$ in the y-axis to form $\triangle A'B'C'$.

b. What is the relationship between the coordinates of the vertices of $\triangle ABC$ and those of $\triangle A'B'C'$?

c. What do you observe about the side lengths and angle measures of the two triangles?

d. *Reflect* $\triangle ABC$ in the x-axis to form $\triangle A'B'C'$. Then repeat parts (b) and (c).

Communicate Your Answer

3. How can you reflect a figure in a coordinate plane?

Name____ Date__________

4.2 Notetaking with Vocabulary

For use after Lesson 4.2

In your own words, write the meaning of each vocabulary term.

reflection

line of reflection

glide reflection

line symmetry

line of symmetry

Core Concepts

Reflections

A **reflection** is a transformation that uses a line like a mirror to reflect a figure. The mirror line is called the **line of reflection**.

A reflection in a line m maps every point P in the plane to a point P', so that for each point on of the following properties is true.

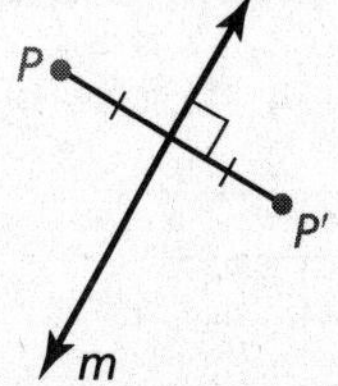

point P not on m

P P' m

point P on m

- If P is not m, then m is the perpendicular bisector of $\overline{PP'}$, or
- If P is on m, then $P = P'$.

Notes:

Core Concepts

Coordinate Rules for Reflections

- If (a, b) is reflected in the x-axis, then its image is the point $(a, -b)$.
- If (a, b) is reflected in the y-axis, then its image is the point $(-a, b)$.
- If (a, b) is reflected in the line $y = x$, then its image is the point (b, a).
- If (a, b) is reflected in the line $y = -x$, then its image is the point $(-b, -a)$.

Notes:

Postulate 4.2 Reflection Postulate

A reflection is a rigid motion.

Extra Practice

In Exercises 1–4, graph $\triangle ABC$ and its image after a reflection in the given line.

1. $A(-1, 5)$, $B(-4, 4)$, $C(-3, 1)$; y-axis

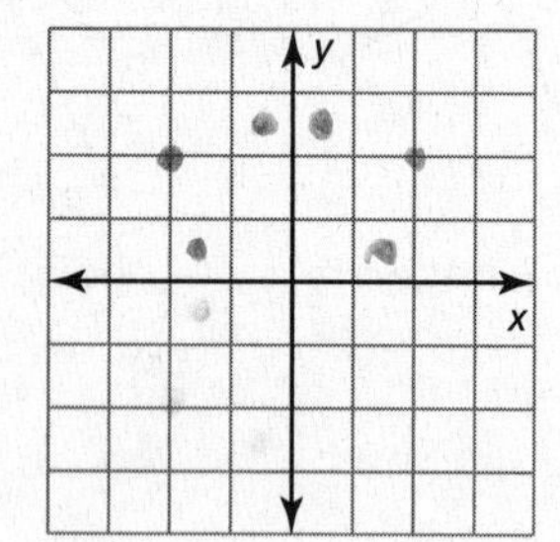

2. $A(0, 2)$, $B(4, 5)$, $C(5, 2)$; x-axis

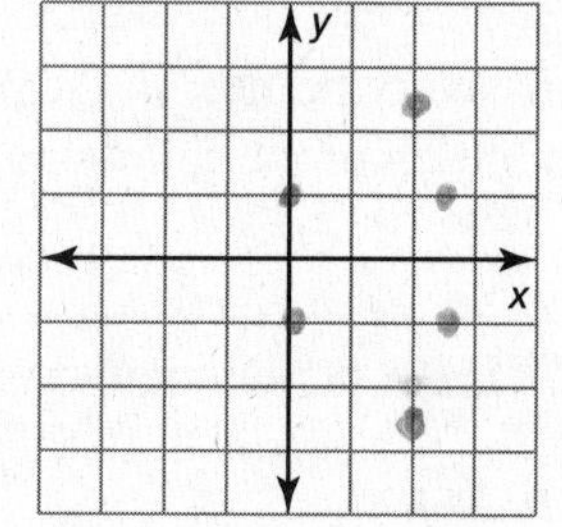

3. $A(2, -1)$, $B(-4, -2)$, $C(-1, -3)$; $y = 1$

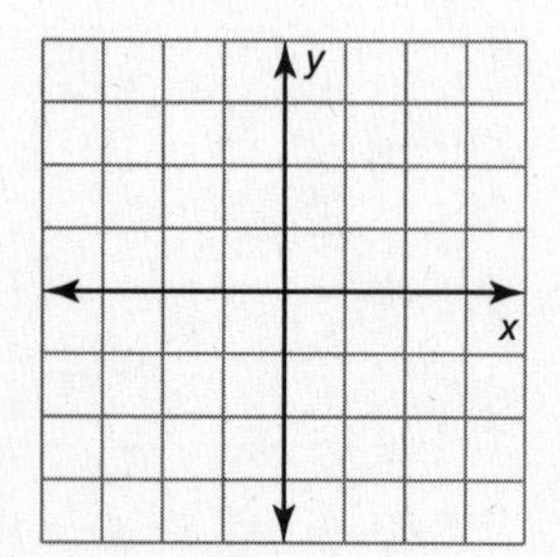

4. $A(-2, 3)$, $B(-2, -2)$, $C(0, -2)$; $x = -3$

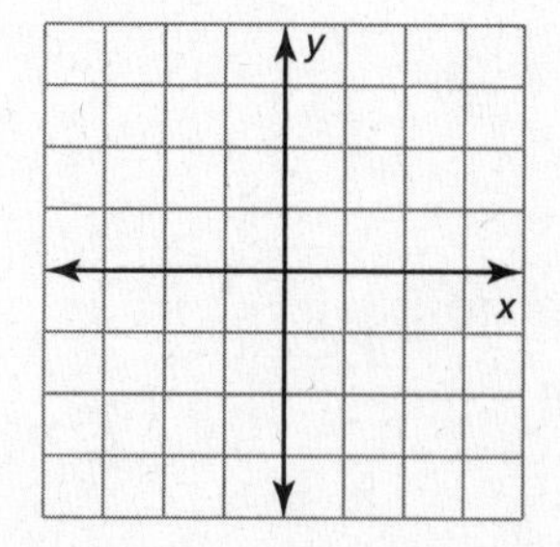

Name____ Date__________

4.2 Notetaking with Vocabulary (continued)

In Exercises 5 and 6, graph the polygon's image after a reflection in the given line.

5. $y = x$

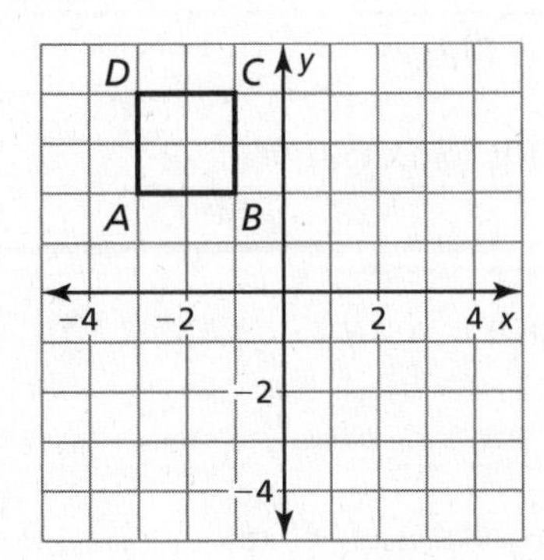

6. $y = -x$

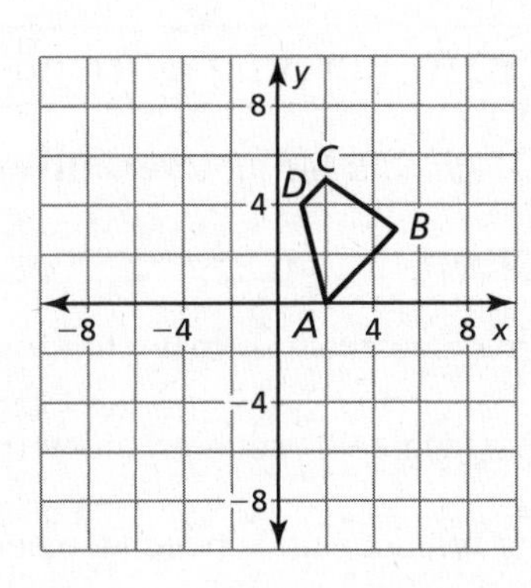

In Exercises 7 and 8, graph $\triangle JKL$ with vertices $J(3, 1)$, $K(4, 2)$, and $L(1, 3)$ and its image after the glide reflection.

7. Translation: $(x, y) \rightarrow (x - 6, y - 1)$

Reflection: in the line $y = -x$

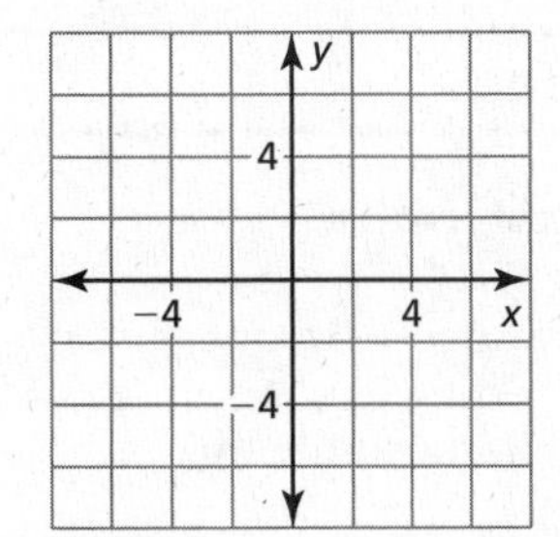

8. Translation: $(x, y) \rightarrow (x, y - 4)$

Reflection: in the line $x = 1$

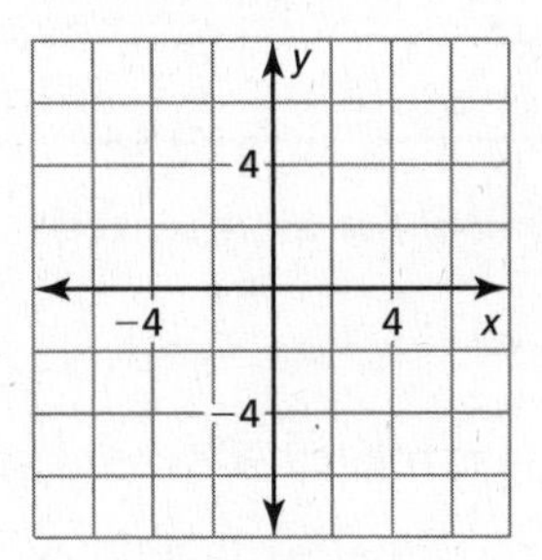

In Exercises 9–12, identify the line symmetry (if any) of the word.

9. MOON **10.** WOW **11.** KID **12.** DOCK

13. You are placing a power strip along wall w that connects to two computers. Where should you place the power strip to minimize the length of the connecting cables?

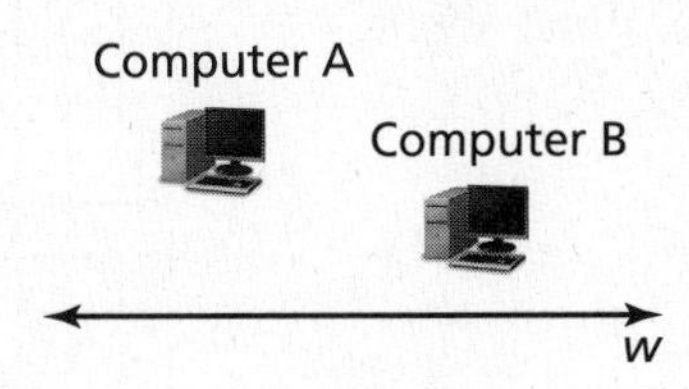

Name ______________________________ Date ________

4.3 Rotations

For use with Exploration 4.3

Essential Question How can you rotate a figure in a coordinate plane?

1 EXPLORATION: Rotating a Triangle in a Coordinate Plane

Go to *BigIdeasMath.com* for an interactive tool to investigate this exploration.

Work with a partner.

a. Use dynamic geometry software to draw any triangle and label it $\triangle ABC$.

b. *Rotate* the triangle 90° counterclockwise about the origin to form $\triangle A'B'C'$.

c. What is the relationship between the coordinates of the vertices of $\triangle ABC$ and those of $\triangle A'B'C'$?

d. What do you observe about the side lengths and angle measures of the two triangles?

2 EXPLORATION: Rotating a Triangle in a Coordinate Plane

Go to *BigIdeasMath.com* for an interactive tool to investigate this exploration.

Work with a partner.

a. The point (x, y) is rotated 90° counterclockwise about the origin. Write a rule to determine the coordinates of the image of (x, y).

b. Use the rule you wrote in part (a) to rotate $\triangle ABC$ 90° counterclockwise about the origin. What are the coordinates of the vertices of the image, $\triangle A'B'C'$?

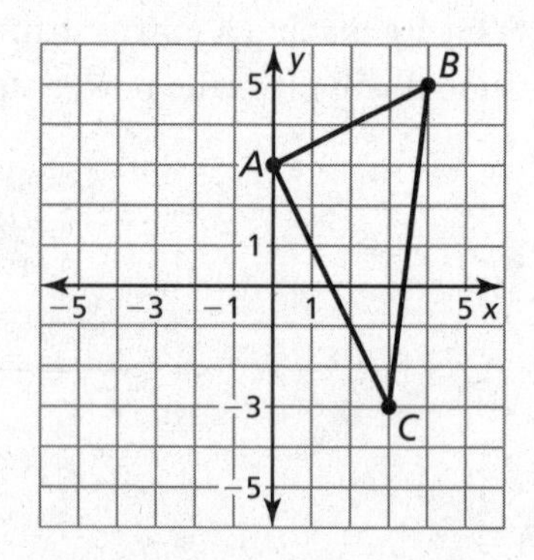

c. Draw $\triangle A'B'C'$. Are its side lengths the same as those of $\triangle ABC$? Justify your answer.

Name______________________________________ Date__________

3 EXPLORATION: Rotating a Triangle in a Coordinate Plane

Work with a partner.

a. The point (x, y) is rotated 180° counterclockwise about the origin. Write a rule to determine the coordinates of the image of (x, y). Explain how you found the rule.

b. Use the rule you wrote in part (a) to rotate $\triangle ABC$ 180° counterclockwise about the origin. What are the coordinates of the vertices of the image, $\triangle A'B'C'$?

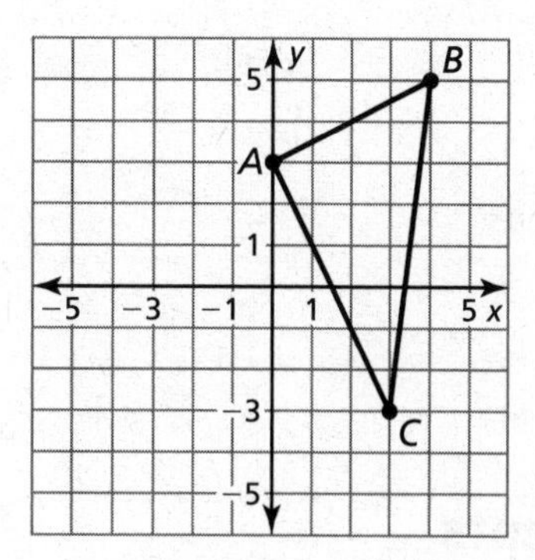

Communicate Your Answer

4. How can you rotate a figure in a coordinate plane?

5. In Exploration 3, rotate $\triangle A'B'C'$ 180° counterclockwise about the origin. What are the coordinates of the vertices of the image, $\triangle A''B''C''$? How are these coordinates related to the coordinates of the vertices of the original triangle, $\triangle ABC$?

Name ______________________________ Date __________

4.3 Notetaking with Vocabulary
For use after Lesson 4.3

In your own words, write the meaning of each vocabulary term.

rotation

center of rotation

angle of rotation

rotational symmetry

center of symmetry

Core Concepts

Rotations

A **rotation** is a transformation is which a figure is turned about a fixed point called the **center of rotation**. Rays drawn from the center of rotation to a point and its image form the **angle of rotation**.

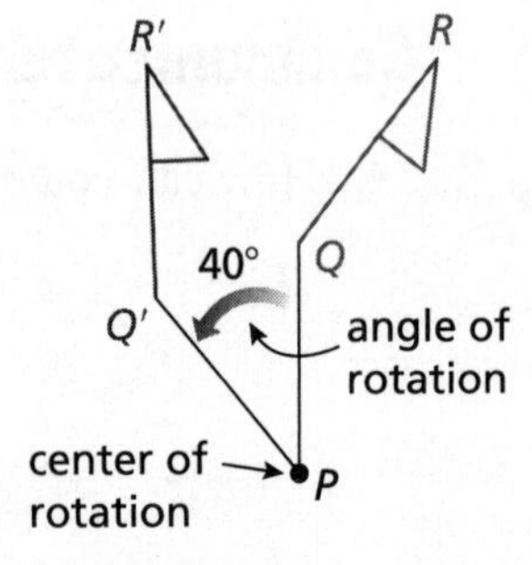

A rotation about a point P through an angle of $x°$ maps every point Q in the plane to a point Q', so that one of the following properties is true.

- If Q is not the center of rotation P, then $QP = Q'P$ and $m\angle QPQ' = x°$, or
- If Q is the center of rotation P, then $Q = Q'$.

Notes:

Name____ Date__________

4.3 Notetaking with Vocabulary (continued)

Coordinate Rules for Rotations about the Origin

When a point (a, b) is rotated counterclockwise about the origin, the following are true.

- For a rotation of 90°, $(a, b) \rightarrow (-b, a)$.
- For a rotation of 180°, $(a, b) \rightarrow (-a, -b)$.
- For a rotation of 270°, $(a, b) \rightarrow (b, -a)$.

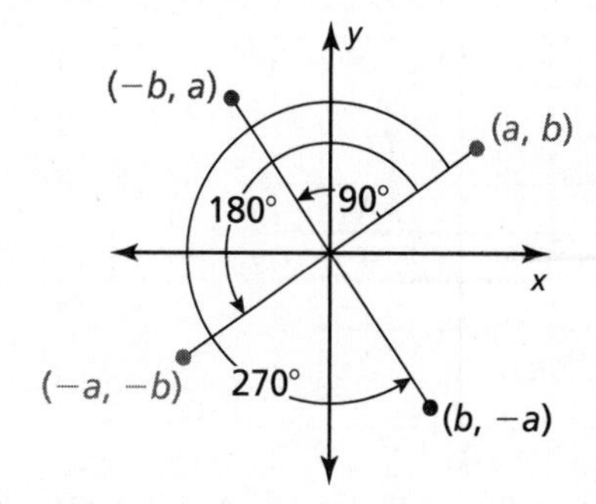

Notes:

Postulate 4.3 Rotation Postulate

A rotation is a rigid motion.

Extra Practice

In Exercises 1–3, graph the image of the polygon after a rotation of the given number of degrees about the origin.

1. 180°

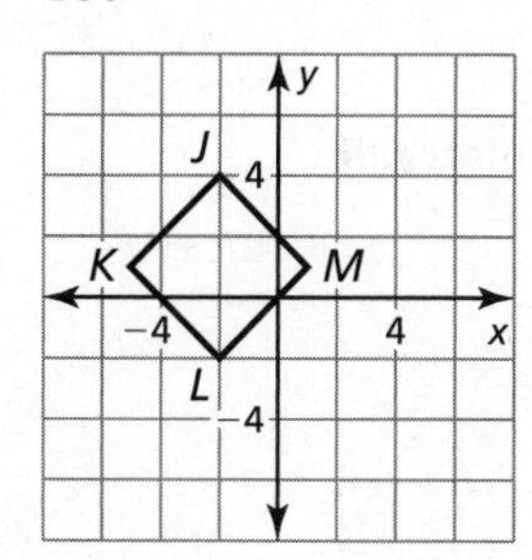

2. 90°

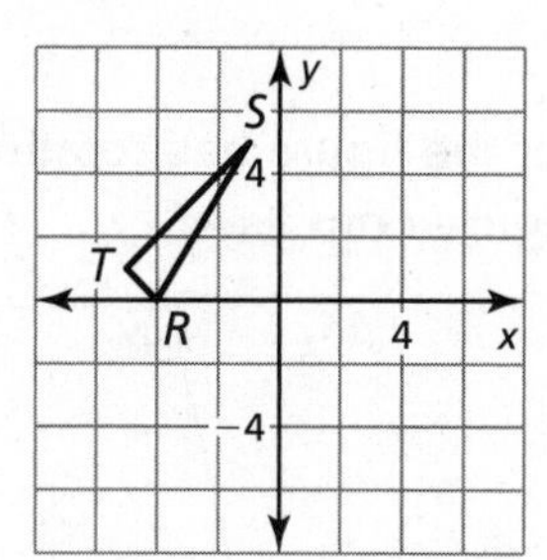

3. 270°

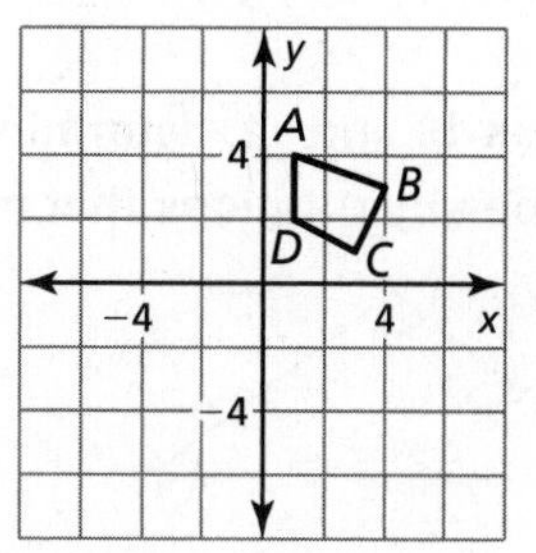

In Exercises 4–7, graph the image of $\overline{MN}$ after the composition.

4. Reflection: x-axis

Rotation: 180° about the origin

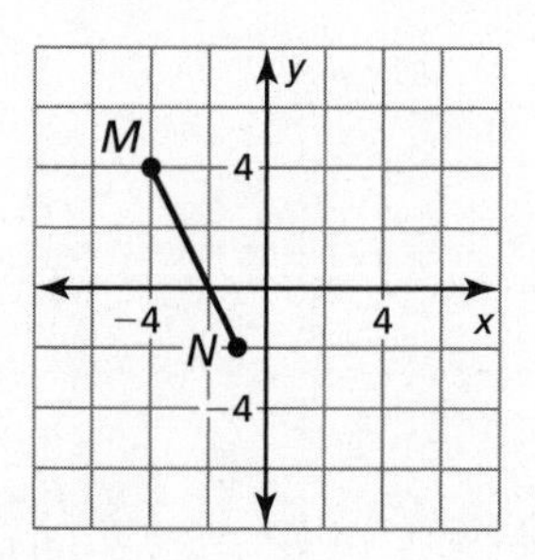

5. Rotation: 90° about the origin

Translation: $(x, y) \rightarrow (x + 2, y - 3)$

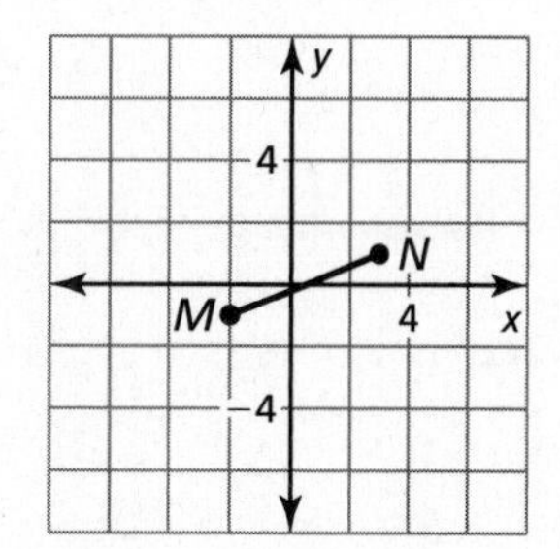

4.3 Notetaking with Vocabulary (continued)

6. Rotation: 270° about the origin

Reflection: in the line $y = x$

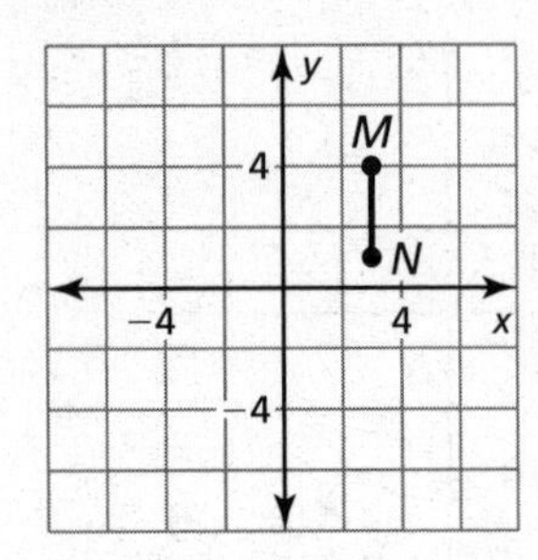

7. Rotation: 90° about the origin

Translation: $(x, y) \rightarrow (x - 5, y)$

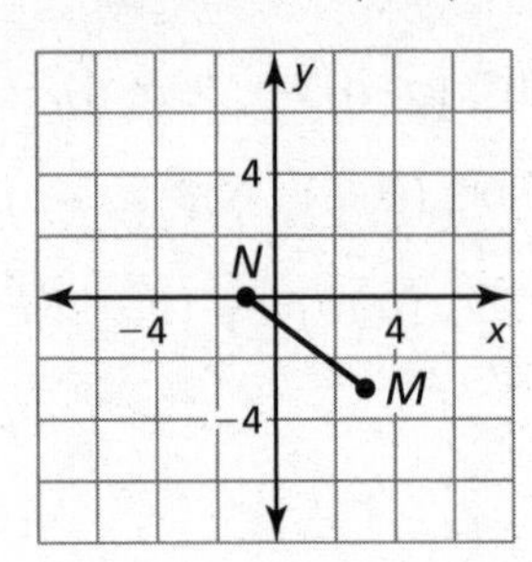

In Exercises 8 and 9, graph △*JKL* with vertices *J*(2, 3), *K*(1, −1), and *L*(−1, 0) and its image after the composition.

8. Rotation: 180° about the origin

Reflection: $x = 2$

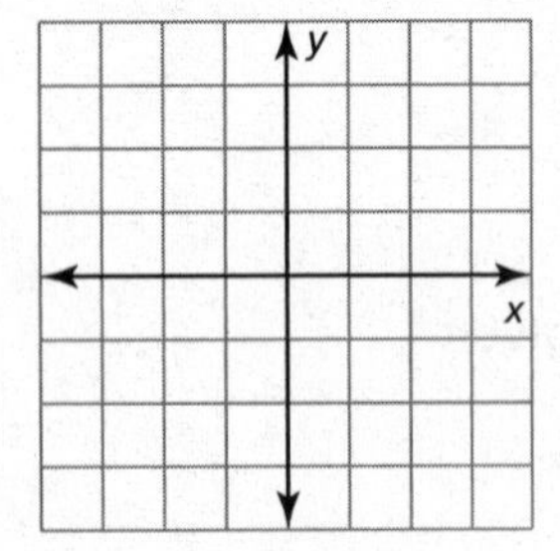

9. Translation: $(x, y) \rightarrow (x - 4, y - 4)$

Rotation: 270° about the origin

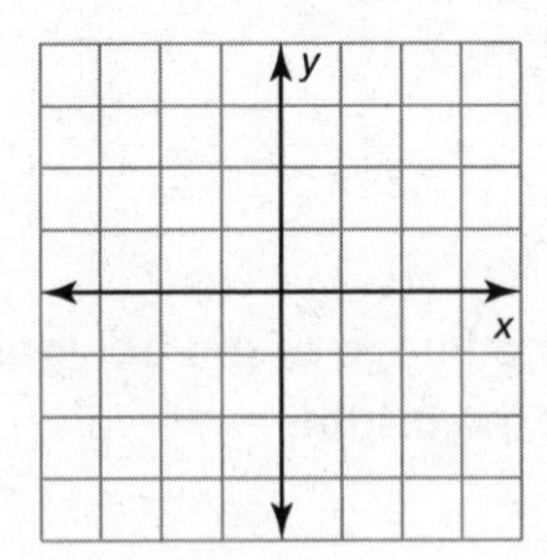

In Exercises 10 and 11, determine whether the figure has rotational symmetry. If so, describe any rotations that map the figure onto itself.

10.

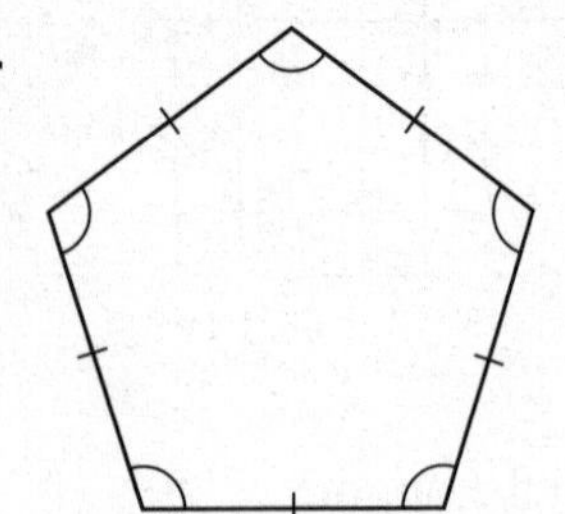

11.

Name______________________________ Date__________

4.4 Congruence and Transformations

For use with Exploration 4.4

Essential Question What conjectures can you make about a figure reflected in two lines?

1 EXPLORATION: Reflections in Parallel Lines

Go to *BigIdeasMath.com* for an interactive tool to investigate this exploration.

Work with a partner. Use dynamic geometry software to draw any scalene triangle and label it $\triangle ABC$.

a. Draw any line $\overleftrightarrow{DE}$. Reflect $\triangle ABC$ in $\overleftrightarrow{DE}$ to form $\triangle A'B'C'$.

b. Draw a line parallel to $\overleftrightarrow{DE}$. Reflect $\triangle A'B'C'$ in the new line to form $\triangle A''B''C''$.

c. Draw the line through point A that is perpendicular to $\overleftrightarrow{DE}$. What do you notice?

d. Find the distance between points A and A''. Find the distance between the two parallel lines. What do you notice?

e. Hide $\triangle A'B'C'$. Is there a single transformation that maps $\triangle ABC$ to $\triangle A''B''C''$. Explain.

f. Make conjectures based on your answers in parts (c)–(e). Test your conjectures by changing $\triangle ABC$ and the parallel lines.

2 EXPLORATION: Reflections in Intersecting Lines

Go to *BigIdeasMath.com* for an interactive tool to investigate this exploration.

Work with a partner. Use dynamic geometry software to draw any scalene triangle and label it $\triangle ABC$.

a. Draw any line $\overleftrightarrow{DE}$. Reflect $\triangle ABC$ in $\overleftrightarrow{DE}$ to form $\triangle A'B'C'$.

b. Draw any line $\overleftrightarrow{DF}$ so that $\angle EDF$ is less than or equal to 90°. Reflect $\triangle A'B'C'$ in $\overleftrightarrow{DF}$ to form $\triangle A''B''C''$.

c. Find the measure of $\angle EDF$. Rotate $\triangle ABC$ counterclockwise about point D twice using the measure of $\angle EDF$.

d. Make a conjecture about a figure reflected in two intersecting lines. Test your conjecture by changing $\triangle ABC$ and the lines.

Communicate Your Answer

3. What conjectures can you make about a figure reflected in two lines?

4. Point Q is reflected in two parallel lines, $\overleftrightarrow{GH}$ and $\overleftrightarrow{JK}$, to form Q' and Q''. The distance from $\overleftrightarrow{GH}$ to $\overleftrightarrow{JK}$ is 3.2 inches. What is the distance QQ''?

Name____ Date__________

4.4 Notetaking with Vocabulary
For use after Lesson 4.4

In your own words, write the meaning of each vocabulary term.

congruent figures

congruence transformation

Theorems

Theorem 4.2 Reflections in Parallel Lines Theorem

If lines k and m are parallel, then a reflection in line k followed by a reflection in line m is the same as a translation.

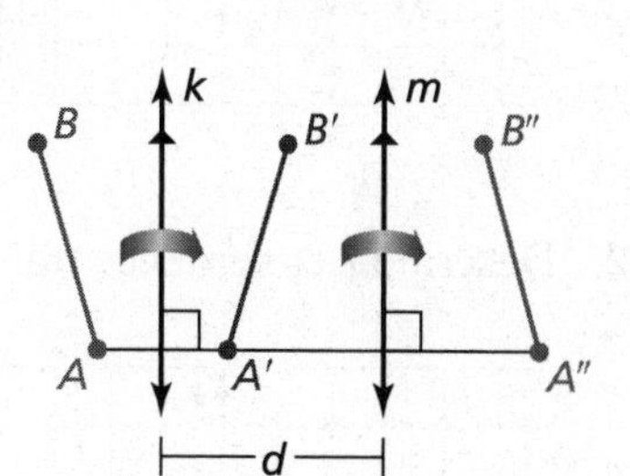

If A'' is the image of A, then

1. AA'' is perpendicular to k and m, and
2. $AA'' = 2d$, where d is the distance between k and m.

Proof Ex. 31. p. 206

Notes:

Theorem 4.3 Reflections in Intersecting Lines Theorem

If lines k and m intersect at point P, then a reflection in line k followed by a reflection in line m is the same as a rotation about point P.

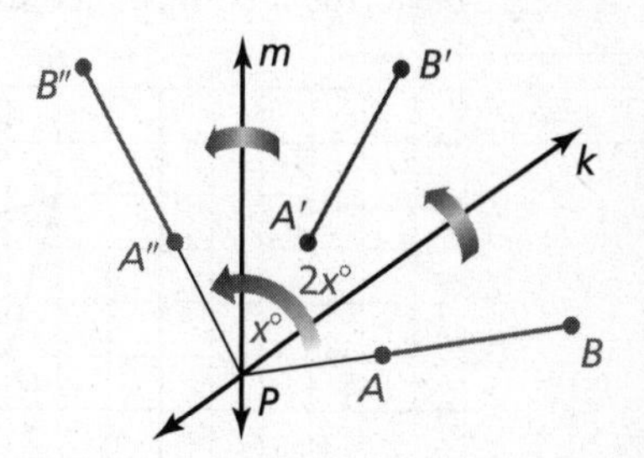

The angle of rotation is $2x°$, where $x°$ is the measure of the acute or right angle formed by lines k and m.

Proof Ex. 31. p. 206

Notes:

Name ______________________________ Date __________

Extra Practice

1. Identify any congruent figures in the coordinate plane. Explain.

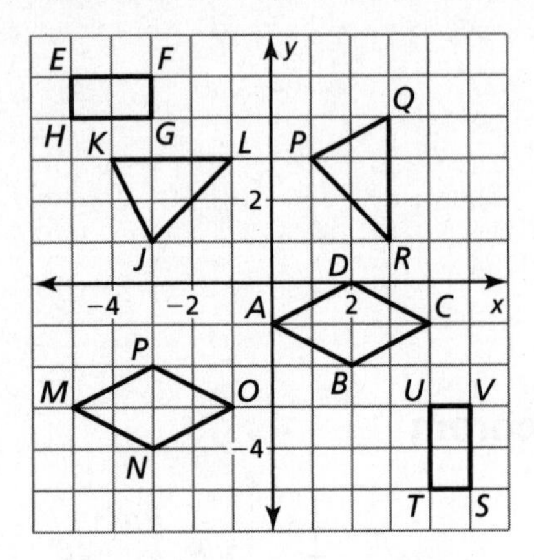

2. Describe a congruence transformation that maps $\triangle PQR$ to $\triangle STU$.

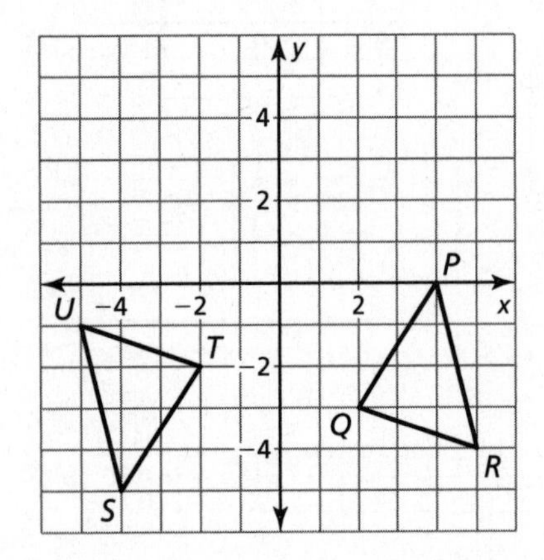

3. Describe a congruence transformation that maps polygon *ABCD* to polygon *EFGH*.

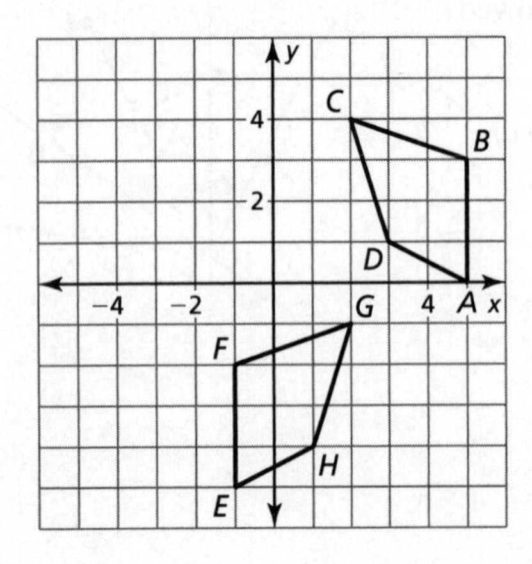

In Exercises 4 and 5, determine whether the polygons with the given vertices are congruent. Use transformations to explain your reasoning.

4. $A(2, 2)$, $B(3, 1)$, $C(1, 1)$ and $D(2, -2)$, $E(3, -1)$, $F(1, -1)$

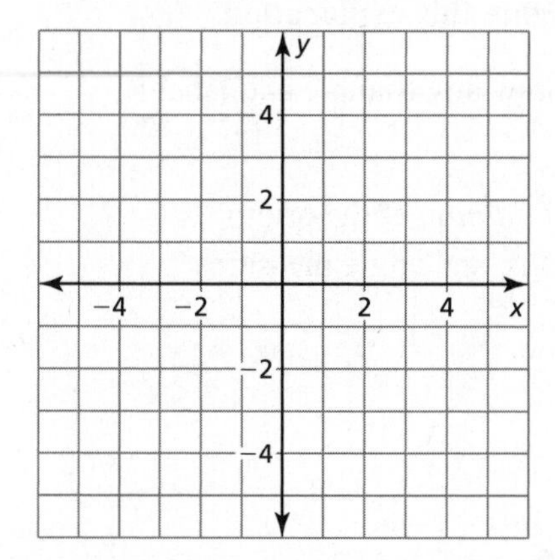

5. $G(3, 3)$, $H(2, 1)$, $I(6, 2)$, $J(6, 3)$ and $K(2, -1)$, $L(-3, -3)$, $M(2, -2)$, $N(2, -1)$

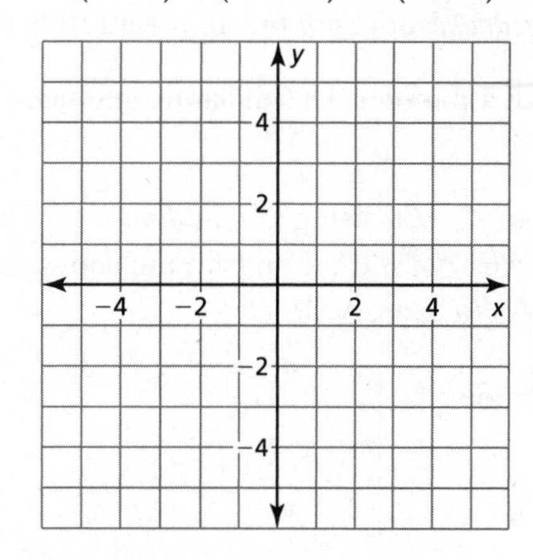

In Exercises 6–9, *k* || *m*, $\overline{UV}$ is reflected in line *k*, and $\overline{U'V'}$ is reflected in line *m*.

6. A translation maps $\overline{UV}$ onto which segment?

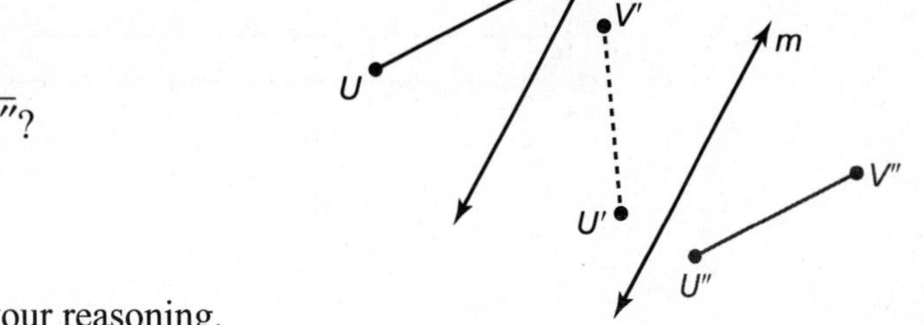

7. Which lines are perpendicular to $\overline{UU''}$?

8. Why is V'' the image of V? Explain your reasoning.

9. If the distance between k and m is 5 inches, what is the length of $\overline{VV''}$?

10. What is the angle of rotation that maps A onto A''?

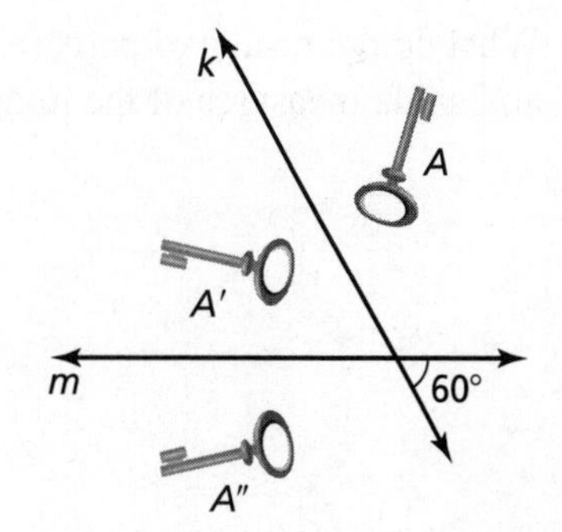

Name ________________________________ Date __________

4.5 Dilations
For use with Exploration 4.5

Essential Question What does it mean to dilate a figure?

1 EXPLORATION: Dilating a Triangle in a Coordinate Plane

Go to *BigIdeasMath.com* for an interactive tool to investigate this exploration.

Work with a partner. Use dynamic geometry software to draw any triangle and label it $\triangle ABC$.

a. *Dilate* $\triangle ABC$ using a *scale factor* of 2 and a *center of dilation* at the origin to form $\triangle A'B'C'$. Compare the coordinates, side lengths, and angle measures of $\triangle ABC$ and $\triangle A'B'C'$.

Sample

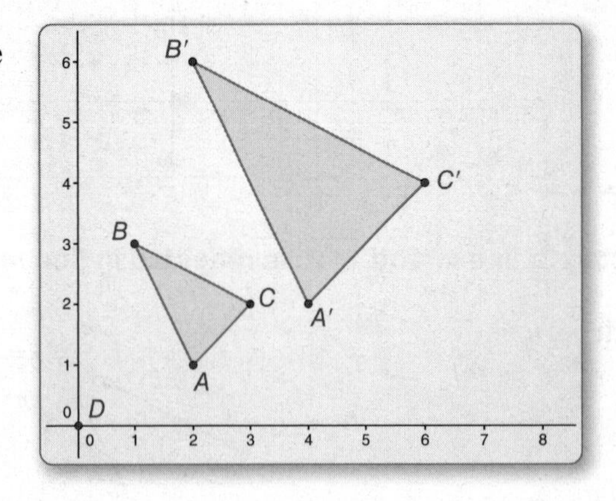

b. Repeat part (a) using a *scale factor* of $\frac{1}{2}$.

c. What do the results of parts (a) and (b) suggest about the coordinates, side lengths, and angle measures of the image of $\triangle ABC$ after a dilation with a scale factor of k?

2 EXPLORATION: Dilating Lines in a Coordinate Plane

Go to *BigIdeasMath.com* for an interactive tool to investigate this exploration.

Work with a partner. Use dynamic geometry software to draw $\overleftrightarrow{AB}$ that passes through the origin and $\overleftrightarrow{AC}$ that does not pass through the origin.

a. *Dilate* $\overleftrightarrow{AB}$ using a *scale factor* of 3 and a *center of dilation* at the origin. Describe the image.

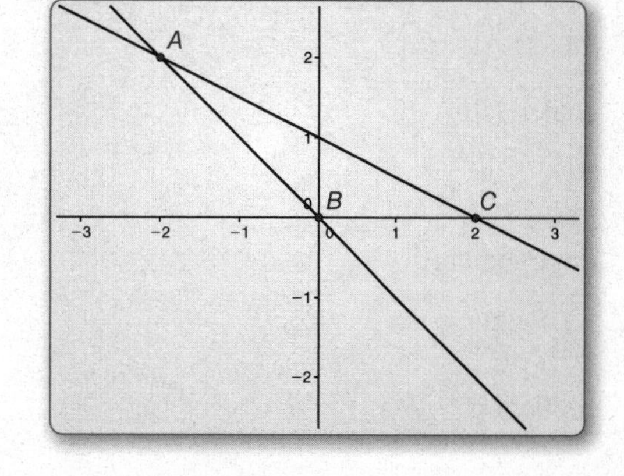

b. *Dilate* $\overleftrightarrow{AC}$ using a *scale factor* of 3 and a *center of dilation* at the origin. Describe the image.

c. Repeat parts (a) and (b) using a scale factor of $\frac{1}{4}$.

d. What do you notice about dilations of lines passing through the center of dilation and dilations of lines not passing through the center of dilation?

Communicate Your Answer

3. What does it mean to dilate a figure?

4. Repeat Exploration 1 using a center of dilation at a point other than the origin.

Name ______________________________ Date __________

4.5 Notetaking with Vocabulary
For use after Lesson 4.5

In your own words, write the meaning of each vocabulary term.

dilation

center of dilation

scale factor

enlargement

reduction

Core Concepts

Dilations

A **dilation** is a transformation in which a figure is enlarged or reduced with respect to a fixed point C called the **center of dilation** and a **scale factor** k, which is the ratio of the lengths of the corresponding sides of the image and the preimage.

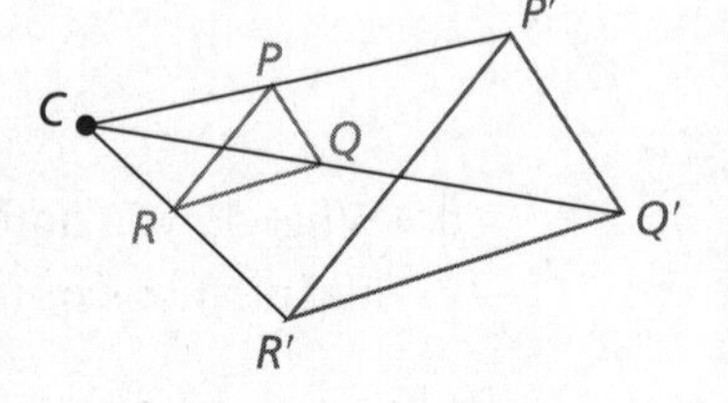

A dilation with center of dilation C and scale factor k maps every point P in a figure to a point P' so that the following are true.

- If P is the center point C, then $P = P'$.
- If P is not the center point C, then the image point P' lies on $\overrightarrow{CP}$. The scale factor k is a positive number such that $k = \dfrac{CP'}{CP}$.
- Angle measures are preserved.

Notes:

Name____ Date__________

4.5 Notetaking with Vocabulary (continued)

Coordinate Rule for Dilations

If $P(x, y)$ is the preimage of a point, then its image after a dilation centered at the origin $(0, 0)$ with scale factor k is the point $P'(kx, ky)$.

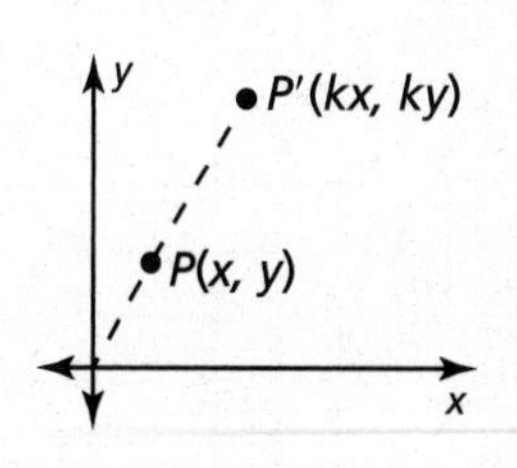

Notes:

Extra Practice

In Exercises 1–3, find the scale factor of the dilation. Then tell whether the dilation is a *reduction* or an *enlargement*.

1.

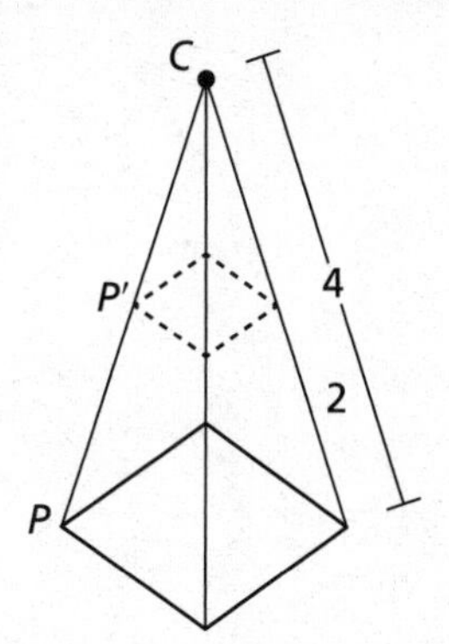

2.

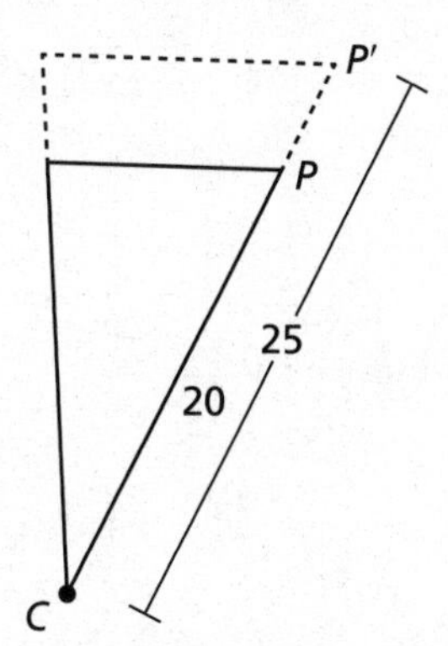

3.

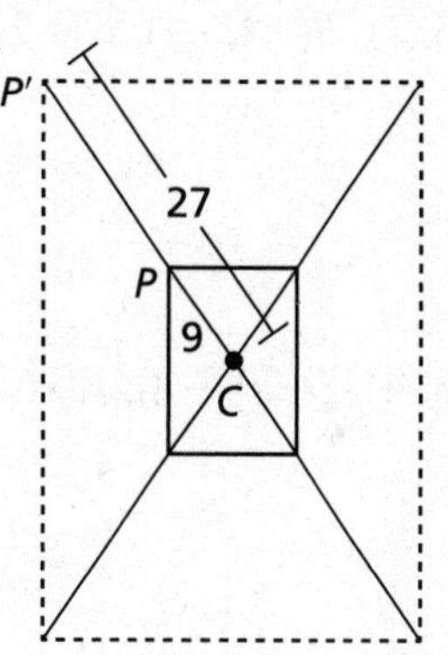

In Exercises 4 and 5, graph the polygon and its image after a dilation with scale factor *k*.

4. $A(-3, 1), B(-4, -1), C(-2, -1); k = 2$

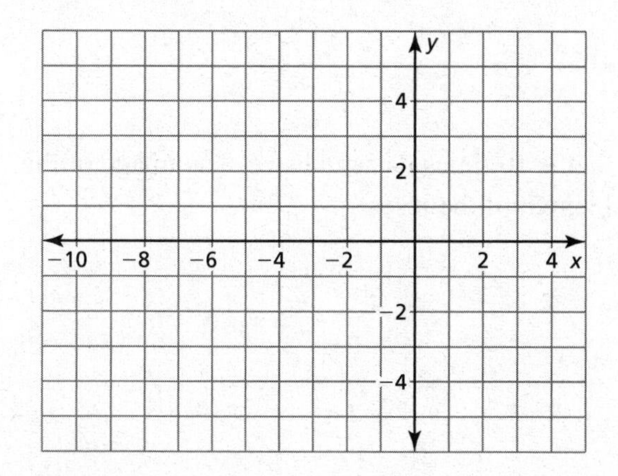

4.5 Notetaking with Vocabulary (continued)

5. $P(-10, 0), Q(-5, 0), R(0, 5), S(-5, 5); k = \frac{1}{5}$

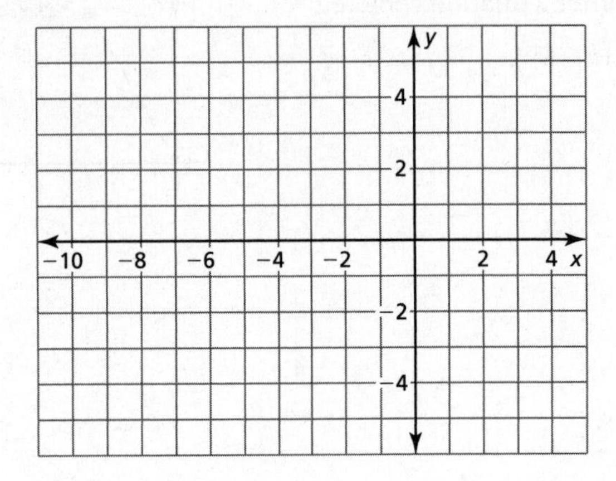

In Exercises 6 and 7, find the coordinates of the image of the polygon after a dilation with scale factor *k*.

6. $A(-3, 1), B(-4, -1), C(-2, -1); k = -6$

7. $P(-8, 4), Q(20, -8), R(16, 4), S(0, 12); k = -0.25$

8. You design a poster on an 8.5-inch by 11-inch paper for a contest at your school. The poster of the winner will be printed on a 34-inch by 44-inch canvas to be displayed. What is the scale factor of this dilation?

9. A biology book shows the image of an insect that is 10 times its actual size. The image of the insect is 8 centimeters long. What is the actual length of the insect?

Name____ Date__________

4.6 Similarity and Transformations

For use with Exploration 4.6

Essential Question When a figure is translated, reflected, rotated, or dilated in the plane, is the image always similar to the original figure?

1 EXPLORATION: Dilations and Similarity

Go to *BigIdeasMath.com* for an interactive tool to investigate this exploration.

Work with a partner.

a. Use dynamic geometry software to draw any triangle and label it $\triangle ABC$.

b. Dilate the triangle using a scale factor of 3. Is the image similar to the original triangle? Justify your answer.

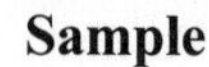

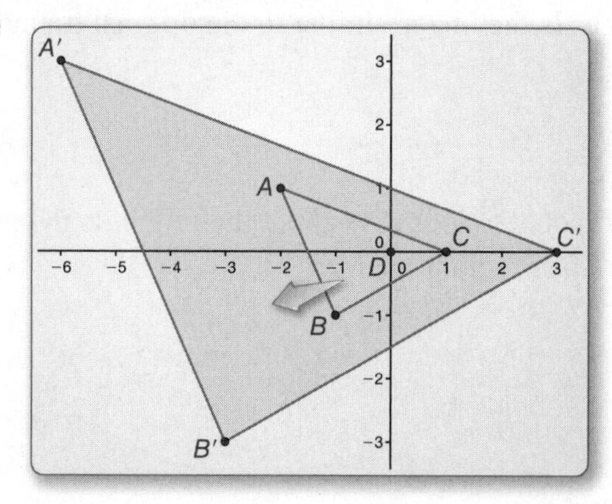

2 EXPLORATION: Rigid Motions and Similarity

Go to *BigIdeasMath.com* for an interactive tool to investigate this exploration.

Work with a partner.

a. Use dynamic geometry software to draw any triangle.

b. Copy the triangle and translate it 3 units left and 4 units up. Is the image similar to the original triangle? Justify your answer.

c. Reflect the triangle in the y-axis. Is the image similar to the original triangle? Justify your answer.

d. Rotate the original triangle 90° counterclockwise about the origin. Is the image similar to the original triangle? Justify your answer.

Communicate Your Answer

3. When a figure is translated, reflected, rotated, or dilated in the plane, is the image always similar to the original figure? Explain your reasoning.

4. A figure undergoes a composition of transformations, which includes translations, reflections, rotations, and dilations. Is the image similar to the original figure? Explain your reasoning.

Name____ Date__________

4.6 Notetaking with Vocabulary
For use after Lesson 4.6

In your own words, write the meaning of each vocabulary term.

similarity transformation

similar figures

Notes:

Name ______________________________ Date __________

Extra Practice

In Exercises 1–3, graph the polygon with the given vertices and its image after the similarity transformation.

1. $A(3, 6)$, $B(2, 5)$, $C(4, 3)$, $D(5, 5)$

   Translation: $(x, y) \rightarrow (x - 5, y - 3)$

   Dilation: $(x, y) \rightarrow (3x, 3y)$

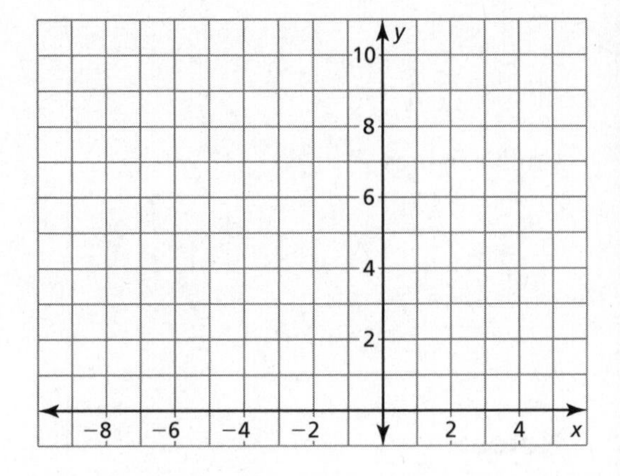

2. $R(12, 8)$, $S(8, 0)$, $T(0, 4)$

   Dilation: $(x, y) \rightarrow \left(\frac{1}{4}x, \frac{1}{4}y\right)$

   Reflection: in the y-axis

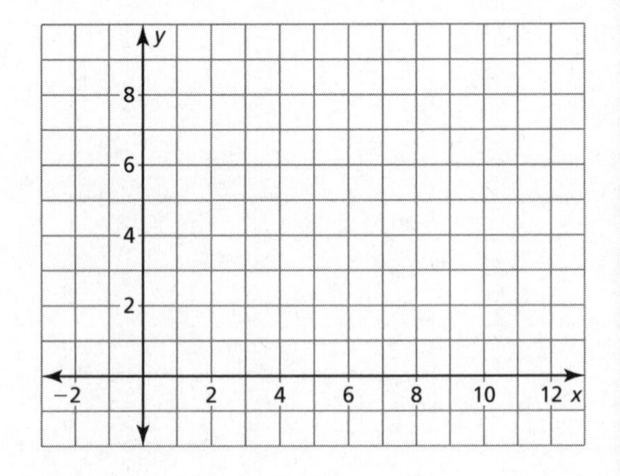

3. $X(9, 6)$, $Y(3, 3)$, $Z(3, 6)$

   Rotation: 90° about the origin

   Dilation: $(x, y) \rightarrow \left(\frac{2}{3}x, \frac{2}{3}y\right)$

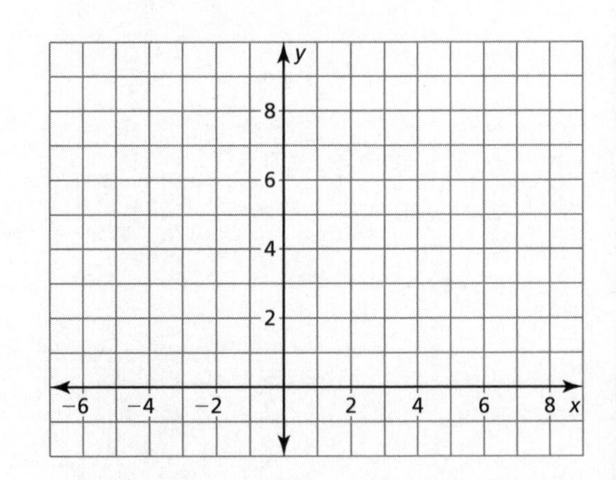

In Exercises 4–6, describe the similarity transformation that maps the preimage to the image.

4.

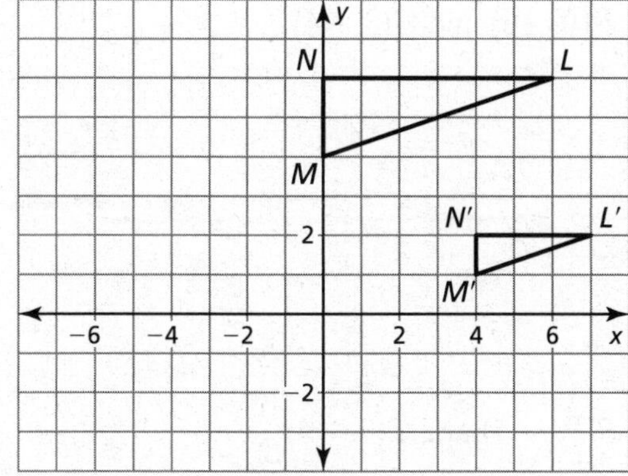

5.

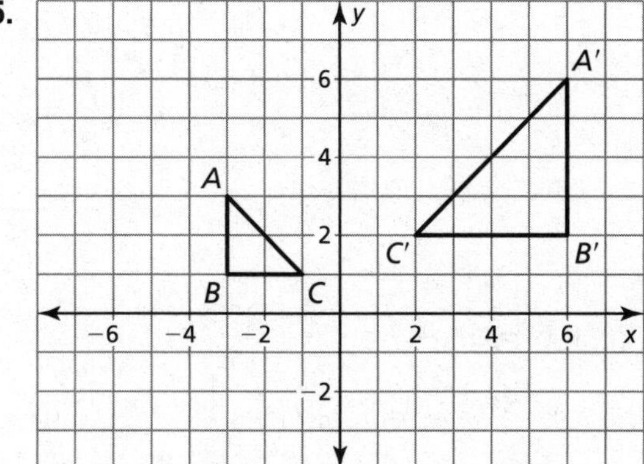

6.

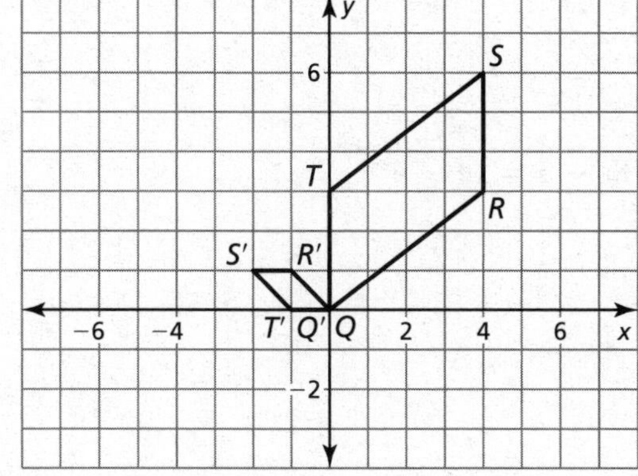

Name ______________________________ Date __________

Chapter 5 Maintaining Mathematical Proficiency

Find the coordinates of the midpoint *M* of the segment with the given endpoints. Then find the distance between the two points.

1. $A(3, 1)$ and $B(5, 5)$

2. $F(0, -6)$ and $G(8, -4)$

3. $P(-2, -7)$ and $B(-4, 5)$

4. $S(10, -5)$ and $T(7, -9)$

Solve the equation.

5. $9x - 6 = 7x$

6. $2r + 6 = 5r - 9$

7. $20 - 3n = 2n + 30$

8. $8t - 5 = 6t - 4$

Name____ Date__________

5.1 Angles of Triangles

For use with Exploration 5.1

Essential Question How are the angle measures of a triangle related?

1 EXPLORATION: Writing a Conjecture

Go to *BigIdeasMath.com* for an interactive tool to investigate this exploration.

Work with a partner.

a. Use dynamic geometry software to draw any triangle and label it $\triangle ABC$.

b. Find the measures of the interior angles of the triangle.

c. Find the sum of the interior angle measures.

d. Repeat parts (a)–(c) with several other triangles. Then write a conjecture about the sum of the measures of the interior angles of a triangle.

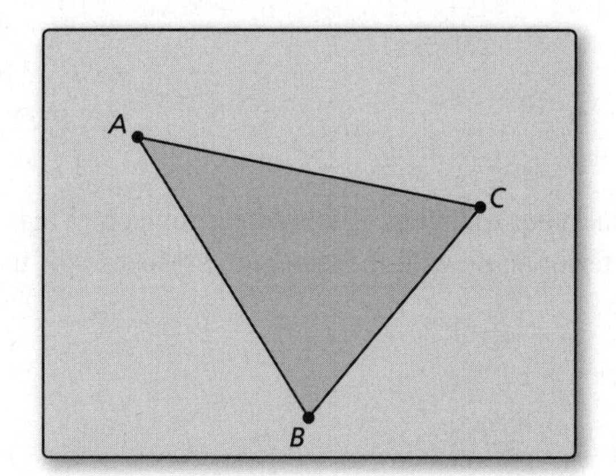

Sample

Angles

$m\angle A = 43.67°$

$m\angle B = 81.87°$

$m\angle C = 54.46°$

Name ____________________ Date ________

2 EXPLORATION: Writing a Conjecture

Go to *BigIdeasMath.com* for an interactive tool to investigate this exploration.

Work with a partner.

a. Use dynamic geometry software to draw any triangle and label it $\triangle ABC$.

b. Draw an exterior angle at any vertex and find its measure.

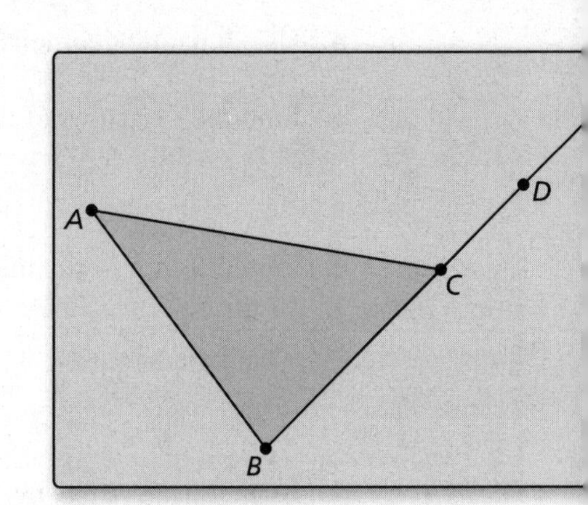

c. Find the measures of the two nonadjacent interior angles of the triangle.

d. Find the sum of the measures of the two nonadjacent interior angles. Compare this sum to the measure of the exterior angle.

Sample
Angles
$m\angle A = 43.67°$
$m\angle B = 81.87°$
$m\angle ACD = 125.54°$

e. Repeat parts (a)–(d) with several other triangles. Then write a conjecture that compares the measure of an exterior angle with the sum of the measures of the two nonadjacent interior angles.

Communicate Your Answer

3. How are the angle measures of a triangle related?

4. An exterior angle of a triangle measures 32°. What do you know about the measures of the interior angles? Explain your reasoning.

Name____ Date__________

5.1 Notetaking with Vocabulary
For use after Lesson 5.1

In your own words, write the meaning of each vocabulary term.

interior angles

exterior angles

corollary to a theorem

Core Concepts

Classifying Triangles by Sides

Scalene Triangle

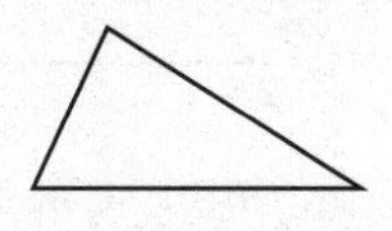

no congruent sides

Isosceles Triangle

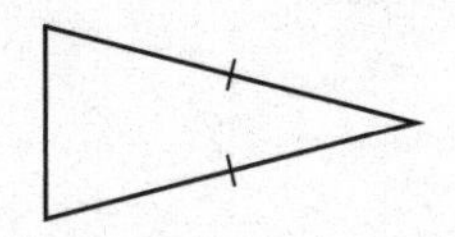

at least 2 congruent sides

Equilateral Triangle

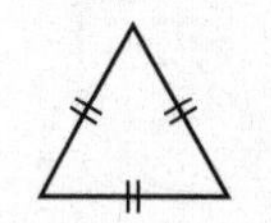

3 congruent sides

Classifying Triangles by Angles

Acute Triangle

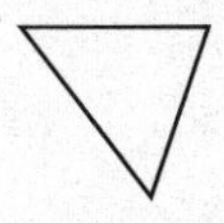

3 acute angles

Right Triangle

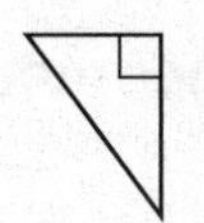

1 right angle

Obtuse Triangle

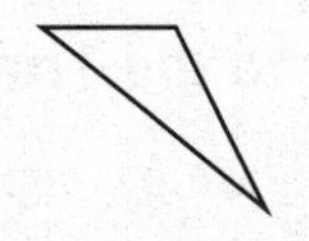

1 obtuse angle

Equiangular Triangle

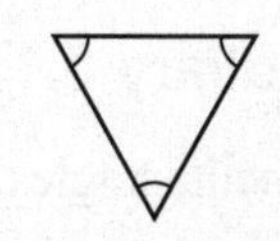

3 congruent angles

Notes:

Name ______________________________ Date __________

5.1 Notetaking with Vocabulary (continued)

Theorems

Theorem 5.1 Triangle Sum Theorem

The sum of the measures of the interior angles of a triangle is 180°.

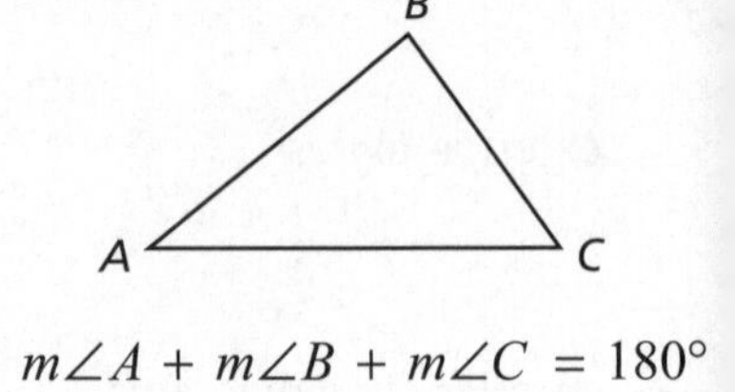

$m\angle A + m\angle B + m\angle C = 180°$

Notes:

Theorem 5.2 Exterior Angle Theorem

The measure of an exterior angle of a triangle is equal to the sum of the measures of the two nonadjacent interior angles.

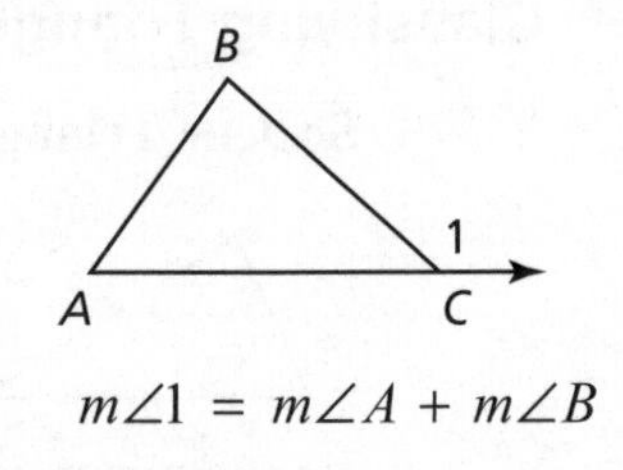

$m\angle 1 = m\angle A + m\angle B$

Notes:

Corollary 5.1 Corollary to the Triangle Sum Theorem

The acute angles of a right triangle are complementary.

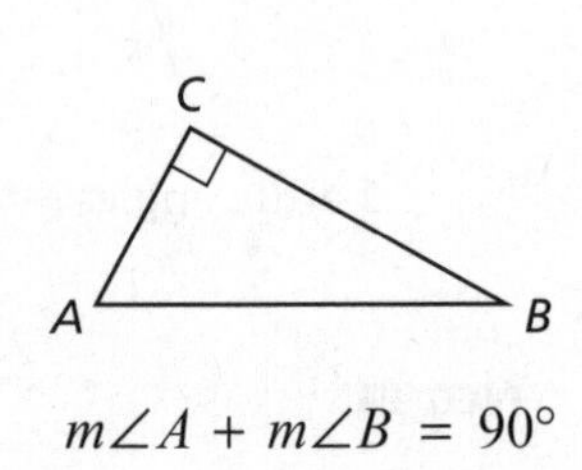

$m\angle A + m\angle B = 90°$

Notes:

Name____ Date__________

5.1 **Notetaking with Vocabulary (continued)**

Extra Practice

In Exercises 1–3, classify the triangle by its sides and by measuring its angles.

1.

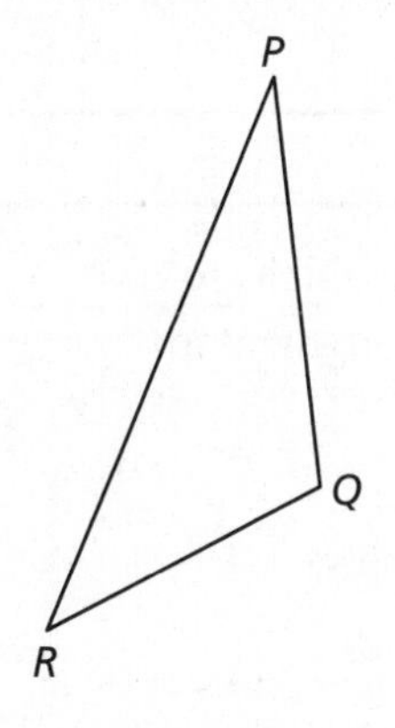

2.

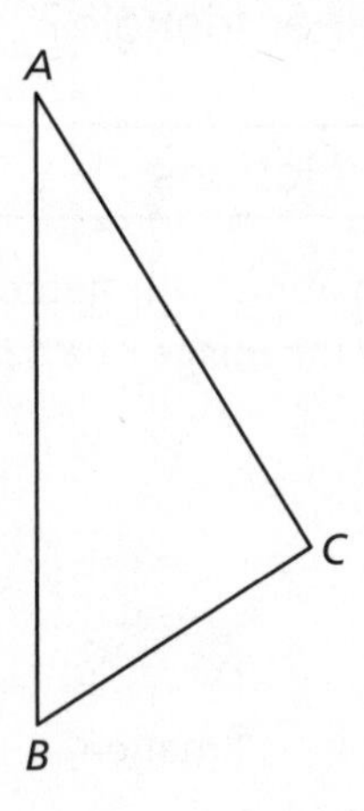

3.

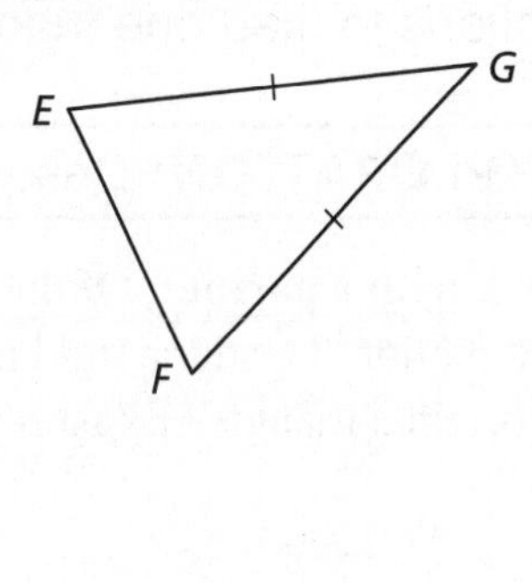

4. Classify $\triangle ABC$ by its sides. Then determine whether it is a right triangle.
$A(6, 6)$, $B(9, 3)$, $C(2, 2)$

In Exercises 5 and 6, find the measure of the exterior angle.

5.

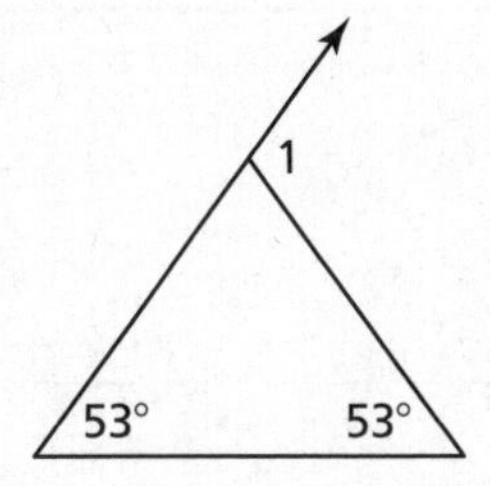

6.

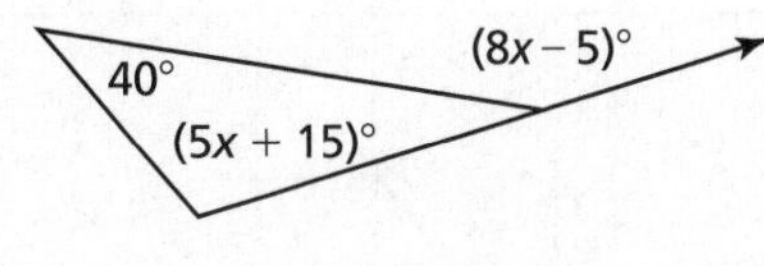

7. In a right triangle, the measure of one acute angle is twice the sum of the measure of the other acute angle and 30. Find the measure of each acute angle in the right triangle.

Name ____ Date __________

5.2 Congruent Polygons

For use with Exploration 5.2

Essential Question Given two congruent triangles, how can you use rigid motions to map one triangle to the other triangle?

1 EXPLORATION: Describing Rigid Motions

Work with a partner. Of the four transformations you studied in Chapter 4, which are rigid motions? Under a rigid motion, why is the image of a triangle always congruent to the original triangle? Explain you reasoning.

Translation

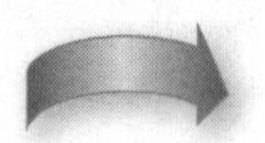
Reflection

Rotation

Dilation

2 EXPLORATION: Finding a Composition of Rigid Motions

Go to *BigIdeasMath.com* for an interactive tool to investigate this exploration.

Work with a partner. Describe a composition of rigid motions that maps $\triangle ABC$ to $\triangle DEF$. Use dynamic geometry software to verify your answer.

a. $\triangle ABC \cong \triangle DEF$

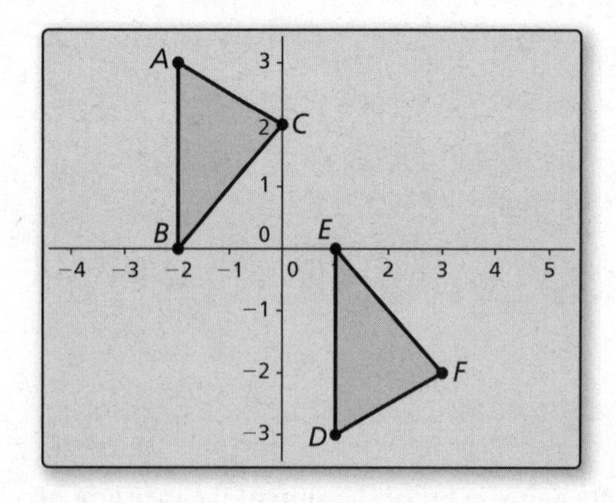

b. $\triangle ABC \cong \triangle DEF$

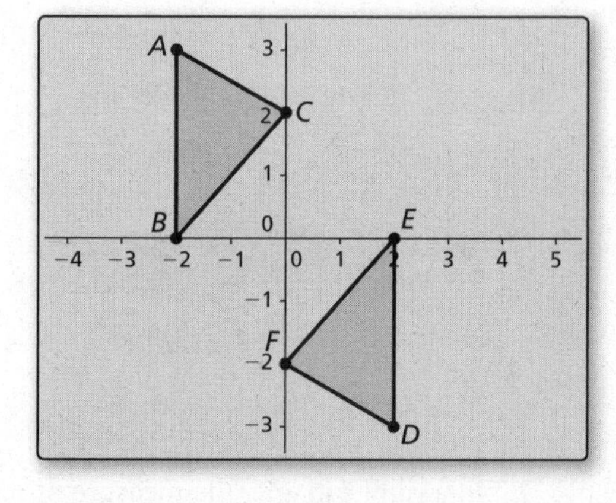

2 EXPLORATION: Finding a Composition of Rigid Motions (continued)

c. $\triangle ABC \cong \triangle DEF$

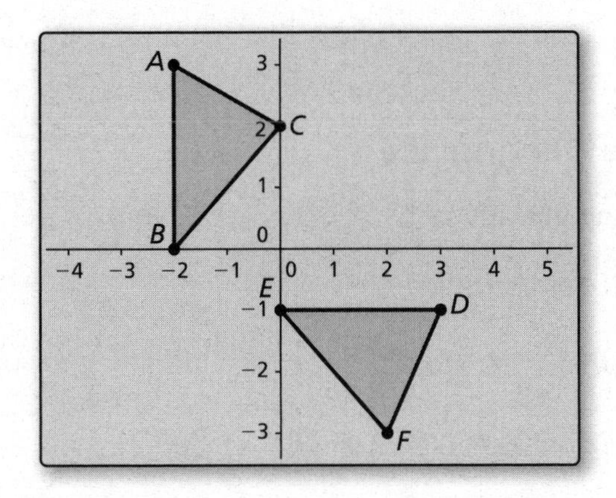

d. $\triangle ABC \cong \triangle DEF$

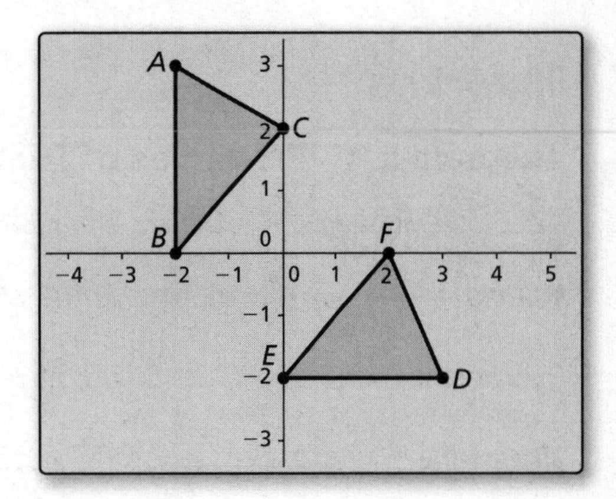

Communicate Your Answer

3. Given two congruent triangles, how can you use rigid motions to map one triangle to the other triangle?

4. The vertices of $\triangle ABC$ are $A(1, 1)$, $B(3, 2)$, and $C(4, 4)$. The vertices of $\triangle DEF$ are $D(2, -1)$, $E(0, 0)$, and $F(-1, 2)$. Describe a composition of rigid motions that maps $\triangle ABC$ to $\triangle DEF$.

Name ______________________________ Date __________

5.2 Notetaking with Vocabulary

For use after Lesson 5.2

In your own words, write the meaning of each vocabulary term.

corresponding parts

Theorems

Theorem 5.3 Properties of Triangle Congruence

Triangle congruence is reflexive, symmetric, and transitive.

Reflexive For any triangle $\triangle ABC$, $\triangle ABC \cong \triangle ABC$.

Symmetric If $\triangle ABC \cong \triangle DEF$, then $\triangle DEF \cong \triangle ABC$.

Transitive If $\triangle ABC \cong \triangle DEF$ and $\triangle DEF \cong \triangle JKL$, then $\triangle ABC \cong \triangle JKL$.

Notes:

Theorem 5.4 Third Angles Theorem

If two angles of one triangle are congruent to two angles of another triangle, then the third angles are also congruent.

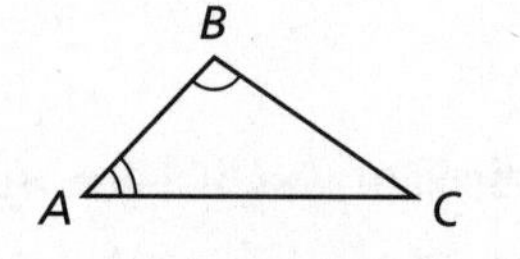

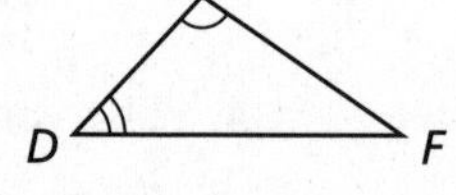

If $\angle A \cong \angle D$ and $\angle B \cong \angle E$, then $\angle C \cong \angle F$.

Notes:

Name____ Date__________

Extra Practice

In Exercises 1 and 2, identify all pairs of congruent corresponding parts. Then write another congruence statement for the polygons.

1. $\triangle PQR \cong \triangle STU$

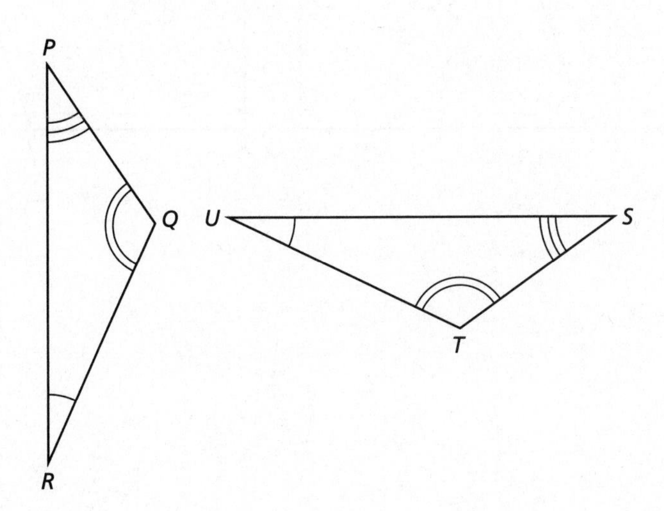

2. $ABCD \cong EFGH$

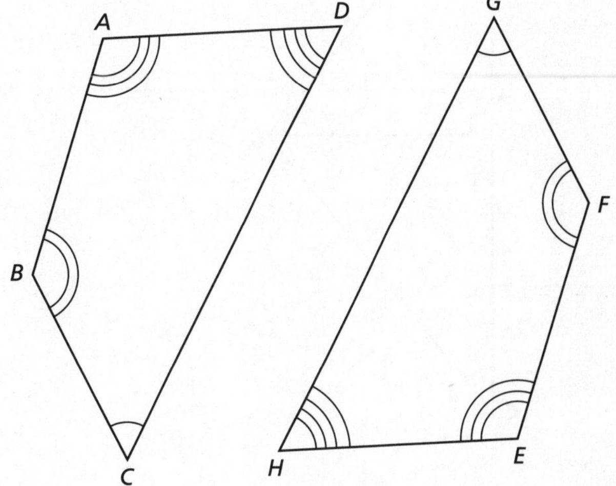

In Exercises 3 and 4, find the values of *x* and *y*.

3. $\triangle XYZ \cong \triangle RST$

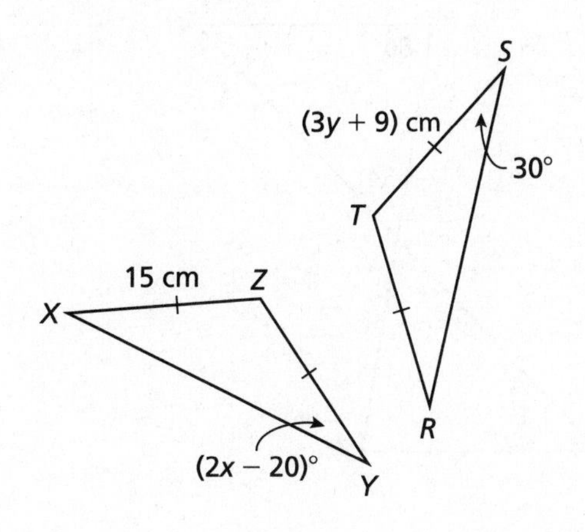

4. $ABCD \cong EFGH$

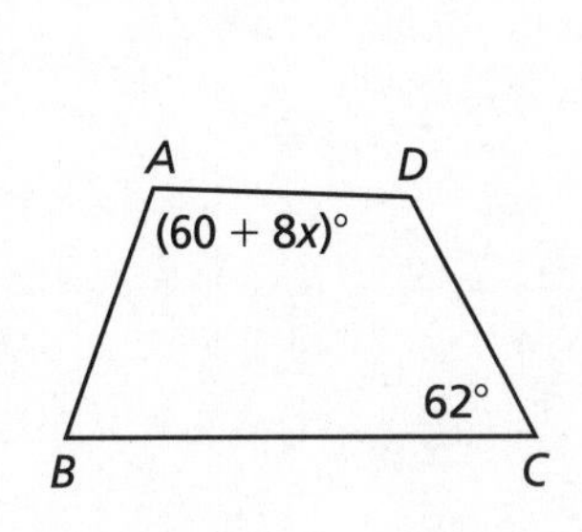

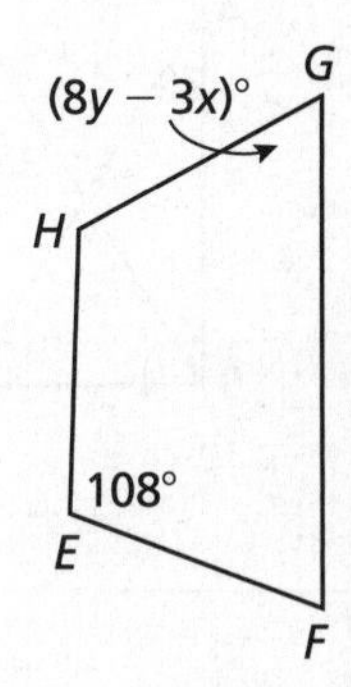

Name ______________________________ Date ________

In Exercises 5 and 6, show that the polygons are congruent. Explain your reasoning.

5.

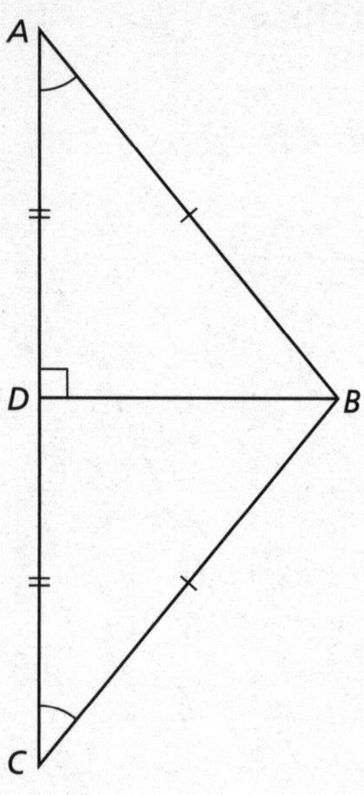

6.

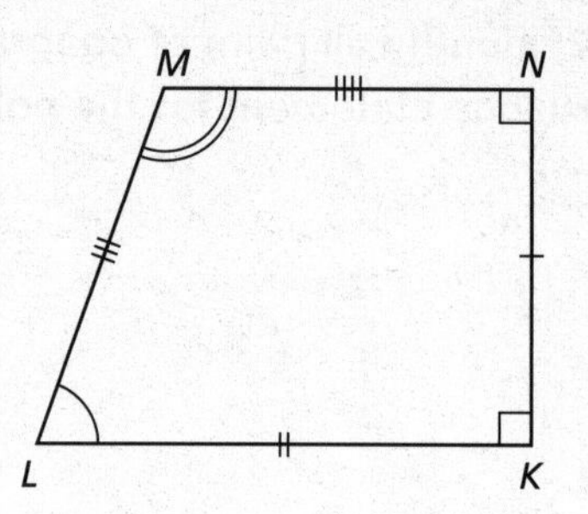

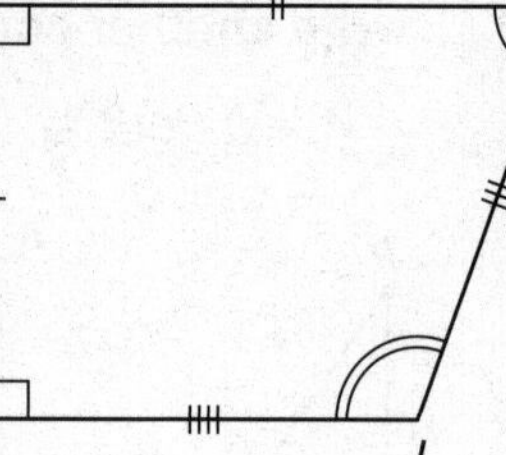

In Exercises 7 and 8, find $m\angle 1$.

7.

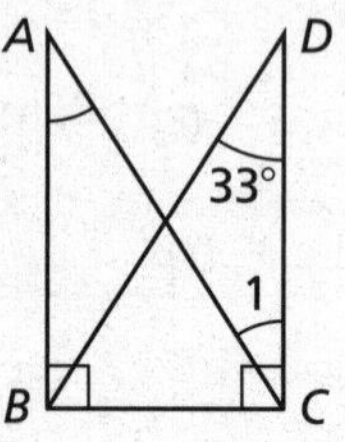

8.

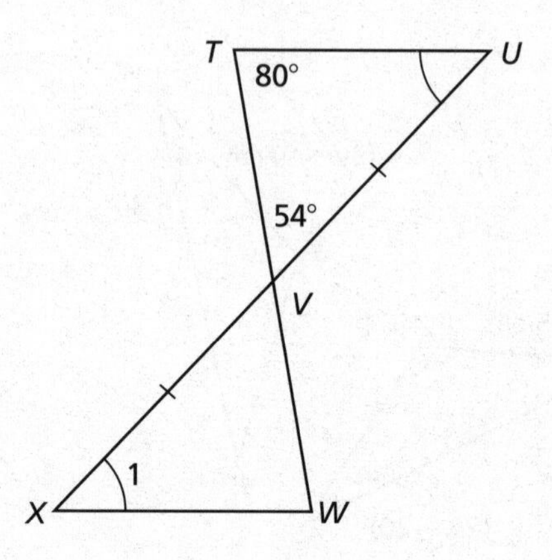

Name____ Date__________

5.3 Proving Triangle Congruence by SAS
For use with Exploration 5.3

Essential Question What can you conclude about two triangles when you know that two pairs of corresponding sides and the corresponding included angles are congruent?

1 EXPLORATION: Drawing Triangles

Go to *BigIdeasMath.com* for an interactive tool to investigate this exploration.

Work with a partner. Use dynamic geometry software.

a. Construct circles with radii of 2 units and 3 units centered at the origin. Construct a 40° angle with its vertex at the origin. Label the vertex A.

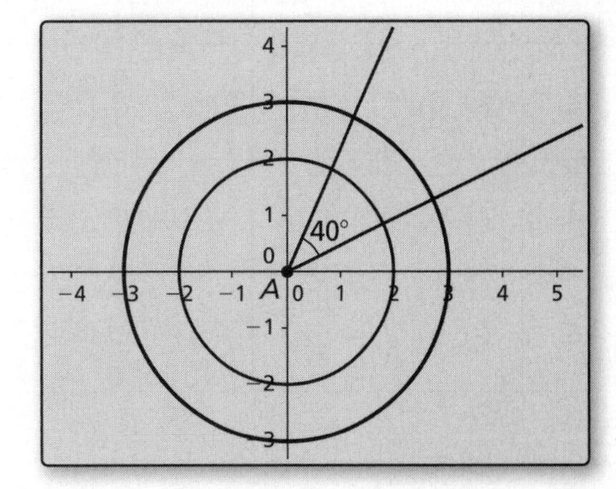

b. Locate the point where one ray of the angle intersects the smaller circle and label this point B. Locate the point where the other ray of the angle intersects the larger circle and label this point C. Then draw $\triangle ABC$.

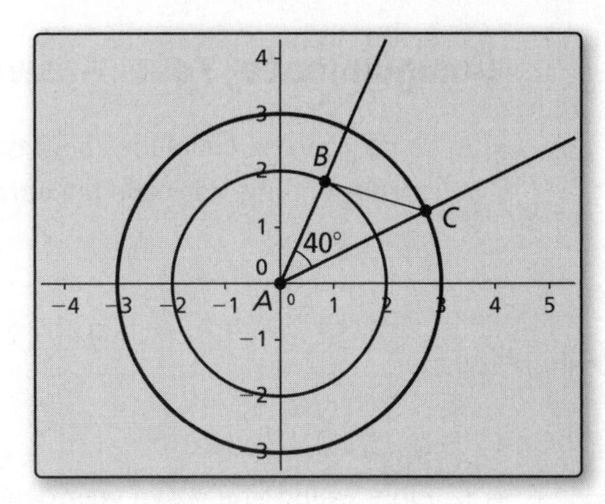

c. Find BC, $m\angle B$, and $m\angle C$.

d. Repeat parts (a)–(c) several times, redrawing the angle in different positions. Keep track of your results by completing the table on the next page. What can you conclude?

1 EXPLORATION: Drawing Triangles (continued)

	A	*B*	*C*	*AB*	*AC*	*BC*	*m∠A*	*m∠B*	*m∠C*
1.	(0, 0)			2	3		40°		
2.	(0, 0)			2	3		40°		
3.	(0, 0)			2	3		40°		
4.	(0, 0)			2	3		40°		
5.	(0, 0)			2	3		40°		

Communicate Your Answer

2. What can you conclude about two triangles when you know that two pairs of corresponding sides and the corresponding included angles are congruent?

3. How would you prove your conclusion in Exploration 1(d)?

Name______________________________ Date__________

5.3 Notetaking with Vocabulary

For use after Lesson 5.3

In your own words, write the meaning of each vocabulary term.

congruent figures

rigid motion

Theorems

Theorem 5.5 Side-Angle-Side (SAS) Congruence Theorem

If two sides and the included angle of one triangle are congruent to two sides and the included angle of a second triangle, then the two triangles are congruent.

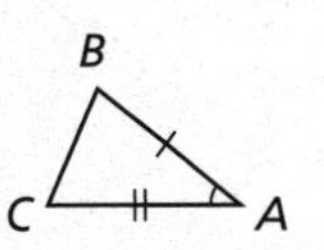

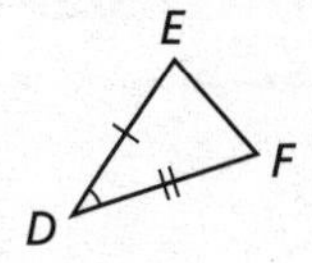

If $\overline{AB} \cong \overline{DE}$, $\angle A \cong \angle D$, and $\overline{AC} \cong \overline{DF}$, then $\triangle ABC \cong \triangle DEF$.

Notes:

Name ______________________________ Date __________

Extra Practice

In Exercises 1 and 2, write a proof.

1. **Given** $\overline{BD} \perp \overline{AC}$, $\overline{AD} \cong \overline{CD}$

   Prove $\triangle ABD \cong \triangle CBD$

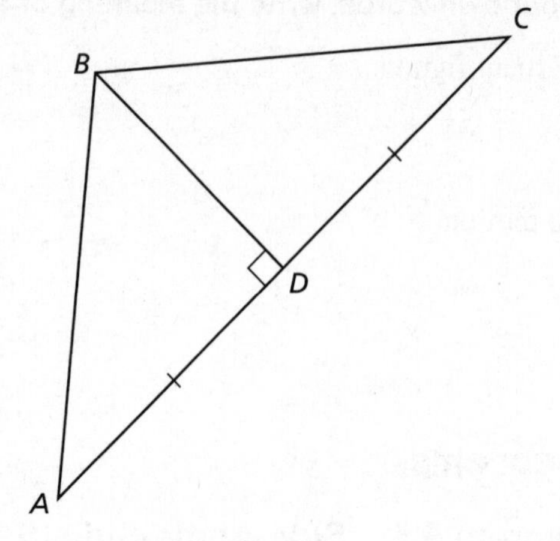

STATEMENTS	REASONS

2. **Given** $\overline{JN} \cong \overline{MN}$, $\overline{NK} \cong \overline{NL}$

   Prove $\triangle JNK \cong \triangle MNL$

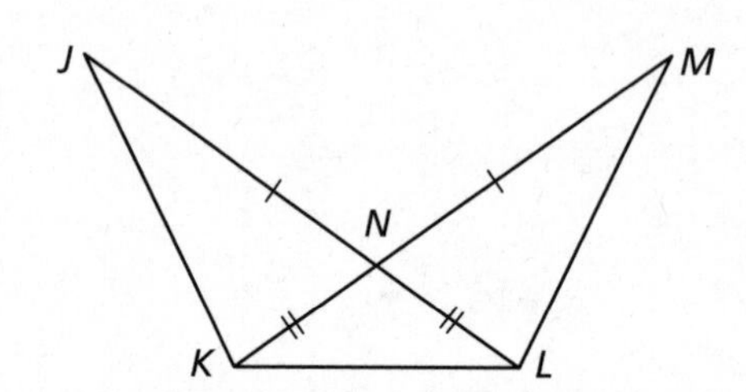

STATEMENTS	REASONS

Name____ Date__________

In Exercises 3 and 4, use the given information to name two triangles that are congruent. Explain your reasoning.

3. $\angle EPF \cong \angle GPH$, and P is the center of the circle.

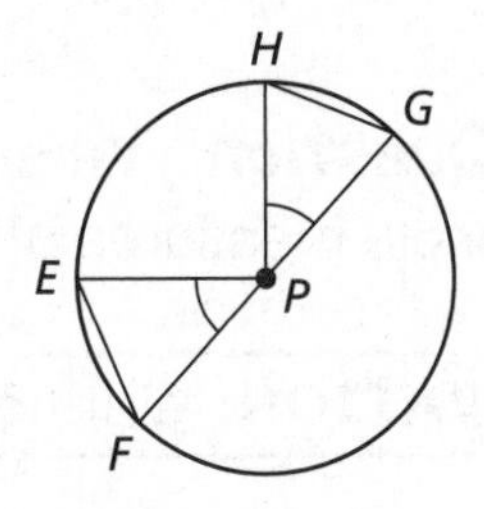

4. *ABCDEF* is a regular hexagon.

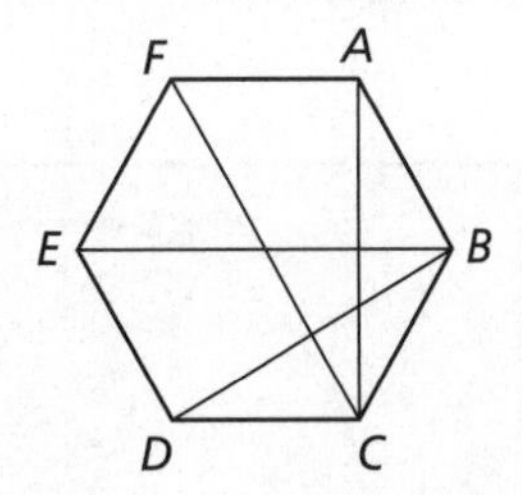

5. A quilt is made of triangles. You know $\overline{PS} \parallel \overline{QR}$ and $\overline{PS} \cong \overline{QR}$. Use the SAS Congruence Theorem (Theorem 5.5) to show that $\triangle PQR \cong \triangle RSP$.

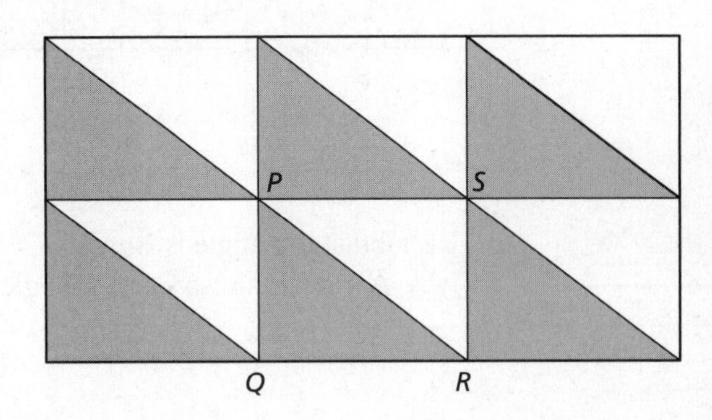

Name ______________________________ Date __________

5.4 Equilateral and Isosceles Triangles

For use with Exploration 5.4

Essential Question What conjectures can you make about the side lengths and angle measures of an isosceles triangle?

1 EXPLORATION: Writing a Conjecture about Isosceles Triangles

Go to *BigIdeasMath.com* for an interactive tool to investigate this exploration.

Work with a partner. Use dynamic geometry software.

a. Construct a circle with a radius of 3 units centered at the origin.

b. Construct $\triangle ABC$ so that B and C are on the circle and A is at the origin.

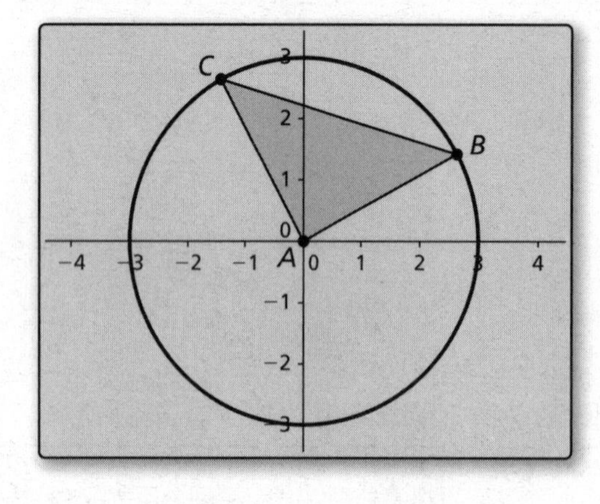

Sample
Points
$A(0, 0)$
$B(2.64, 1.42)$
$C(-1.42, 2.64)$
Segments
$AB = 3$
$AC = 3$
$BC = 4.24$
Angles
$m\angle A = 90°$
$m\angle B = 45°$
$m\angle C = 45°$

c. Recall that a triangle is *isosceles* if it has at least two congruent sides. Explain why $\triangle ABC$ is an isosceles triangle.

d. What do you observe about the angles of $\triangle ABC$?

e. Repeat parts (a)–(d) with several other isosceles triangles using circles of different radii. Keep track of your observations by completing the table on the next page. Then write a conjecture about the angle measures of an isosceles triangle.

Name____ Date__________

1 EXPLORATION: Writing a Conjecture about Isosceles Triangles (continued)

		A	B	C	AB	AC	BC	$m\angle A$	$m\angle B$	$m\angle C$
Sample	1.	(0, 0)	(2.64, 1.42)	(−1.42, 2.64)	3	3	4.24	90°	45°	45°
	2.	(0, 0)								
	3.	(0, 0)								
	4.	(0, 0)								
	5.	(0, 0)								

f. Write the converse of the conjecture you wrote in part (e). Is the converse true?

Communicate Your Answer

2. What conjectures can you make about the side lengths and angle measures of an isosceles triangle?

3. How would you prove your conclusion in Exploration 1(e)? in Exploration 1(f)?

Name ______________________________ Date ________

5.4 Notetaking with Vocabulary

For use after Lesson 5.4

In your own words, write the meaning of each vocabulary term.

legs

vertex angle

base

base angles

Theorems

Theorem 5.6 Base Angles Theorem

If two sides of a triangle are congruent, then the angles opposite them are congruent.

If $\overline{AB} \cong \overline{AC}$, then $\angle B \cong \angle C$.

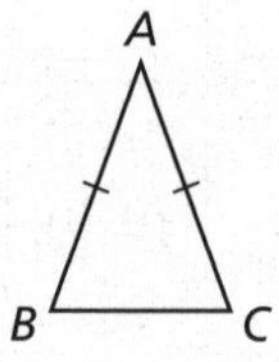

Theorem 5.7 Converse of the Base Angles Theorem

If two angles of a triangle are congruent, then the sides opposite them are congruent.

If $\angle B \cong \angle C$, then $\overline{AB} \cong \overline{AC}$.

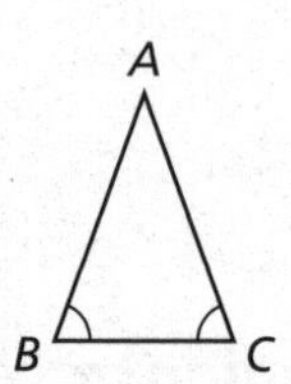

Notes:

Name____ Date__________

5.4 Notetaking with Vocabulary (continued)

Corollaries

Corollary 5.2 Corollary to the Base Angles Theorem

If a triangle is equilateral, then it is equiangular.

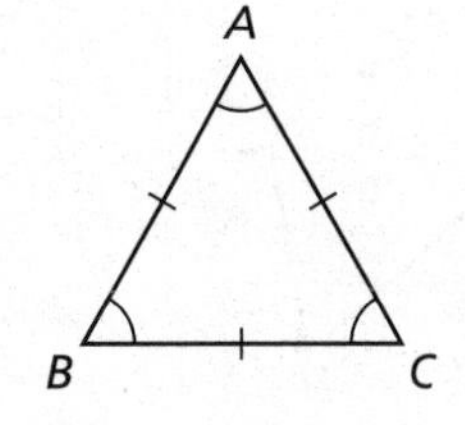

Corollary 5.3 Corollary to the Converse of the Base Angles Theorem

If a triangle is equiangular, then it is equilateral.

Notes:

Extra Practice

In Exercises 1–4, complete the statement. State which theorem you used.

1. If $\overline{NJ} \cong \overline{NM}$, then $\angle$_______ $\cong \angle$_______.

2. If $\overline{LM} \cong \overline{LN}$, then $\angle$_______ $\cong \angle$_______.

3. If $\angle NKM \cong \angle NMK$, then _______ $\cong$ _______.

4. If $\angle LJN \cong \angle LNJ$, then _______ $\cong$ _______.

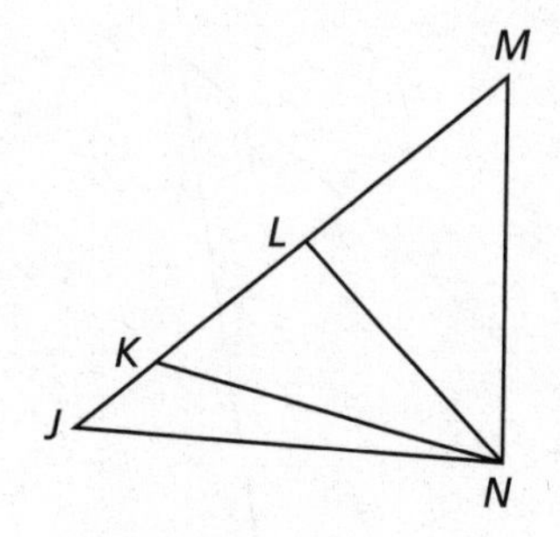

5.4 Notetaking with Vocabulary (continued)

In Exercises 5 and 6, find the value of *x*.

5.

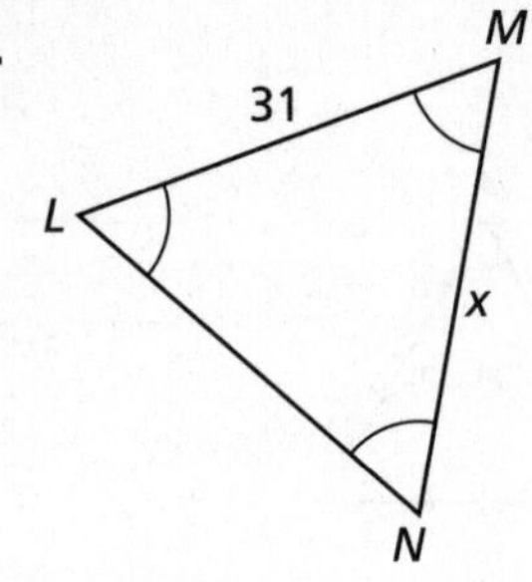

6.

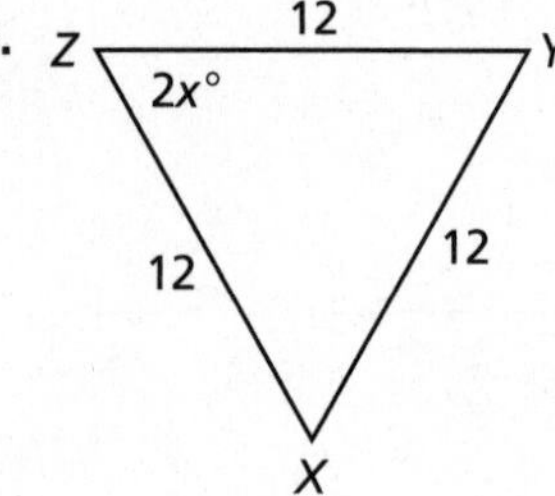

In Exercises 7 and 8, find the values of *x* and *y*.

7.

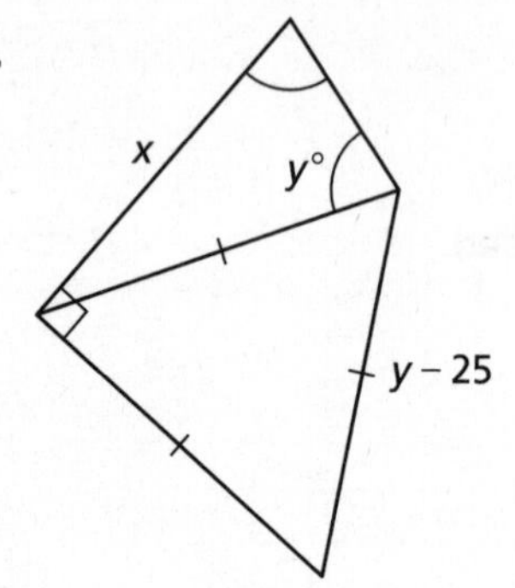

8.

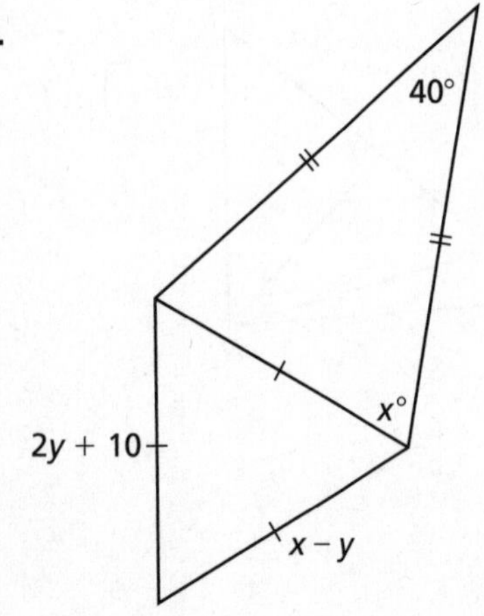

Name________________________________ Date__________

5.5 Proving Triangle Congruence by SSS
For use with Exploration 5.5

Essential Question What can you conclude about two triangles when you know the corresponding sides are congruent?

1 EXPLORATION: Drawing Triangles

Go to *BigIdeasMath.com* for an interactive tool to investigate this exploration.

Work with a partner. Use dynamic geometry software.

a. Construct circles with radii of 2 units and 3 units centered at the origin. Label the origin A. Then draw $\overline{BC}$ of length 4 units.

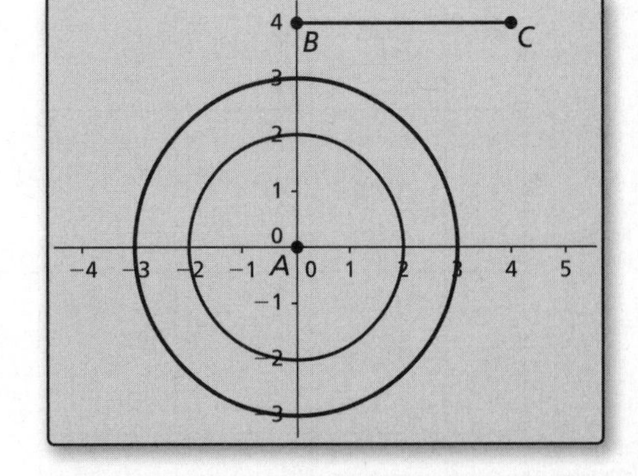

b. Move $\overline{BC}$ so that B is on the smaller circle and C is on the larger circle. Then draw $\triangle ABC$.

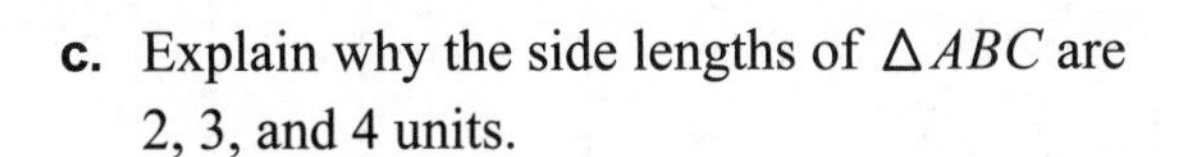

c. Explain why the side lengths of $\triangle ABC$ are 2, 3, and 4 units.

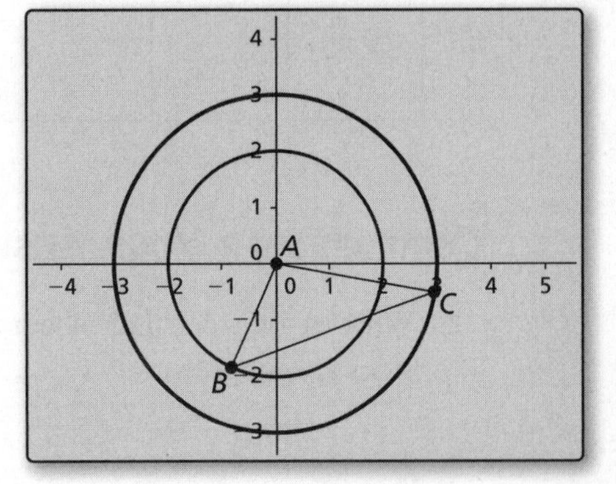

d. Find $m\angle A$, $m\angle B$, and $m\angle C$.

e. Repeat parts (b) and (d) several times, moving $\overline{BC}$ to different locations. Keep track of your results by completing the table on the next page. What can you conclude?

Name ______________________________ Date __________

1 EXPLORATION: Drawing Triangles (continued)

	A	*B*	*C*	*AB*	*AC*	*BC*	$m\angle A$	$m\angle B$	$m\angle C$
1.	(0, 0)			2	3	4			
2.	(0, 0)			2	3	4			
3.	(0, 0)			2	3	4			
4.	(0, 0)			2	3	4			
5.	(0, 0)			2	3	4			

Communicate Your Answer

2. What can you conclude about two triangles when you know the corresponding sides are congruent?

3. How would you prove your conclusion in Exploration 1(e)?

Name____ Date__________

5.5 Notetaking with Vocabulary
For use after Lesson 5.5

In your own words, write the meaning of each vocabulary term.

legs

hypotenuse

Theorems

Theorem 5.8 Side-Side-Side (SSS) Congruence Theorem

If three sides of one triangle are congruent to three sides of a second triangle, then the two triangles are congruent.

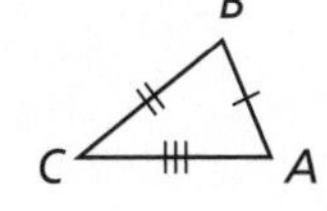

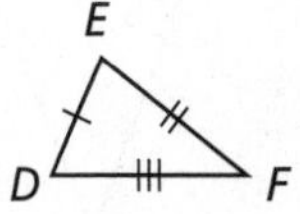

If $\overline{AB} \cong \overline{DE}$, $\overline{BC} \cong \overline{EF}$, and $\overline{AC} \cong \overline{DF}$, then $\triangle ABC \cong \triangle DEF$.

Notes:

Theorem 5.9 Hypotenuse-Leg (HL) Congruence Theorem

If the hypotenuse and a leg of a right triangle are congruent to the hypotenuse and a leg of a second right triangle, then the two triangles are congruent.

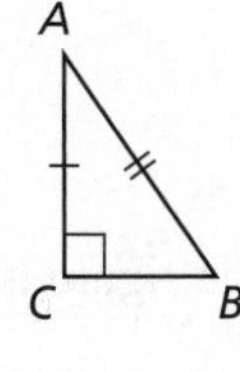

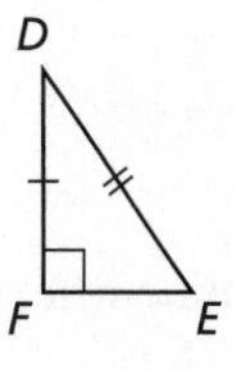

If $\overline{AB} \cong \overline{DE}$, $\overline{AC} \cong \overline{DF}$, and $m\angle C = m\angle F = 90°$, then $\triangle ABC \cong \triangle DEF$.

Notes:

Name ______________________________ Date __________

5.5 **Notetaking with Vocabulary (continued)**

Extra Practice

In Exercises 1–4, decide whether the congruence statement is true. Explain your reasoning.

1. $\triangle ABC \cong \triangle EDC$

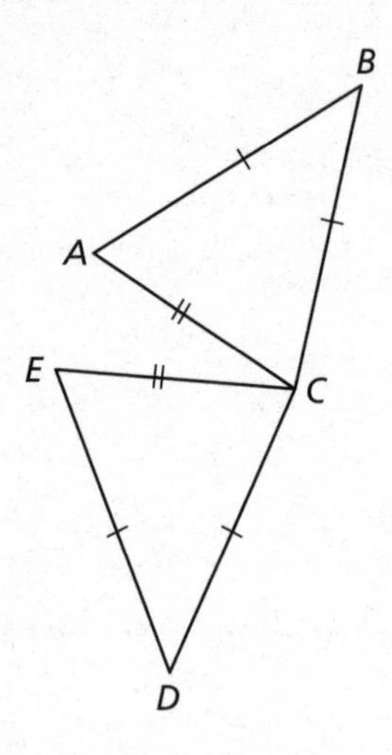

2. $\triangle KGH \cong \triangle HJK$

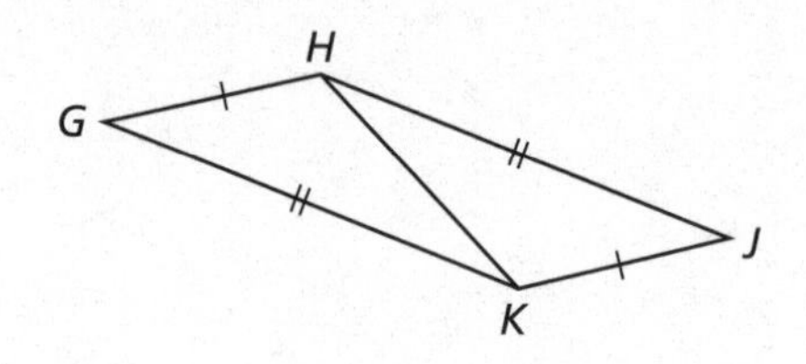

3. $\triangle UVW \cong \triangle XYZ$

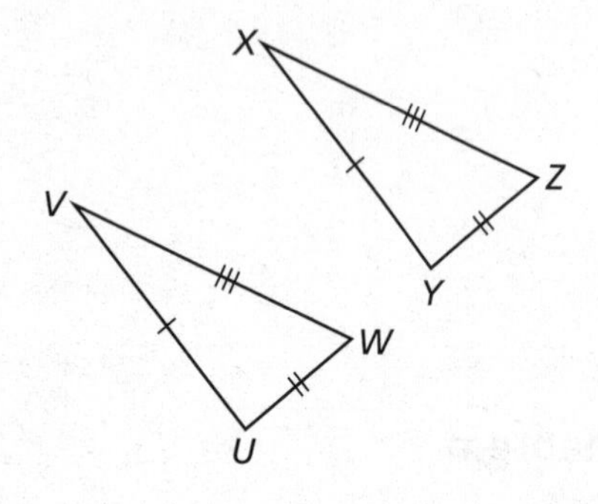

4. $\triangle RST \cong \triangle RPQ$

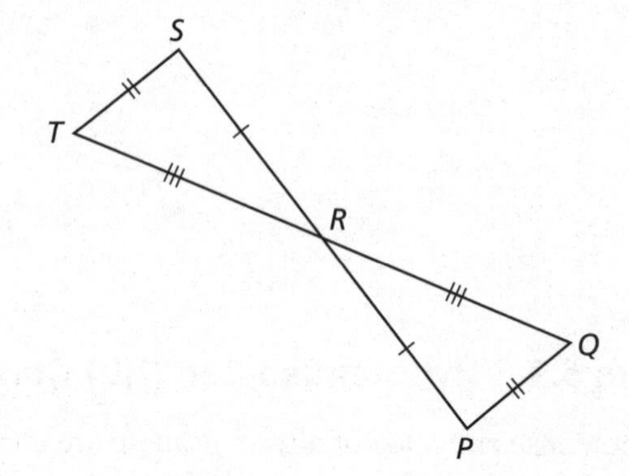

5. Determine whether the figure is stable. Explain your reasoning.

Name______________________________ Date__________

5.5 Notetaking with Vocabulary (continued)

6. Redraw the triangles so they are side by side with corresponding parts in the same position. Then write a proof.

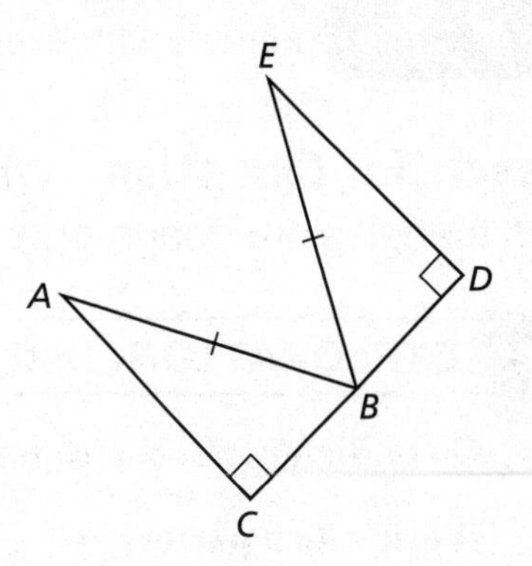

Given B is the midpoint of $\overline{CD}$,
$\overline{AB} \cong \overline{EB}, \angle C$ and $\angle D$ are right angles.

Prove $\triangle ABC \cong \triangle EBD$

STATEMENTS	REASONS

7. Write a proof.

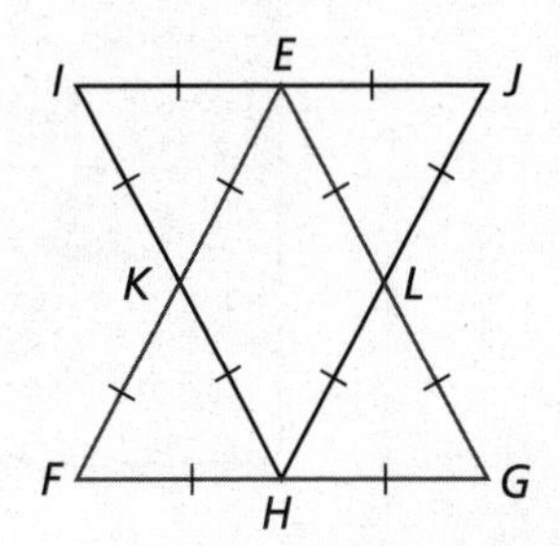

Given $\overline{IE} \cong \overline{EJ} \cong \overline{JL} \cong \overline{LH} \cong \overline{HK} \cong \overline{KI} \cong \overline{EK} \cong \overline{KF} \cong \overline{FH} \cong \overline{HG} \cong \overline{GL} \cong \overline{LE}$

Prove $\triangle EFG \cong \triangle HIJ$

STATEMENTS	REASONS

Name ______________________________ Date __________

5.6 Proving Triangle Congruence by ASA and AAS
For use with Exploration 5.6

Essential Question What information is sufficient to determine whether two triangles are congruent?

1 EXPLORATION: Determining Whether SSA Is Sufficient

Go to *BigIdeasMath.com* for an interactive tool to investigate this exploration.

Work with a partner.

a. Use dynamic geometry software to construct $\triangle ABC$. Construct the triangle so that vertex B is at the origin, $\overline{AB}$ has a length of 3 units, and $\overline{BC}$ has a length of 2 units.

b. Construct a circle with a radius of 2 units centered at the origin. Locate point D where the circle intersects $\overline{AC}$. Draw $\overline{BD}$.

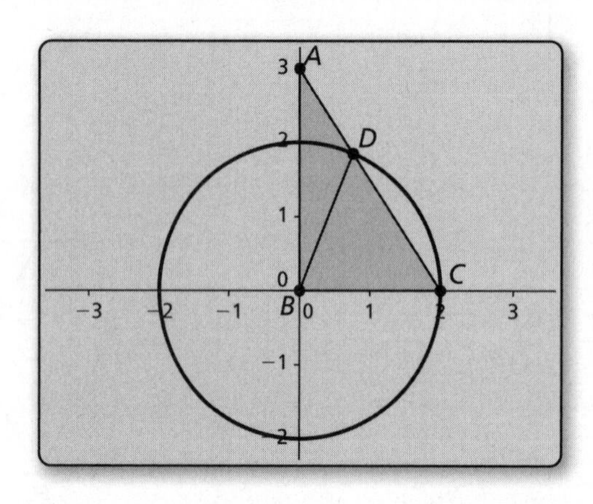

Sample

Points
$A(0, 3)$
$B(0, 0)$
$C(2, 0)$
$D(0.77, 1.85)$
Segments
$AB = 3$
$AC = 3.61$
$BC = 2$
$AD = 1.38$
Angle
$m\angle A = 33.69°$

c. $\triangle ABC$ and $\triangle ABD$ have two congruent sides and a nonincluded congruent angle. Name them.

d. Is $\triangle ABC \cong \triangle ABD$? Explain your reasoning.

e. Is SSA sufficient to determine whether two triangles are congruent? Explain your reasoning.

Name______________________________ Date__________

2 EXPLORATION: Determining Valid Congruence Theorems

Go to *BigIdeasMath.com* for an interactive tool to investigate this exploration.

Work with a partner. Use dynamic geometry software to determine which of the following are valid triangle congruence theorems. For those that are not valid, write a counterexample. Explain your reasoning.

Possible Congruence Theorem	Valid or not valid?
SSS	
SSA	
SAS	
AAS	
ASA	
AAA	

Communicate Your Answer

3. What information is sufficient to determine whether two triangles are congruent?

4. Is it possible to show that two triangles are congruent using more than one congruence theorem? If so, give an example.

Name ______________________________ Date __________

5.6 Notetaking with Vocabulary
For use after Lesson 5.6

In your own words, write the meaning of each vocabulary term.

congruent figures

rigid motion

Theorems

Theorem 5.10 Angle-Side-Angle (ASA) Congruence Theorem

If two angles and the included side of one triangle are congruent to two angles and the included side of a second triangle, then the two triangles are congruent.

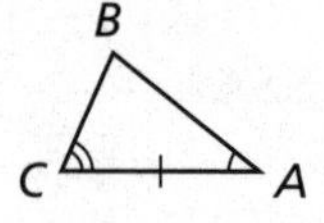

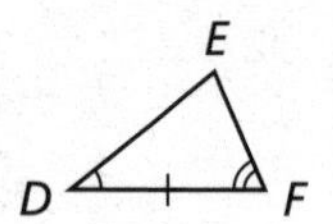

If $\angle A \cong \angle D$, $\overline{AC} \cong \overline{DF}$, and $\angle C \cong \angle F$, then $\triangle ABC \cong \triangle DEF$.

Notes:

Theorem 5.11 Angle-Angle-Side (AAS) Congruence Theorem

If two angles and a non-included side of one triangle are congruent to two angles and the corresponding non-included side of a second triangle, then the two triangles are congruent.

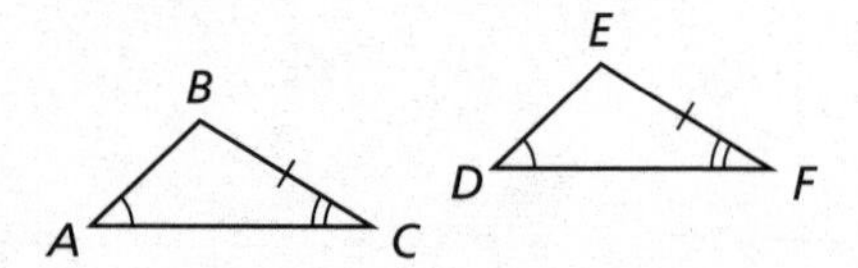

If $\angle A \cong \angle D$, $\angle C \cong \angle F$, and $\overline{BC} \cong \overline{EF}$, then $\triangle ABC \cong \triangle DEF$.

Notes:

Extra Practice

In Exercises 1–4, decide whether enough information is given to prove that the triangles are congruent. If so, state the theorem you would use.

1. $\triangle GHK, \triangle JKH$

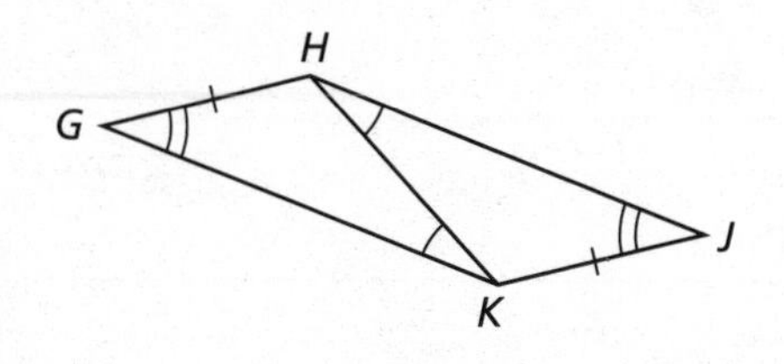

2. $\triangle ABC, \triangle DEC$

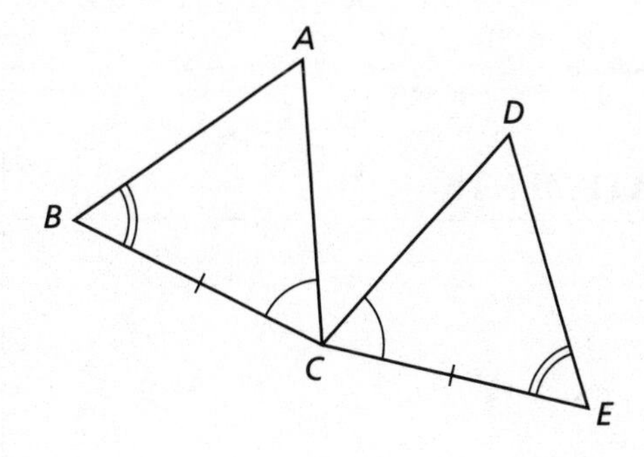

3. $\triangle JKL, \triangle MLK$

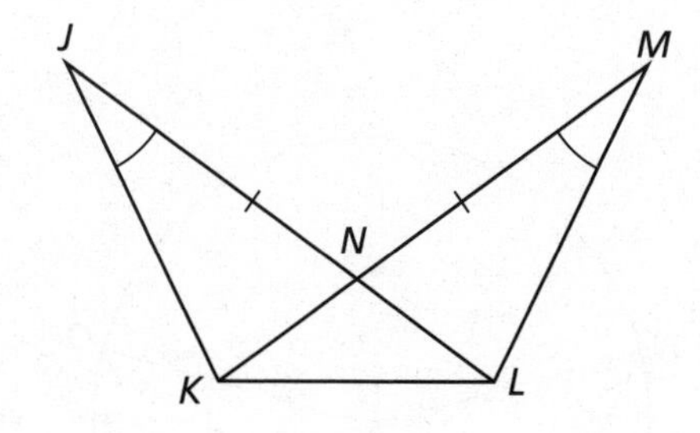

4. $\triangle RST, \triangle UVW$

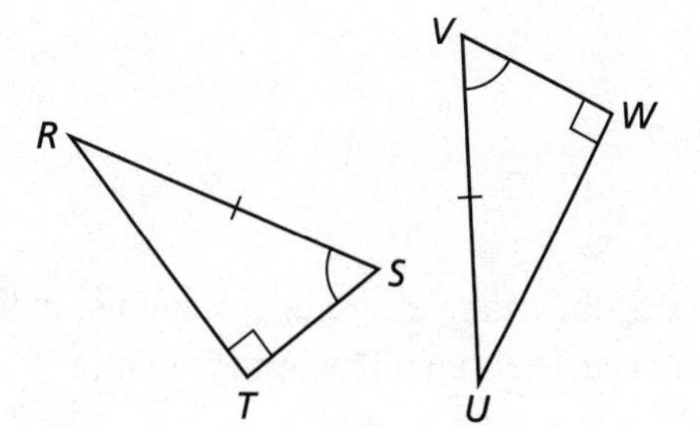

In Exercises 5 and 6, decide whether you can use the given information to prove that $\triangle LMN \cong \triangle PQR$. Explain your reasoning.

5. $\angle M \cong \angle Q, \angle N \cong \angle R, \overline{NL} \cong \overline{RP}$

6. $\angle L \cong \angle R, \angle M \cong \angle Q, \overline{LM} \cong \overline{PQ}$

Name ______________________________ Date __________

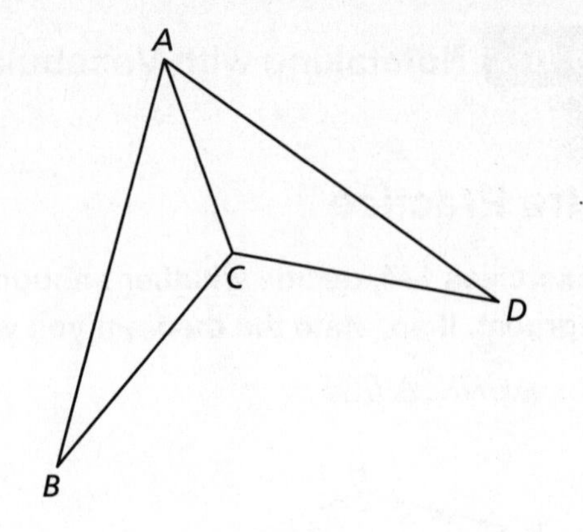

7. Prove that the triangles are congruent using the ASA Congruence Theorem (Theorem 5.10).

Given $\overline{AC}$ bisects $\angle DAB$ and $\angle DCB$.

Prove $\triangle ABC \cong \triangle ADC$

STATEMENTS	REASONS

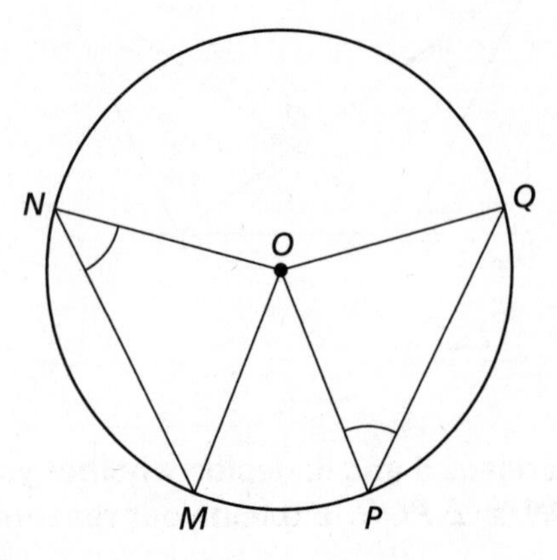

8. Prove that the triangles are congruent using the AAS Congruence Theorem (Theorem 5.11).

Given O is the center of the circle and $\angle N \cong \angle P$.

Prove $\triangle MNO \cong \triangle PQO$

STATEMENTS	REASONS

Name____ Date__________

5.7 Using Congruent Triangles

For use with Exploration 5.7

Essential Question How can you use congruent triangles to make an indirect measurement?

1 EXPLORATION: Measuring the Width of a River

Work with a partner. The figure shows how a surveyor can measure the width of a river by making measurements on only one side of the river.

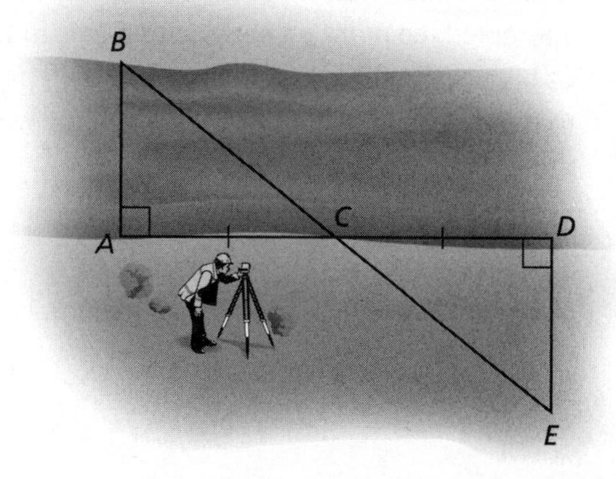

a. Study the figure. Then explain how the surveyor can find the width of the river.

b. Write a proof to verify that the method you described in part (a) is valid.

Given $\angle A$ is a right angle, $\angle D$ is a right angle, $\overline{AC} \cong \overline{CD}$

c. Exchange proofs with your partner and discuss the reasoning used.

Name ______________________________ Date __________

2 EXPLORATION: Measuring the Width of a River

Work with a partner. It was reported that one of Napoleon's officers estimated the width of a river as follows. The officer stood on the bank of the river and lowered the visor on his cap until the farthest thing visible was the edge of the bank on the other side. He then turned and noted the point on his side that was in line with the tip of his visor and his eye. The officer then paced the distance to this point and concluded that distance was the width of the river.

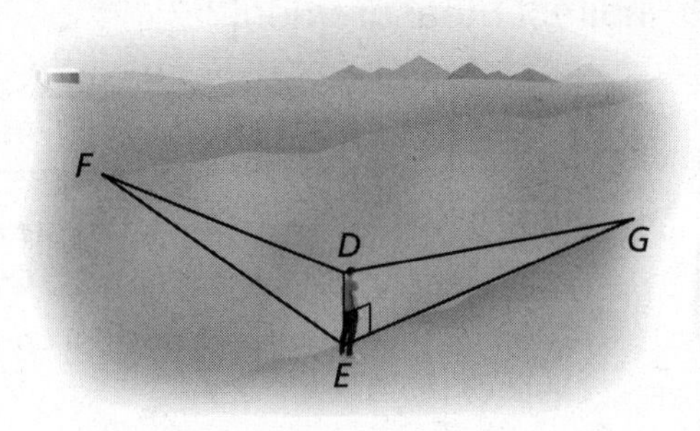

a. Study the figure. Then explain how the officer concluded that the width of the river is EG.

b. Write a proof to verify that the conclusion the officer made is correct.

Given $\angle DEG$ is a right angle, $\angle DEF$ is a right angle, $\angle EDG \cong \angle EDF$

c. Exchange proofs with your partner and discuss the reasoning used.

Communicate Your Answer

3. How can you use congruent triangles to make an indirect measurement?

4. Why do you think the types of measurements described in Explorations 1 and 2 are called *indirect* measurements?

Name____ Date__________

5.7 Notetaking with Vocabulary
For use after Lesson 5.7

In your own words, write the meaning of each vocabulary term.

congruent figures

corresponding parts

construction

Notes:

Extra Practice

In Exercises 1–3, explain how to prove that the statement is true.

1. $\overline{UV} \cong \overline{XV}$

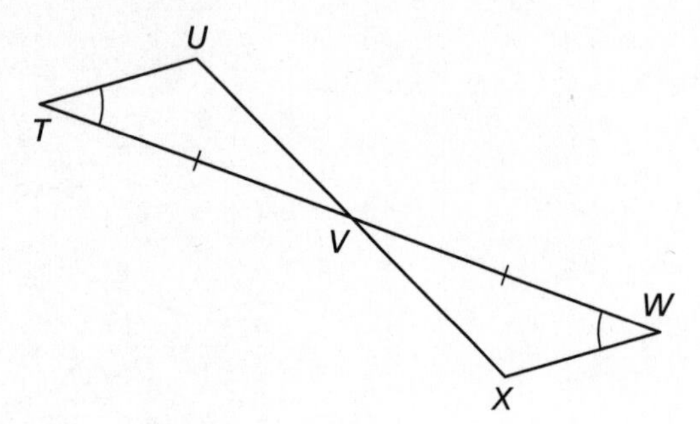

2. $\overline{TS} \cong \overline{VR}$

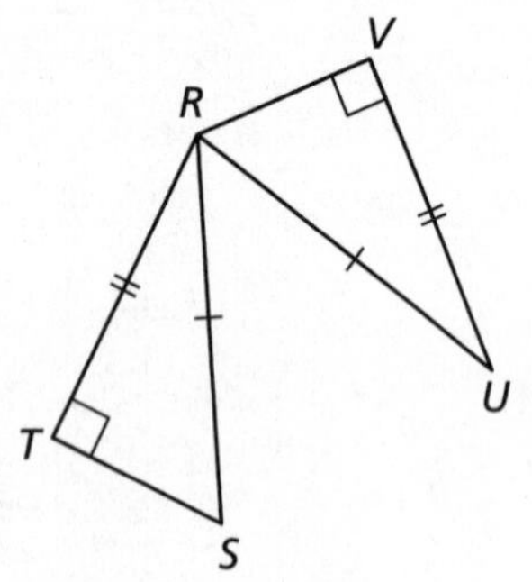

3. $\angle JLK \cong \angle MLN$

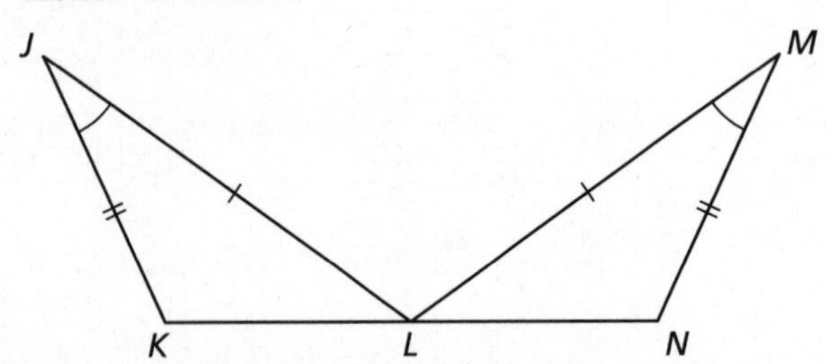

In Exercises 4 and 5, write a plan to prove that $\angle 1 \cong \angle 2$.

4.

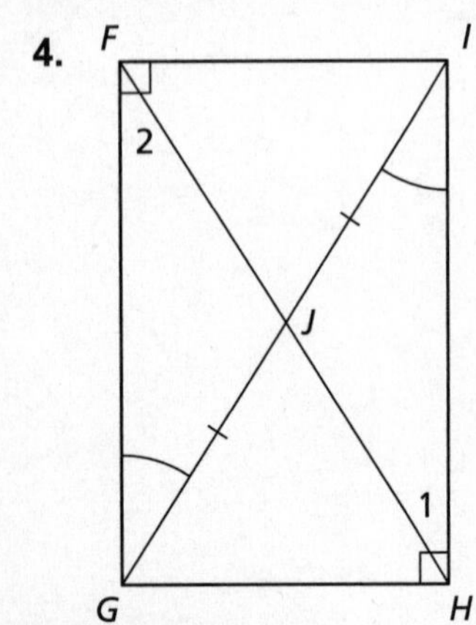

Name____ Date__________

5.

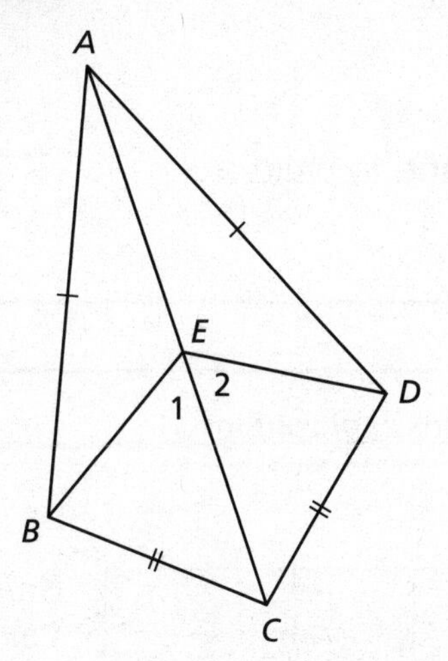

6. Write a proof to verify that the construction is valid.

Ray bisects an angle

Plan for Proof Show that $\triangle ABD \cong \triangle ACD$ by the SSS Congruence Theorem (Thm. 5.8). Use corresponding parts of congruent triangles to show that $\angle BAD \cong \angle CAD$.

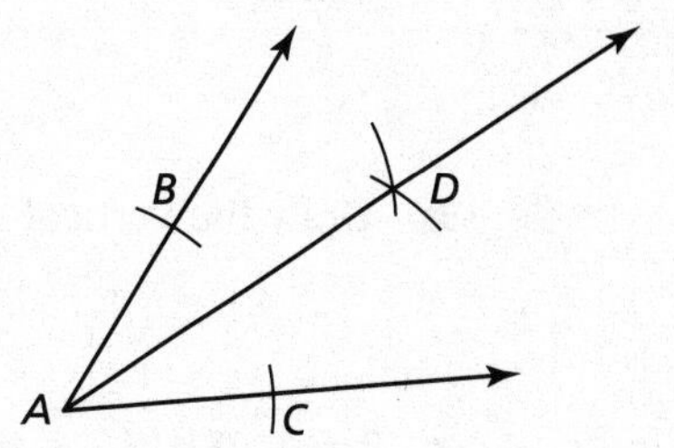

STATEMENTS	REASONS

Name ______________________________ Date __________

5.8 Coordinate Proofs

For use with Exploration 5.8

Essential Question How can you use a coordinate plane to write a proof?

1 EXPLORATION: Writing a Coordinate Proof

Go to *BigIdeasMath.com* for an interactive tool to investigate this exploration.

Work with a partner.

a. Use dynamic geometry software to draw $\overline{AB}$ with endpoints $A(0, 0)$ and $B(6, 0)$.

b. Draw the vertical line $x = 3$.

c. Draw $\triangle ABC$ so that C lies on the line $x = 3$.

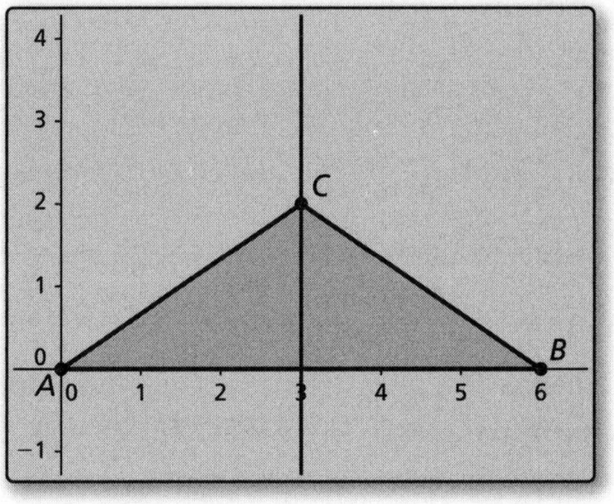

Sample
Points
$A(0, 0)$
$B(6, 0)$
$C(3, y)$
Segments
$AB = 6$
Line
$x = 3$

d. Use your drawing to prove that $\triangle ABC$ is an isosceles triangle.

2 EXPLORATION: Writing a Coordinate Proof

Go to *BigIdeasMath.com* for an interactive tool to investigate this exploration.

Work with a partner.

a. Use dynamic geometry software to draw $\overline{AB}$ with endpoints $A(0, 0)$ and $B(6, 0)$.

b. Draw the vertical line $x = 3$.

c. Plot the point $C(3, 3)$ and draw $\triangle ABC$. Then use your drawing to prove that $\triangle ABC$ is an isosceles right triangle.

2 EXPLORATION: Writing a Coordinate Proof (continued)

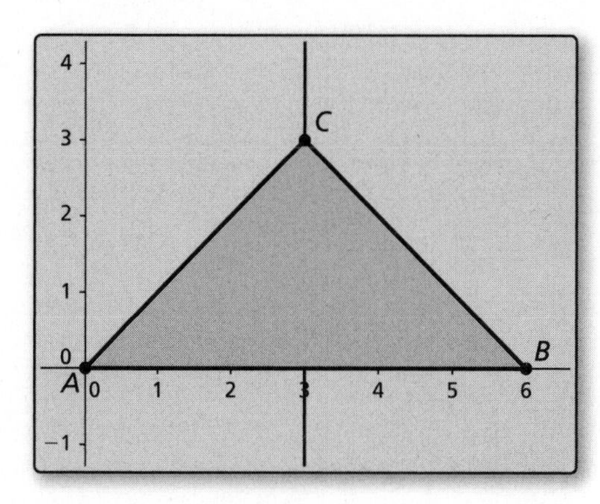

Sample
Points
$A(0, 0)$
$B(6, 0)$
$C(3, 3)$
Segments
$AB = 6$
$BC = 4.24$
$AC = 4.24$
Line
$x = 3$

d. Change the coordinates of C so that C lies below the x-axis and ΔABC is an isosceles right triangle.

e. Write a coordinate proof to show that if C lies on the line $x = 3$ and ΔABC is an isosceles right triangle, then C must be the point (3, 3) or the point found in part (d).

Communicate Your Answer

3. How can you use a coordinate plane to write a proof?

4. Write a coordinate proof to prove that ΔABC with vertices $A(0, 0)$, $B(6, 0)$, and $C\left(3, 3\sqrt{3}\right)$ is an equilateral triangle.

Name ______________________________ Date __________

5.8 Notetaking with Vocabulary

For use after Lesson 5.8

In your own words, write the meaning of each vocabulary term.

coordinate proof

Notes:

Extra Practice

In Exercises 1 and 2, place the figure in a coordinate plane in a convenient way. Assign coordinates to each vertex. Explain the advantages of your placement.

1. an obtuse triangle with height of 3 units and base of 2 units

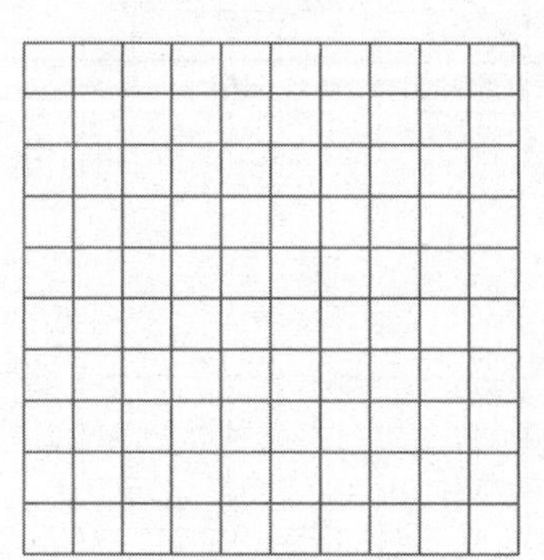

2. a rectangle with length of $2w$

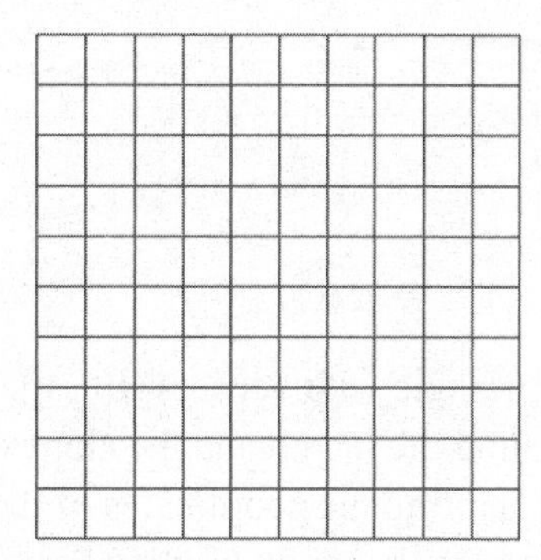

In Exercises 3 and 4, write a plan for the proof.

3. Given Coordinates of vertices of $\triangle OPR$ and $\triangle QRP$

Proof $\triangle OPR \cong \triangle QRP$

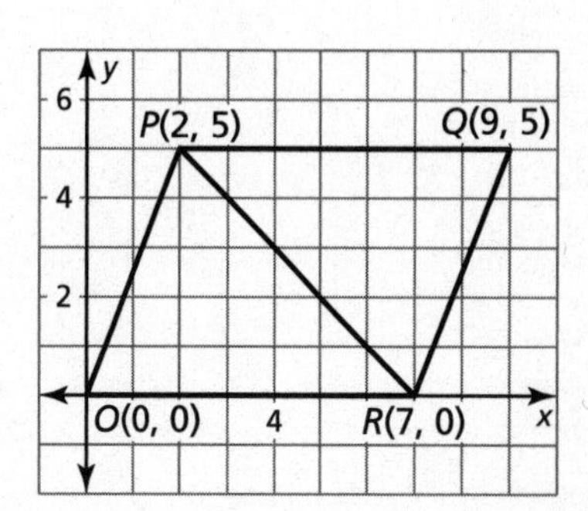

4. Given Coordinates of vertices of $\triangle OAB$ and $\triangle CDB$

Prove B is the midpoint of $\overline{AD}$ and $\overline{OC}$.

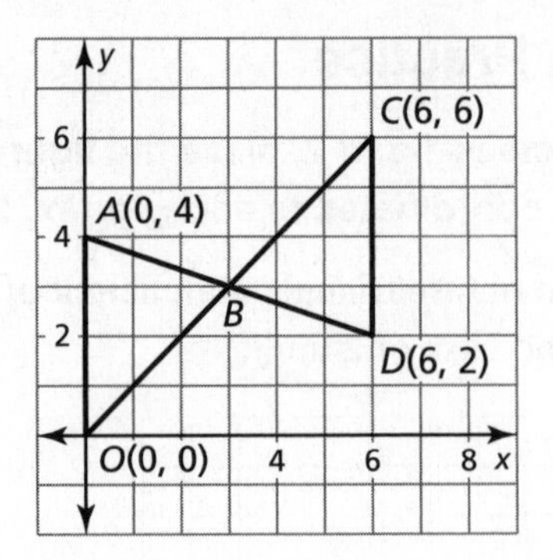

5. Graph the triangle with vertices $A(0, 0)$, $B(3m, m)$, and $C(0, 3m)$. Find the length and the slope of each side of the triangle. Then find the coordinates of the midpoint of each side. Is the triangle a right triangle? isosceles? Explain. (Assume all variables are positive.)

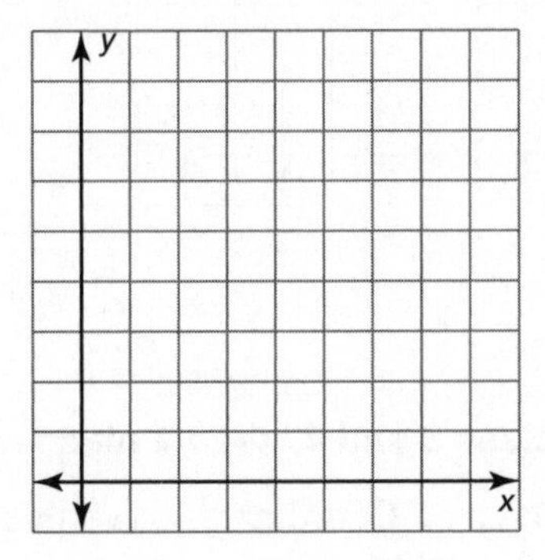

6. Write a coordinate proof.

Given Coordinates of vertices of $\triangle OEF$ and $\triangle OGF$

Prove $\triangle OEF \cong \triangle OGF$

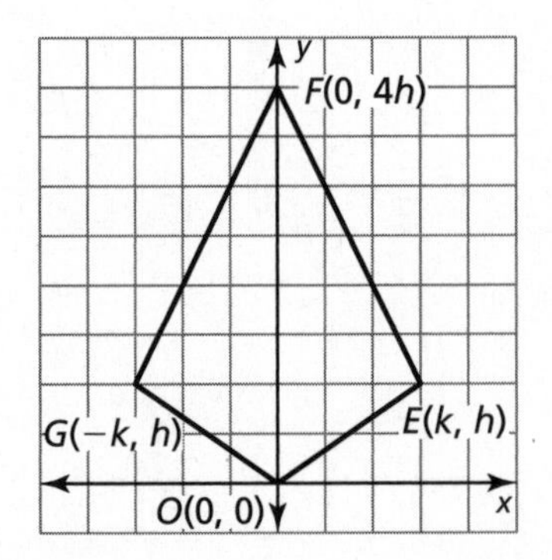

Name____ Date__________

Maintaining Mathematical Proficiency

Write an equation of the line passing through point *P* that is perpendicular to the given line.

1. $P(5, 2), y = 2x + 6$

2. $P(4, 2), y = 6x - 3$

3. $P(-1, -2), y = -3x + 6$

4. $P(-8, 3), y = 3x - 1$

5. $P(6, 7), y = x - 5$

6. $P(3, 7), y = \frac{1}{4}x + 4$

Write the sentence as an inequality.

7. A number g is at least 4 and no more than 12.

8. A number r is more than 2 and less than 7.

9. A number q is less than or equal to 6 or greater than 1.

10. A number p is fewer than 17 or no less than 5.

11. A number k is greater than or equal to –4 and less than 1.

Name ________________________________ Date __________

6.1 Perpendicular and Angle Bisectors

For use with Exploration 6.1

Essential Question What conjectures can you make about a point on the perpendicular bisector of a segment and a point on the bisector of an angle?

1 EXPLORATION: Points on a Perpendicular Bisector

Go to *BigIdeasMath.com* for an interactive tool to investigate this exploration.

Work with a partner. Use dynamic geometry software.

a. Draw any segment and label it $\overline{AB}$. Construct the perpendicular bisector of $\overline{AB}$.

b. Label a point C that is on the perpendicular bisector of $\overline{AB}$ but is not on $\overline{AB}$.

c. Draw $\overline{CA}$ and $\overline{CB}$ and find their lengths. Then move point C to other locations on the perpendicular bisector and note the lengths of $\overline{CA}$ and $\overline{CB}$.

d. Repeat parts (a)–(c) with other segments. Describe any relationship(s) you notice.

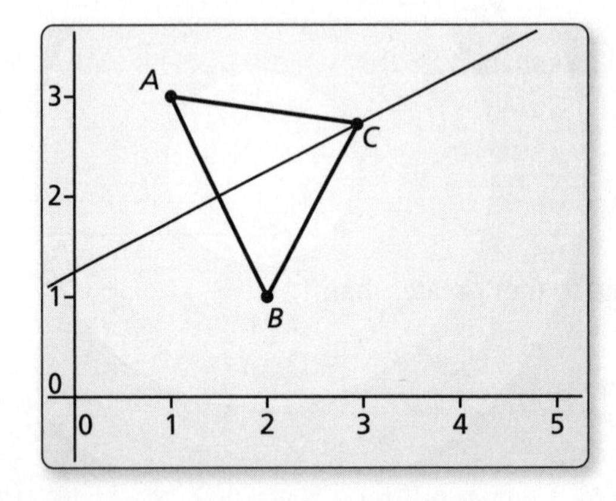

Sample
Points
$A(1, 3)$
$B(2, 1)$
$C(2.95, 2.73)$
Segments
$AB = 2.24$
$CA = ?$
$CB = ?$
Line
$-x + 2y = 2.5$

2 EXPLORATION: Points on an Angle Bisector

Go to *BigIdeasMath.com* for an interactive tool to investigate this exploration.

Work with a partner. Use dynamic geometry software.

a. Draw two rays $\overrightarrow{AB}$ and $\overrightarrow{AC}$ to form $\angle BAC$. Construct the bisector of $\angle BAC$.

b. Label a point D on the bisector of $\angle BAC$.

2 EXPLORATION: Points on an Angle Bisector (continued)

c. Construct and find the lengths of the perpendicular segments from D to the sides of $\angle BAC$. Move point D along the angle bisector and note how the lengths change.

d. Repeat parts (a)–(c) with other angles. Describe any relationship(s) you notice.

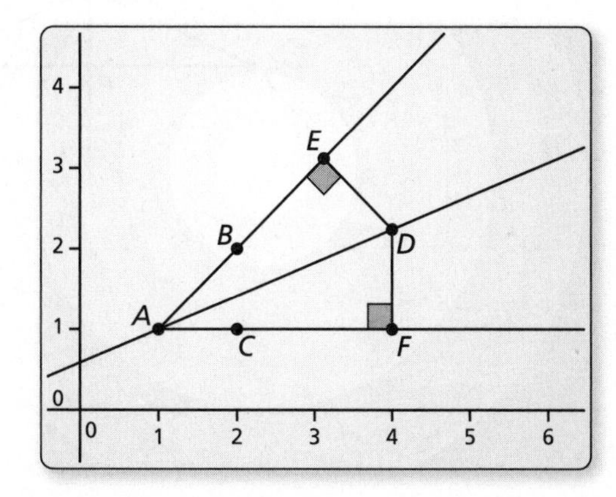

Sample
Points
$A(1, 1)$
$B(2, 2)$
$C(2, 1)$
$D(4, 2.24)$
Rays
$AB = -x + y = 0$
$AC = y = 1$
Line
$-0.38x + 0.92y = 0.54$

Communicate Your Answer

3. What conjectures can you make about a point on the perpendicular bisector of a segment and a point on the bisector of an angle?

4. In Exploration 2, what is the distance from point D to $\overrightarrow{AB}$ when the distance from D to $\overrightarrow{AC}$ is 5 units? Justify your answer.

Name ______________________________ Date __________

6.1 Notetaking with Vocabulary
For use after Lesson 6.1

In your own words, write the meaning of each vocabulary term.

equidistant

Theorems

Theorem 6.1 Perpendicular Bisector Theorem

In a plane, if a point lies on the perpendicular bisector of a segment, then it is equidistant from the endpoints of the segment.

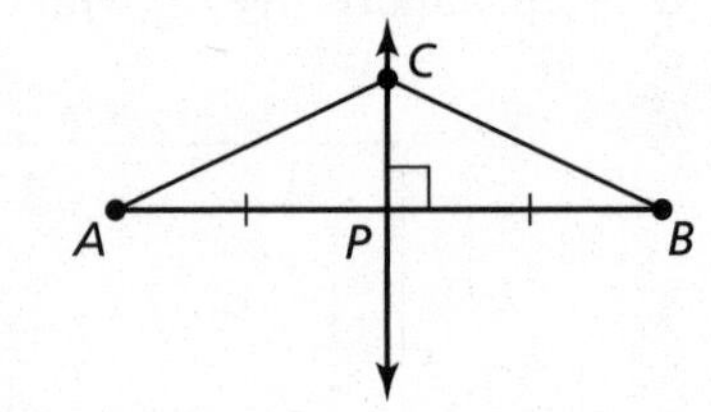

If $\overleftrightarrow{CP}$ is the $\perp$ bisector of $\overline{AB}$, then $CA = CB$.

Notes:

Theorem 6.2 Converse of the Perpendicular Bisector Theorem

In a plane, if a point is equidistant from the endpoints of a segment, then it lies on the perpendicular bisector of the segment.

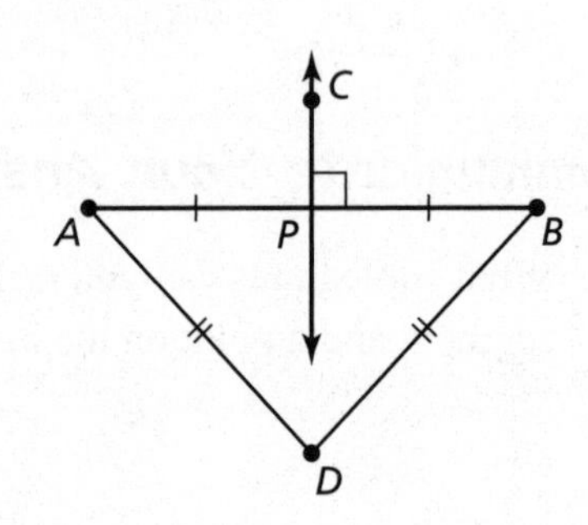

If $DA = DB$, then point D lies on the $\perp$ bisector of $\overline{AB}$.

Notes:

Name______________________________ Date__________

Theorem 6.3 Angle Bisector Theorem

If a point lies on the bisector of an angle, then it is equidistant from the two sides of the angle.

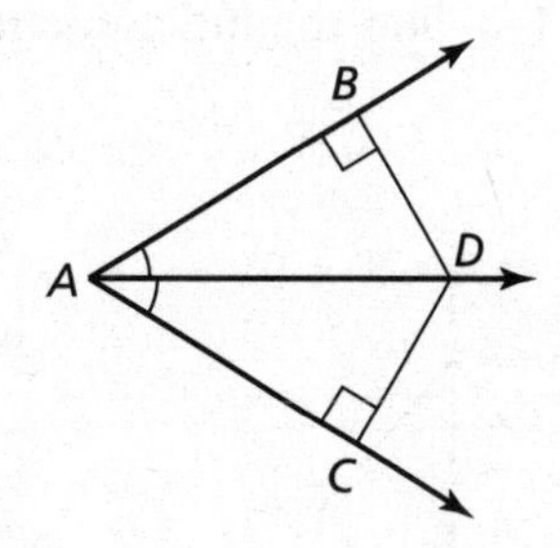

If $\overrightarrow{AD}$ bisects $\angle BAC$ and $\overline{DB} \perp \overrightarrow{AB}$ and $\overline{DC} \perp \overrightarrow{AC}$, then $DB = DC$.

Notes:

Theorem 6.4 Converse of the Angle Bisector Theorem

If a point is in the interior of an angle and is equidistant from the two sides of the angle, then it lies on the bisector of the angle.

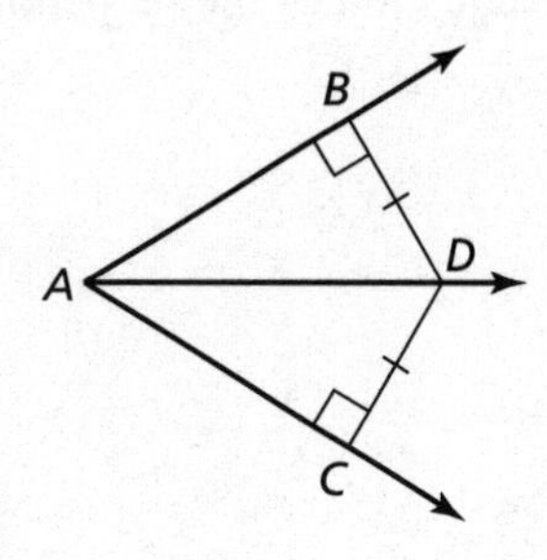

If $\overline{DB} \perp \overrightarrow{AB}$ and $\overline{DC} \perp \overrightarrow{AC}$ and $DB = DC$, then $\overrightarrow{AD}$ bisects $\angle BAC$.

Notes:

Name ______________________________ Date __________

Extra Practice

In Exercises 1–3, find the indicated measure. Explain your reasoning.

1. AB

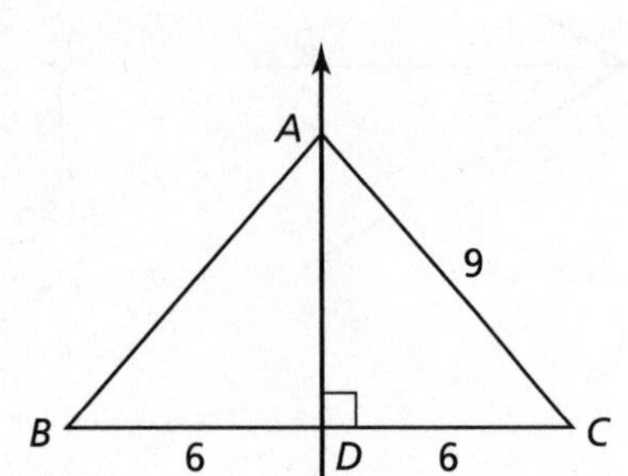

2. EG

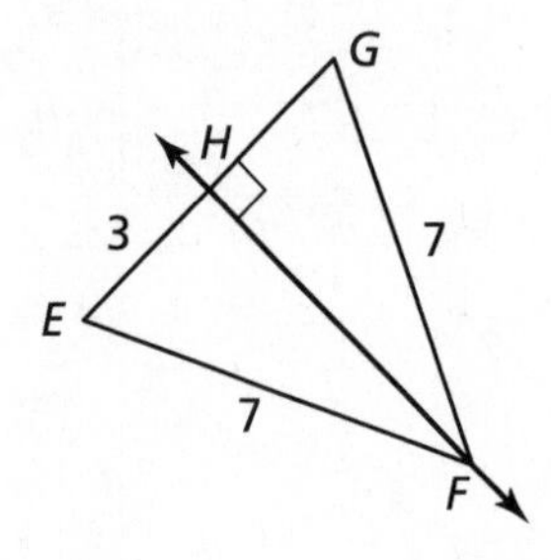

3. SU

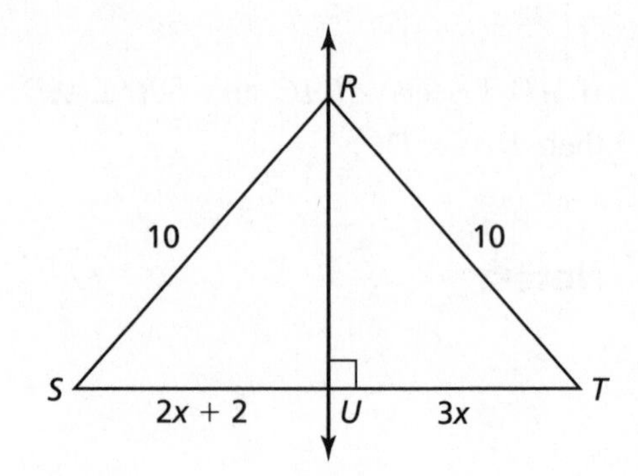

4. Find the equation of the perpendicular bisector of AB.

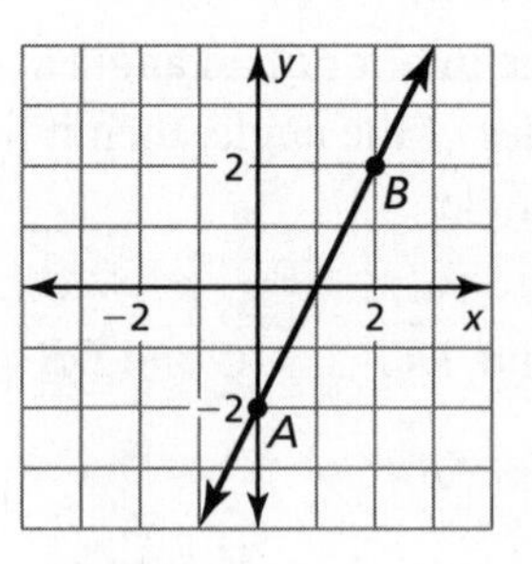

In Exercises 5–7, find the indicated measure. Explain your reasoning.

5. $m\angle CAB$

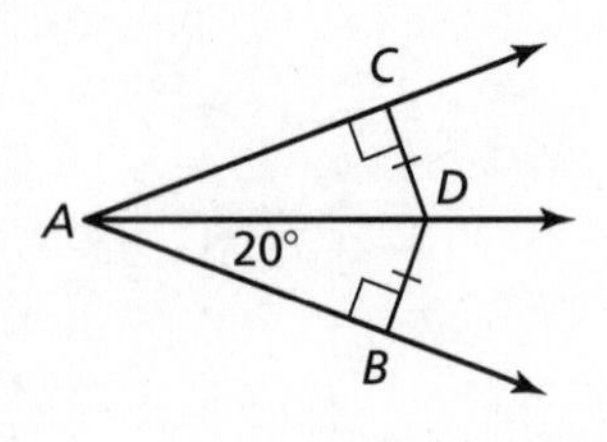

6. DC

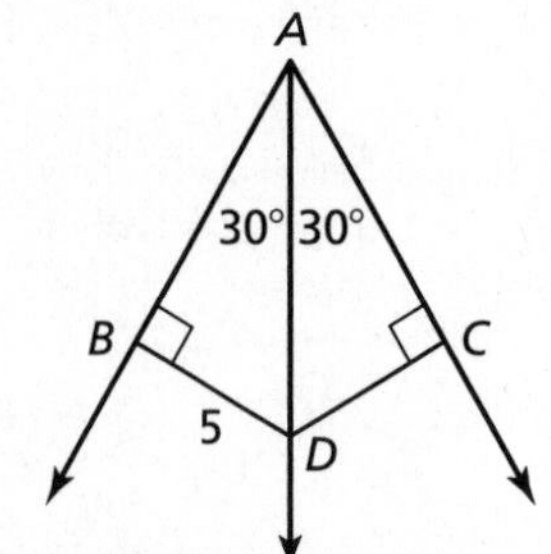

7. BD

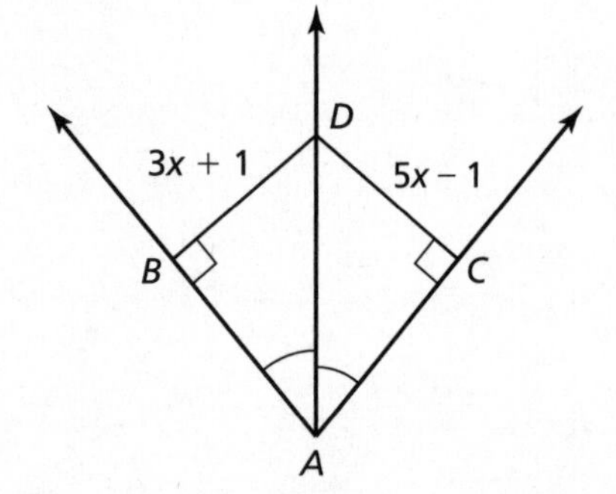

Name____ Date__________

6.2 Bisectors of Triangles

For use with Exploration 6.2

Essential Question What conjectures can you make about the perpendicular bisectors and the angle bisectors of a triangle?

1 EXPLORATION: Properties of the Perpendicular Bisectors of a Triangle

Go to *BigIdeasMath.com* for an interactive tool to investigate this exploration.

Work with a partner. Use dynamic geometry software. Draw any $\triangle ABC$.

a. Construct the perpendicular bisectors of all three sides of $\triangle ABC$. Then drag the vertices to change $\triangle ABC$. What do you notice about the perpendicular bisectors?

b. Label a point D at the intersection of the perpendicular bisectors.

c. Draw the circle with center D through vertex A of $\triangle ABC$. Then drag the vertices to change $\triangle ABC$. What do you notice?

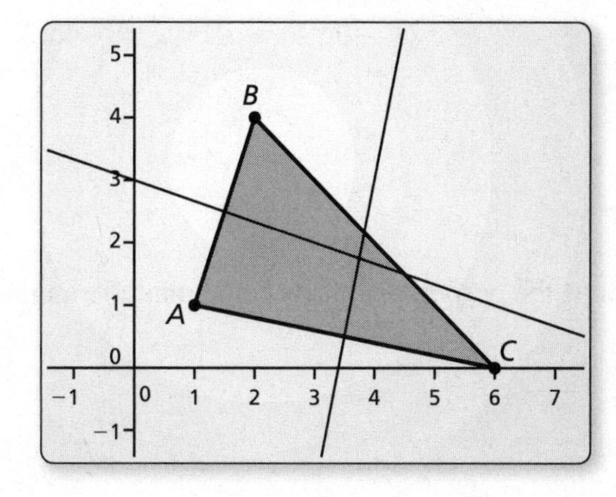

Sample
Points
$A(1, 1)$
$B(2, 4)$
$C(6, 0)$
Segments
$BC = 5.66$
$AC = 5.10$
$AB = 3.16$
Lines
$x + 3y = 9$
$-5x + y = -17$

2 EXPLORATION: Properties of the Angle Bisectors of a Triangle

Go to *BigIdeasMath.com* for an interactive tool to investigate this exploration.

Work with a partner. Use dynamic geometry software. Draw any $\triangle ABC$.

a. Construct the angle bisectors of all three angles of $\triangle ABC$. Then drag the vertices to change $\triangle ABC$. What do you notice about the angle bisectors?

2 EXPLORATION: Properties of the Angle Bisectors of a Triangle (continued)

b. Label a point D at the intersection of the angle bisectors.

c. Find the distance between D and $\overline{AB}$. Draw the circle with center D and this distance as a radius. Then drag the vertices to change $\triangle ABC$. What do you notice?

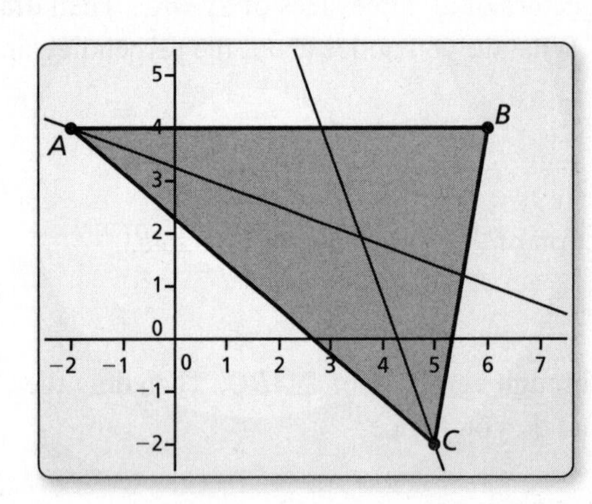

Sample

Points

$A(-2, 4)$

$B(6, 4)$

$C(5, -2)$

Segments

$BC = 6.08$

$AC = 9.22$

$AB = 8$

Lines

$0.35x + 0.94y = 3.06$

$-0.94x - 0.34y = -4.02$

Communicate Your Answer

3. What conjectures can you make about the perpendicular bisectors and the angle bisectors of a triangle?

Name____ Date__________

6.2 Notetaking with Vocabulary

For use after Lesson 6.2

In your own words, write the meaning of each vocabulary term.

concurrent

point of concurrency

circumcenter

incenter

Theorems

Theorem 6.5 Circumcenter Theorem

The circumcenter of a triangle is equidistant from the vertices of the triangle.

If $\overline{PD}$, $\overline{PE}$, and $\overline{PF}$ are perpendicular bisectors, then $PA = PB = PC$.

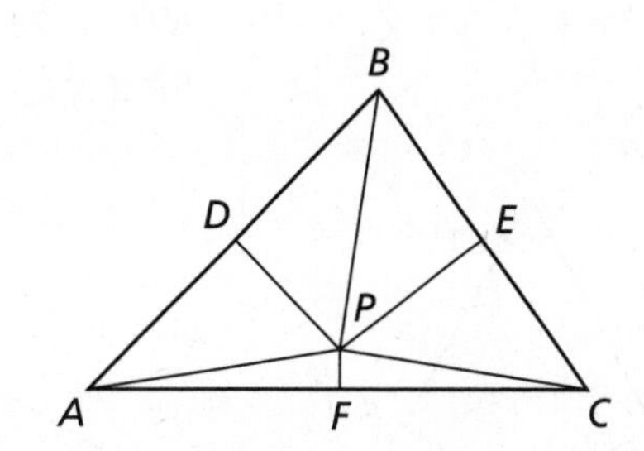

Notes:

Theorem 6.6 Incenter Theorem

The incenter of a triangle is equidistant from the sides of the triangle.

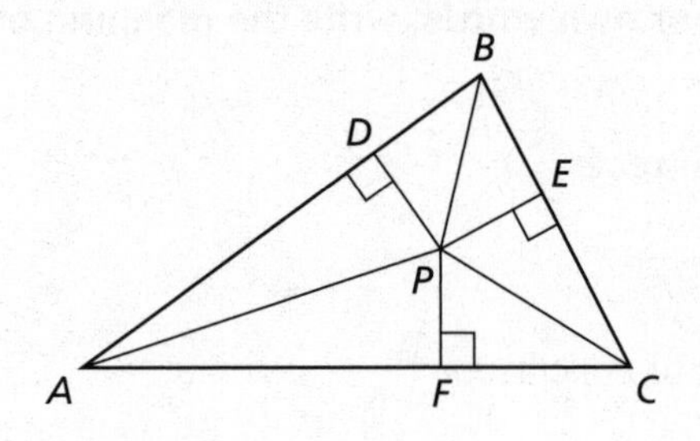

If $\overline{AP}$, $\overline{BP}$, and $\overline{CP}$ are angle bisectors of $\triangle ABC$, then $PD = PE = PF$.

Notes:

Extra Practice

In Exercises 1–3, *N* is the incenter of $\triangle ABC$. Use the given information to find the indicated measure.

1. $ND = 2x - 5$
   $NE = -2x + 7$
   Find NF.

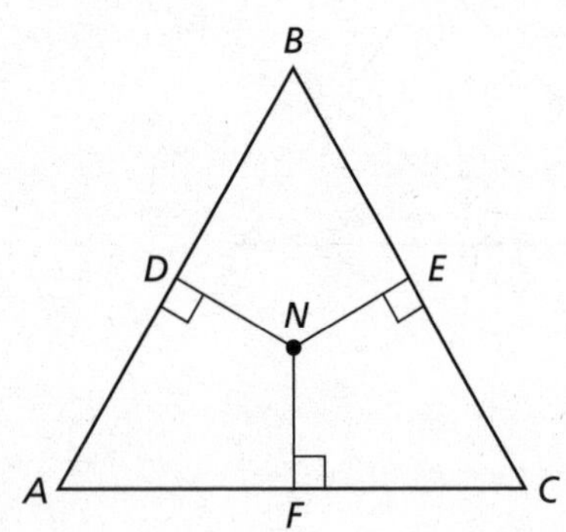

2. $NG = x - 1$
   $NH = 2x - 6$
   Find NJ.

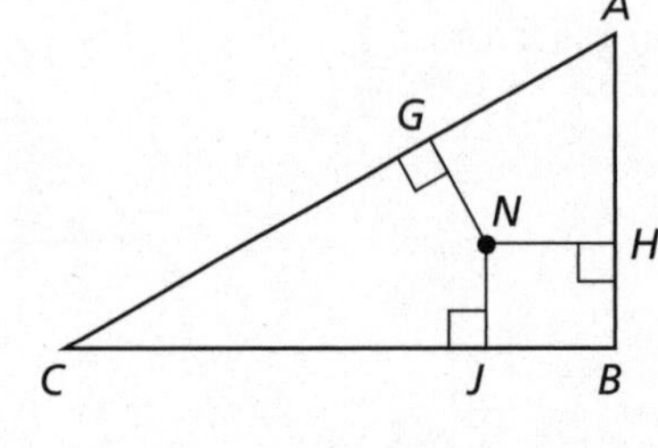

3. $NK = x + 10$
   $NL = -2x + 1$
   Find NM.

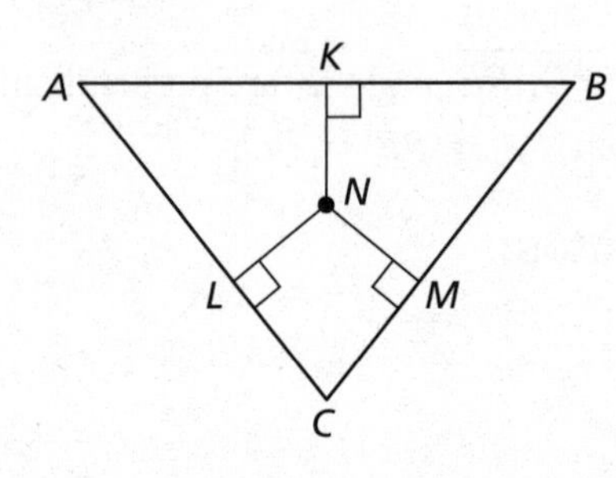

Name____ Date__________

In Exercises 4–7, find the indicated measure.

4. PA

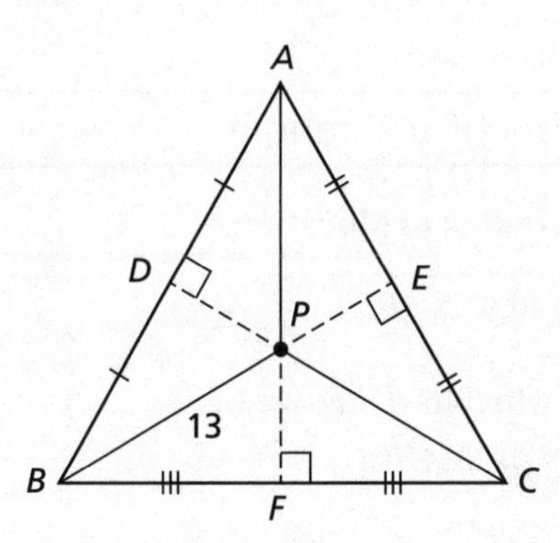

5. PS

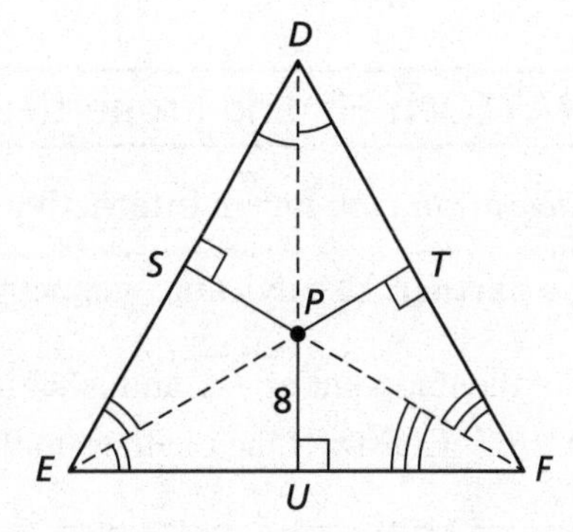

6. GE

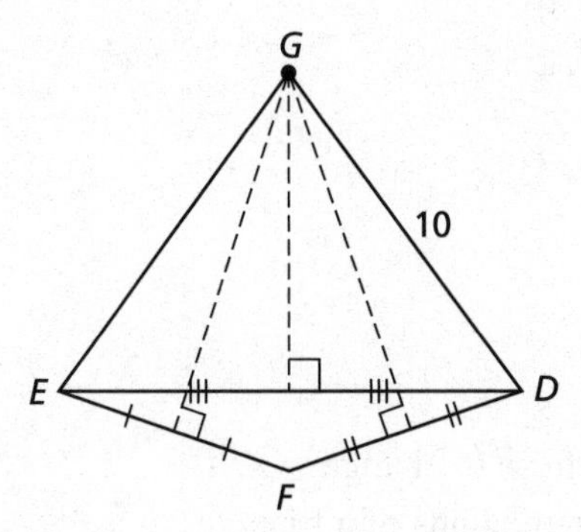

7. NF

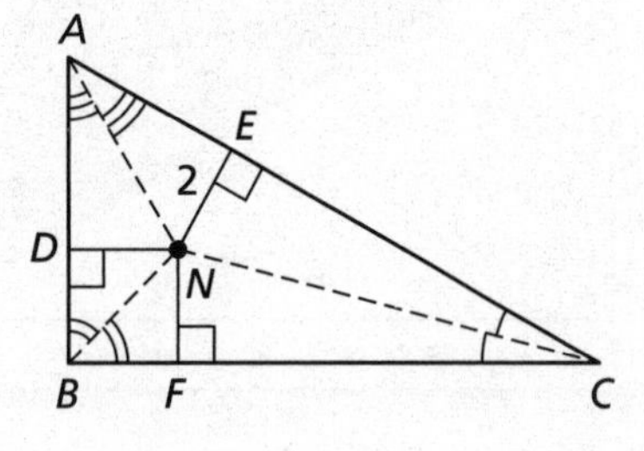

In Exercises 8–10, find the coordinates of the circumcenter of the triangle with the given vertices.

8. $A(-2,-2), B(-2,4), C(6,4)$ **9.** $D(3,5), E(3,1), F(9,5)$ **10.** $J(4,-7), K(4,-3), L(-6,-3)$

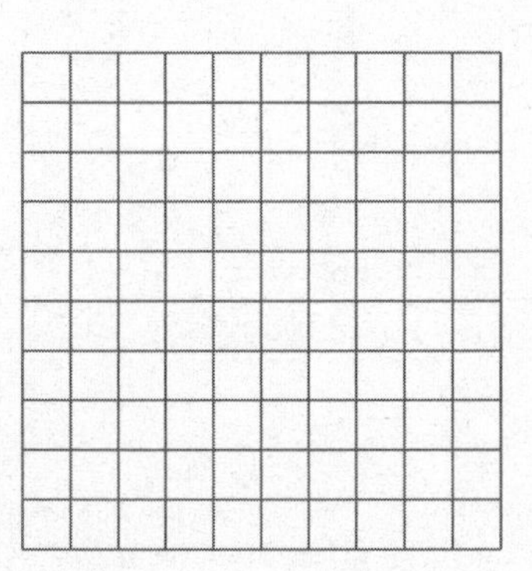

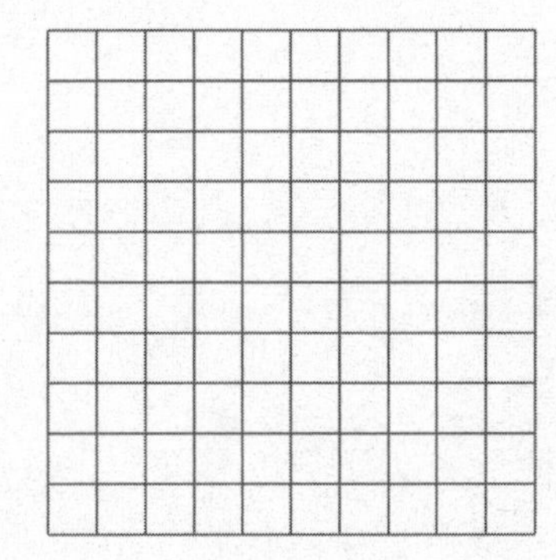

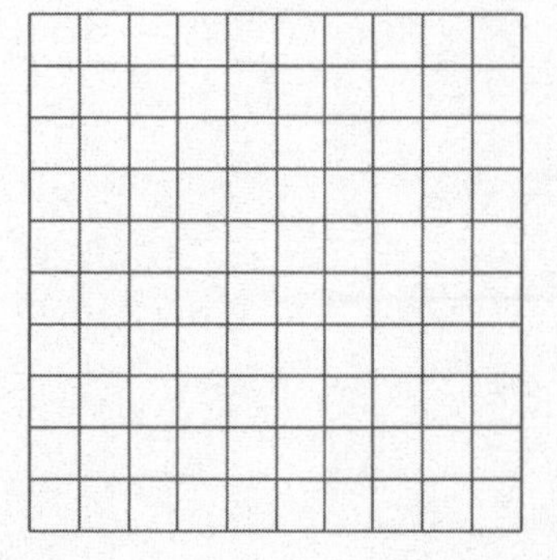

Name ________________________________ Date __________

6.3 Medians and Altitudes of Triangles

For use with Exploration 6.3

Essential Question What conjectures can you make about the medians and altitudes of a triangle?

1 EXPLORATION: Finding Properties of the Medians of a Triangle

Go to *BigIdeasMath.com* for an interactive tool to investigate this exploration.

Work with a partner. Use dynamic geometry software. Draw any $\triangle ABC$.

a. Plot the midpoint of $\overline{BC}$ and label it D. Draw $\overline{AD}$, which is a *median* of $\triangle ABC$. Construct the medians to the other two sides of $\triangle ABC$.

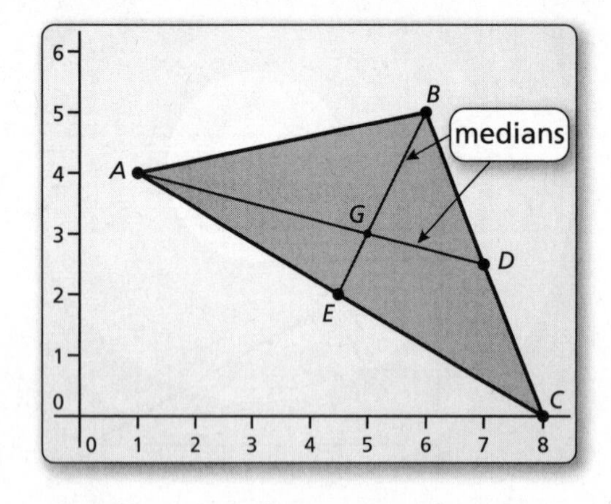

Sample
Points
$A(1, 4)$
$B(6, 5)$
$C(8, 0)$
$D(7, 2.5)$
$E(4.5, 2)$
$G(5, 3)$

b. What do you notice about the medians? Drag the vertices to change $\triangle ABC$. Use your observations to write a conjecture about the medians of a triangle.

c. In the figure above, point G divides each median into a shorter segment and a longer segment. Find the ratio of the length of each longer segment to the length of the whole median. Is this ratio always the same? Justify your answer.

2 EXPLORATION: Finding Properties of the Altitudes of a Triangle

Go to *BigIdeasMath.com* for an interactive tool to investigate this exploration.

Work with a partner. Use dynamic geometry software. Draw any $\triangle ABC$.

a. Construct the perpendicular segment from vertex A to $\overline{BC}$. Label the endpoint D. $\overline{AD}$ is an *altitude* of $\triangle ABC$.

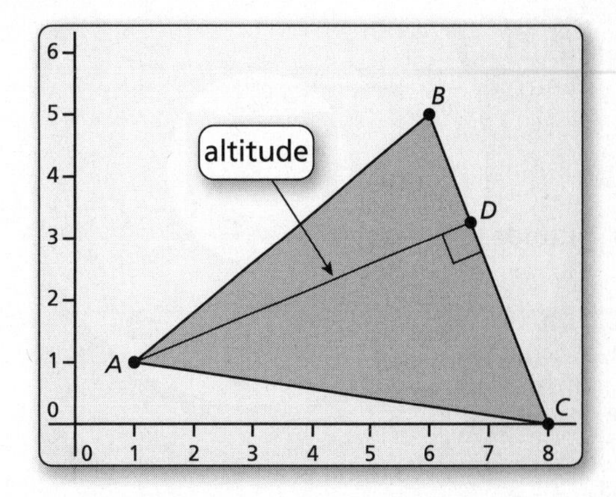

b. Construct the altitudes to the other two sides of $\triangle ABC$. What do you notice?

c. Write a conjecture about the altitudes of a triangle. Test your conjecture by dragging the vertices to change $\triangle ABC$.

Communicate Your Answer

3. What conjectures can you make about the medians and altitudes of a triangle?

4. The length of median $\overline{RU}$ in $\triangle RST$ is 3 inches. The point of concurrency of the three medians of $\triangle RST$ divides $\overline{RU}$ into two segments. What are the lengths of these two segments?

Name ______________________________ Date __________

6.3 Notetaking with Vocabulary
For use after Lesson 6.3

In your own words, write the meaning of each vocabulary term.

median of a triangle

centroid

altitude of a triangle

orthocenter

Theorems

Theorem 6.7 Centroid Theorem

The centroid of a triangle is two-thirds of the distance from each vertex to the midpoint of the opposite side.

The medians of $\triangle ABC$ meet at point P, and

$AP = \frac{2}{3}AE$, $BP = \frac{2}{3}BF$, and $CP = \frac{2}{3}CD$.

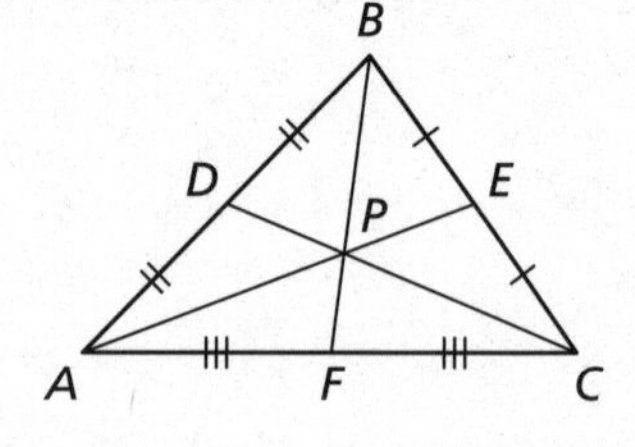

Notes:

Core Concepts

Orthocenter

The lines containing the altitudes of a triangle are concurrent. This point of concurrency is the **orthocenter** of the triangle.

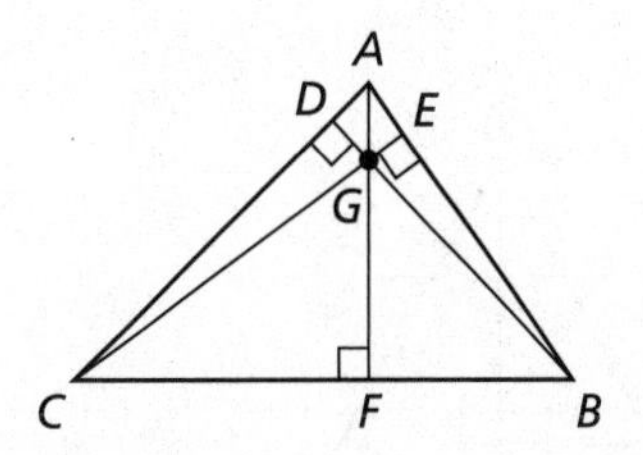

The lines containing $\overline{AF}$, $\overline{BD}$, and $\overline{CE}$ meet at the orthocenter G of $\triangle ABC$.

Notes:

Extra Practice

In Exercises 1–3, point *P* is the centroid of △*LMN*. Find *PN* and *QP*.

1. $QN = 33$

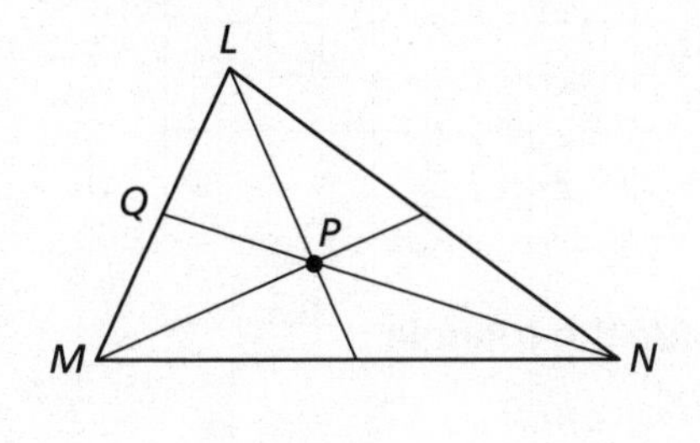

2. $QN = 45$

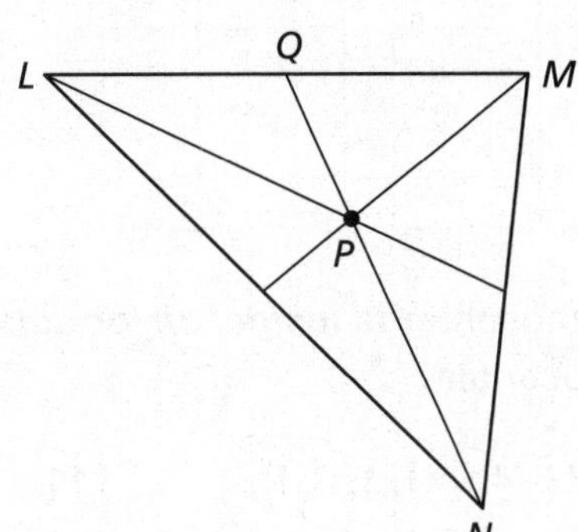

3. $QN = 39$

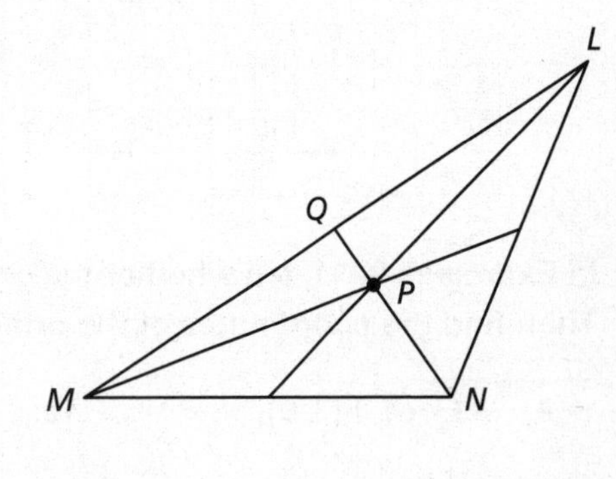

Name ______________________________ Date __________

6.3 Notetaking with Vocabulary (continued)

In Exercises 4 and 5, point *D* is the centroid of △*ABC*. Find *CD* and *CE*.

4. $DE = 7$

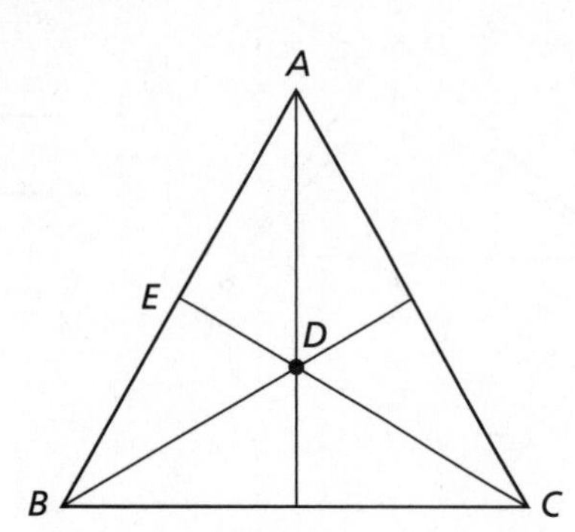

5. $DE = 12$

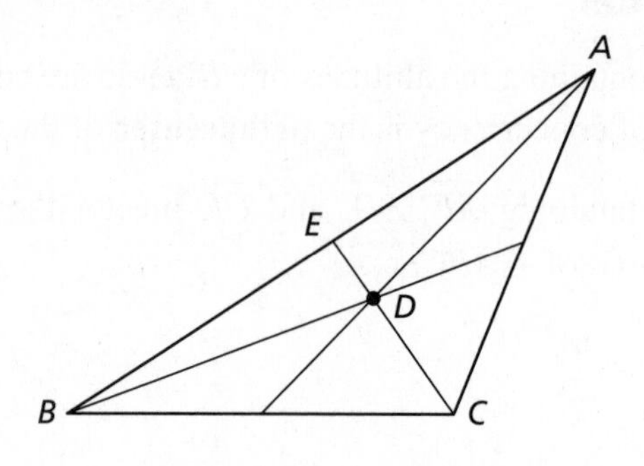

In Exercises 6–8, find the coordinates of the centroid of the triangle with the given vertices.

6. $A(-2, -1), B(1, 8), C(4, -1)$

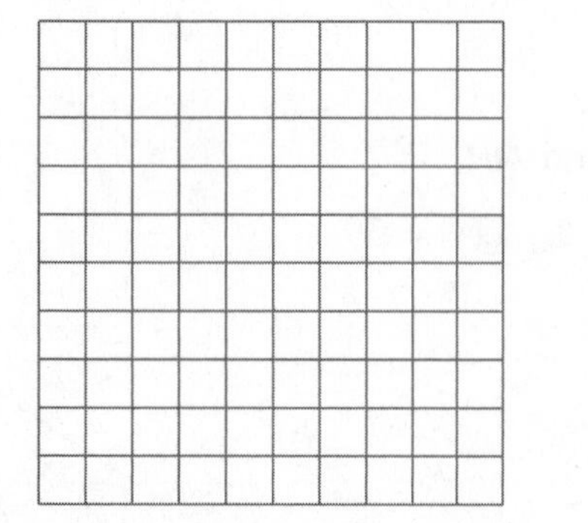

7. $D(-5, 4), E(-3, -2), F(-1, 4)$

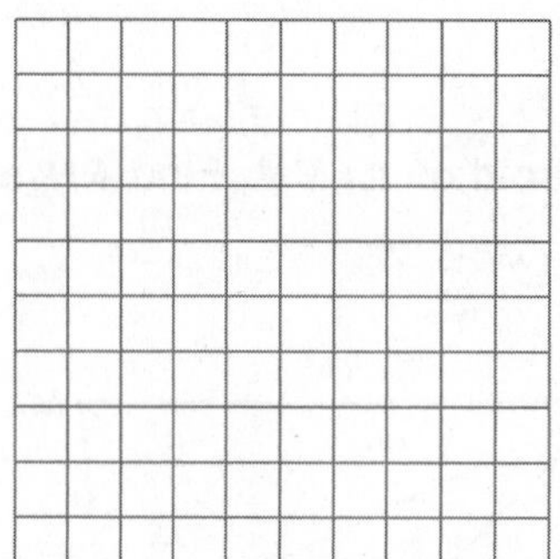

8. $J(8, 7), K(20, 5), L(8, 3)$

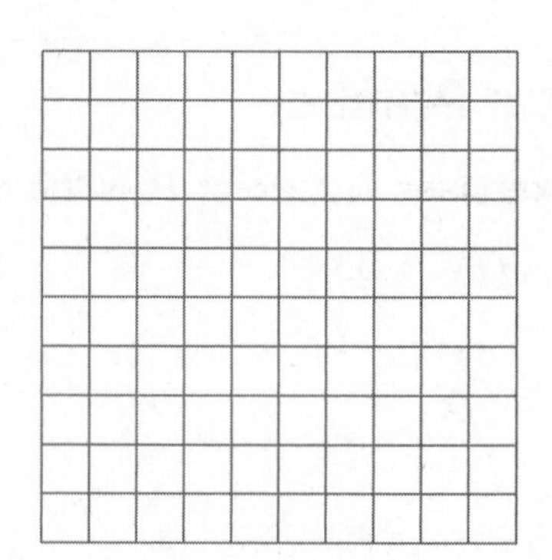

In Exercises 9–11, tell whether the orthocenter is *inside, on,* or *outside* the triangle. Then find the coordinates of the orthocenter.

9. $X(3, 6), Y(3, 0), Z(11, 0)$

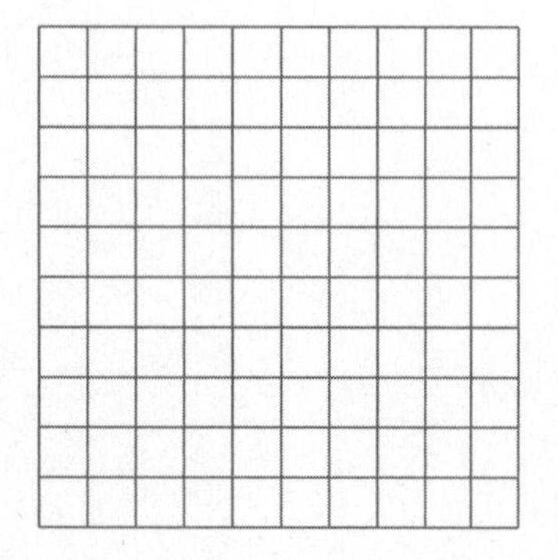

10. $L(-4, -4), M(1, 1), N(6, -4)$

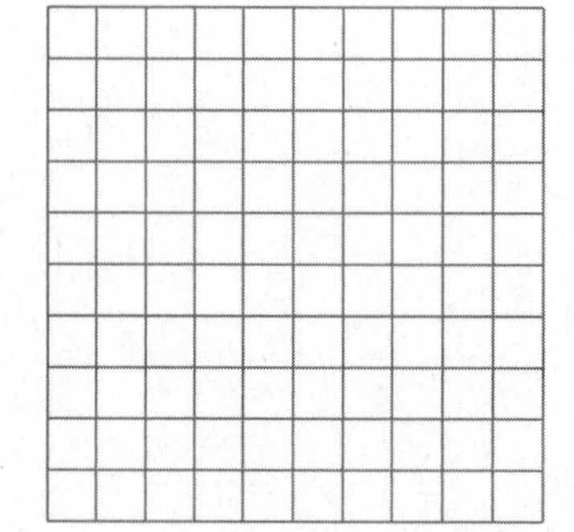

11. $P(3, 4), Q(11, 4), R(9, -2)$

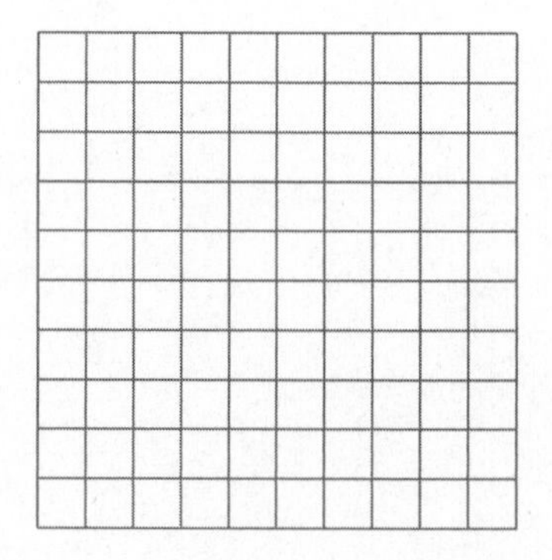

Name____ Date__________

The Triangle Midsegment Theorem

For use with Exploration 6.4

Essential Question How are the midsegments of a triangle related to the sides of the triangle?

1 EXPLORATION: Midsegments of a Triangle

Go to *BigIdeasMath.com* for an interactive tool to investigate this exploration.

Work with a partner. Use dynamic geometry software. Draw any $\triangle ABC$.

a. Plot midpoint D of $\overline{AB}$ and midpoint E of $\overline{BC}$. Draw $\overline{DE}$, which is a *midsegment* of $\triangle ABC$.

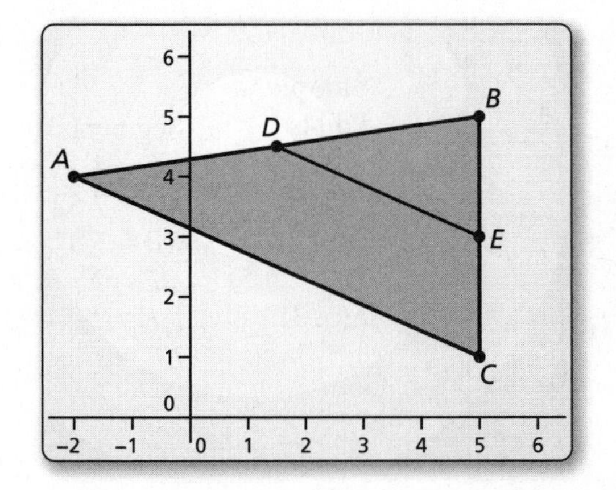

Sample
Points
$A(-2, 4)$
$B(5, 5)$
$C(5, 1)$
$D(1.5, 4.5)$
$E(5, 3)$
Segments
$BC = 4$
$AC = 7.62$
$AB = 7.07$
$DE = ?$

b. Compare the slope and length of $\overline{DE}$ with the slope and length of $\overline{AC}$.

c. Write a conjecture about the relationships between the midsegments and sides of a triangle. Test your conjecture by drawing the other midsegments of $\triangle ABC$, dragging vertices to change $\triangle ABC$, and noting whether the relationships hold.

2 EXPLORATION: Midsegments of a Triangle

Go to *BigIdeasMath.com* for an interactive tool to investigate this exploration.

Work with a partner. Use dynamic geometry software. Draw any $\triangle ABC$.

a. Draw all three midsegments of $\triangle ABC$.

b. Use the drawing to write a conjecture about the triangle formed by the midsegments of the original triangle.

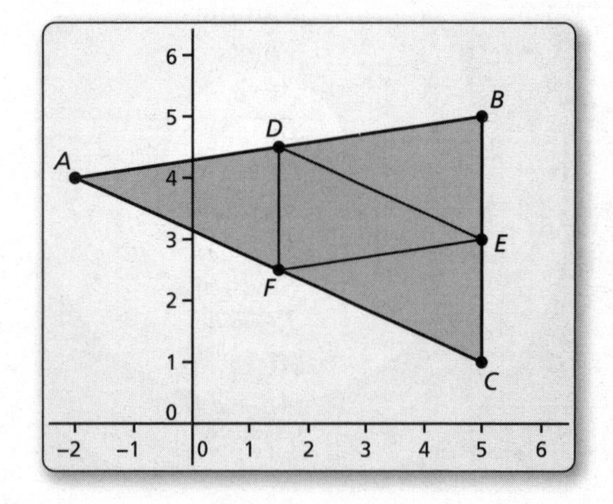

Sample

Points	Segments
$A(-2, 4)$	$BC = 4$
$B(5, 5)$	$AC = 7.62$
$C(5, 1)$	$AB = 7.07$
$D(1.5, 4.5)$	$DE = ?$
$E(5, 3)$	$DF = ?$
	$EF = ?$

Communicate Your Answer

3. How are the midsegments of a triangle related to the sides of the triangle?

4. In $\triangle RST$, $\overline{UV}$ is the midsegment connecting the midpoints of $\overline{RS}$ and $\overline{ST}$. Given $UV = 12$, find RT.

Name____ Date__________

6.4 Notetaking with Vocabulary

For use after Lesson 6.4

In your own words, write the meaning of each vocabulary term.

midsegment of a triangle

Theorems

Theorem 6.8 Triangle Midsegment Theorem

The segment connecting the midpoints of two sides of a triangle is parallel to the third side and is half as long as that side.

$\overline{DE}$ is a midsegment of $\triangle ABC$, $\overline{DE} \parallel \overline{AC}$, and $DE = \frac{1}{2}AC$.

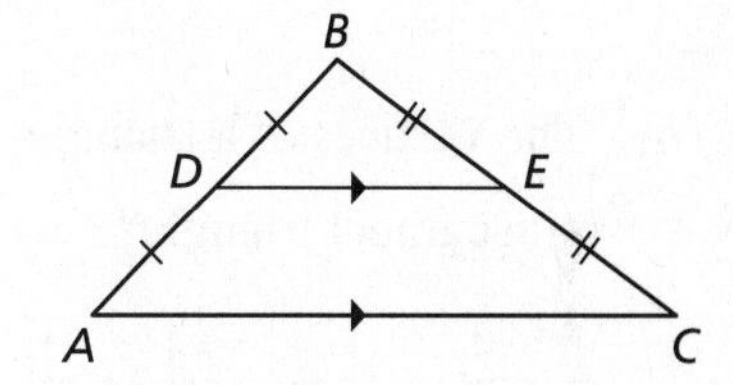

Notes:

Name ____ Date __________

Extra Practice

In Exercises 1–3, *DE* is a midsegment of $\triangle ABC$. Find the value of *x*.

1.

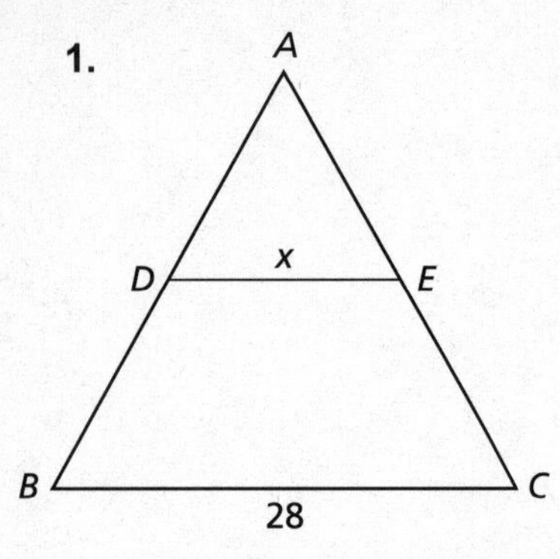

2.

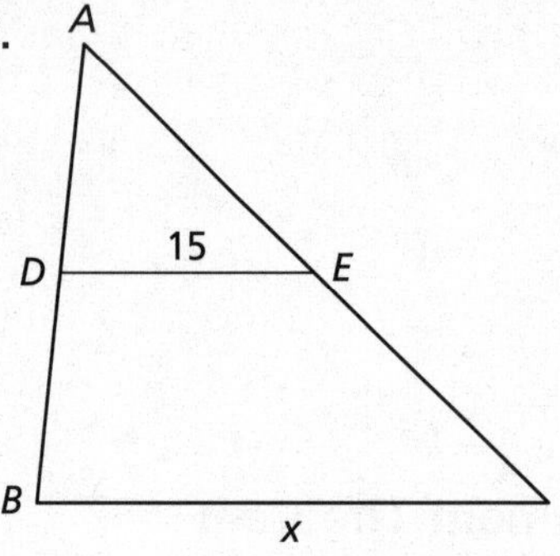

3.

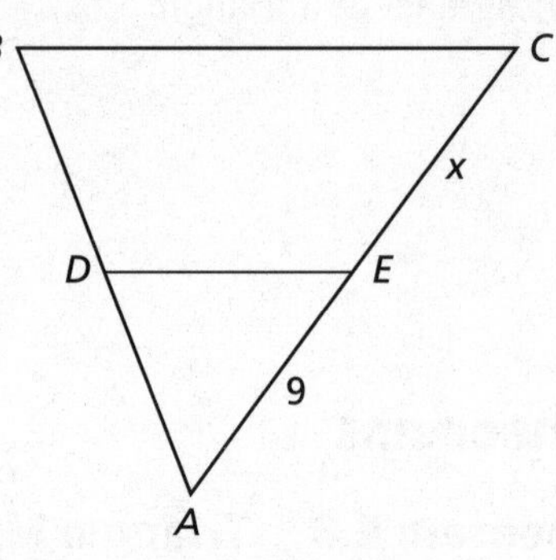

4. The vertices of a triangle are $A(-5, 6)$, $B(3, 8)$, and $C(1, -4)$. What are the vertices of the midsegment triangle?

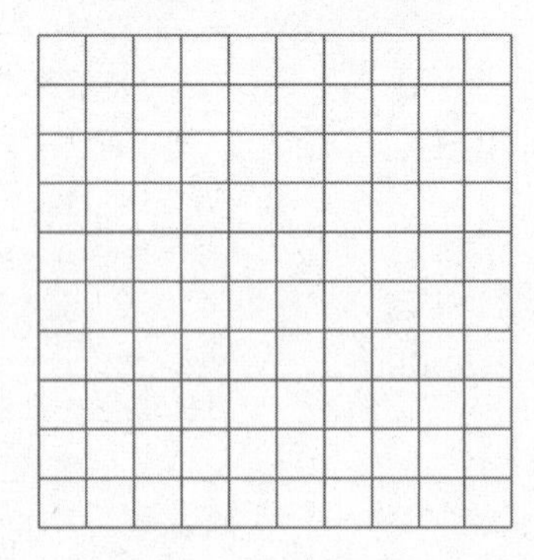

5. What is the perimeter of $\triangle DEF$?

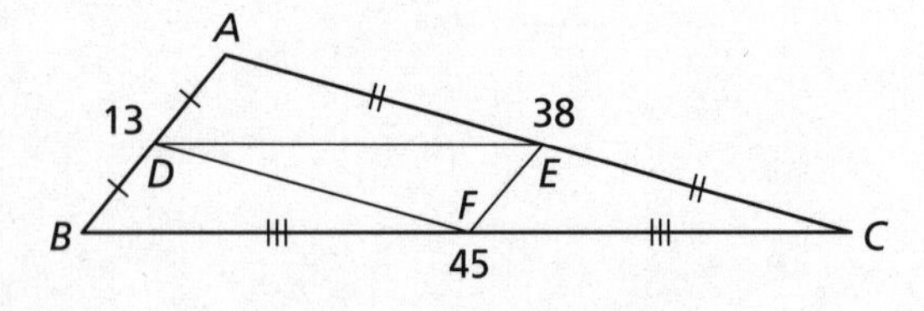

6. In the diagram, $\overline{DE}$ is a midsegment of $\triangle ABC$, and $\overline{FG}$ is a midsegment of $\triangle ADE$. Find *FG*.

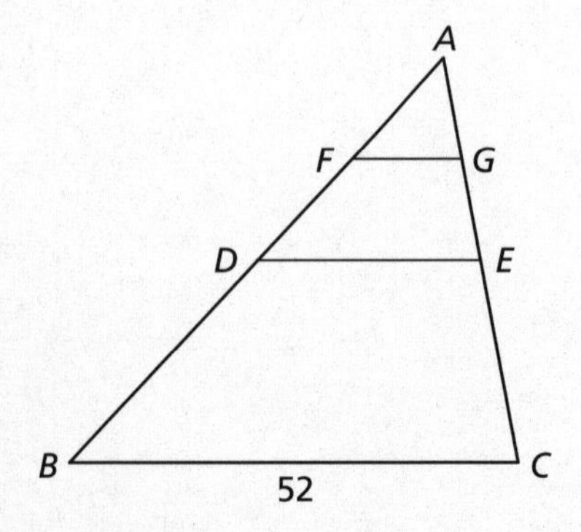

7. The area of $\triangle ABC$ is 48 cm^2. $\overline{DE}$ is a midsegment of $\triangle ABC$. What is the area of $\triangle ADE$?

8. The diagram below shows a triangular wood shed. You want to install a shelf halfway up the 8-foot wall that will be built between the two walls.

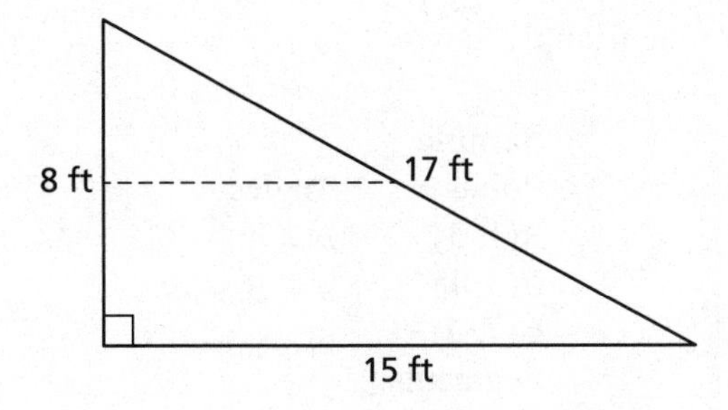

a. How long will the shelf be?

b. How many feet should you measure from the ground along the slanting wall to find where to attach the opposite end of the shelf so that it will be level?

Name ______________________________ Date __________

6.5 Indirect Proof and Inequalities in One Triangle
For use with Exploration 6.5

Essential Question How are the sides related to the angles of a triangle? How are any two sides of a triangle related to the third side?

1 EXPLORATION: Comparing Angle Measures and Side Lengths

Go to *BigIdeasMath.com* for an interactive tool to investigate this exploration.

Work with a partner. Use dynamic geometry software. Draw any scalene $\triangle ABC$.

a. Find the side lengths and angle measures of the triangle.

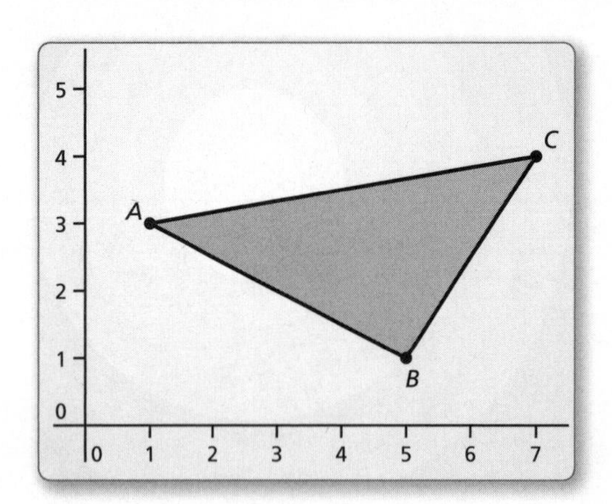

Sample

Points	Angles
$A(1, 3)$	$m\angle A = ?$
$B(5, 1)$	$m\angle B = ?$
$C(7, 4)$	$m\angle C = ?$

Segments
$BC = ?$
$AC = ?$
$AB = ?$

b. Order the side lengths. Order the angle measures. What do you observe?

c. Drag the vertices of $\triangle ABC$ to form new triangles. Record the side lengths and angle measures in the following table. Write a conjecture about your findings.

BC	AC	AB	$m\angle A$	$m\angle B$	$m\angle C$

Name____ Date__________

2 EXPLORATION: A Relationship of the Side Lengths of a Triangle

Go to *BigIdeasMath.com* for an interactive tool to investigate this exploration.

Work with a partner. Use dynamic geometry software. Draw any $\triangle ABC$.

a. Find the side lengths of the triangle.

b. Compare each side length with the sum of the other two side lengths.

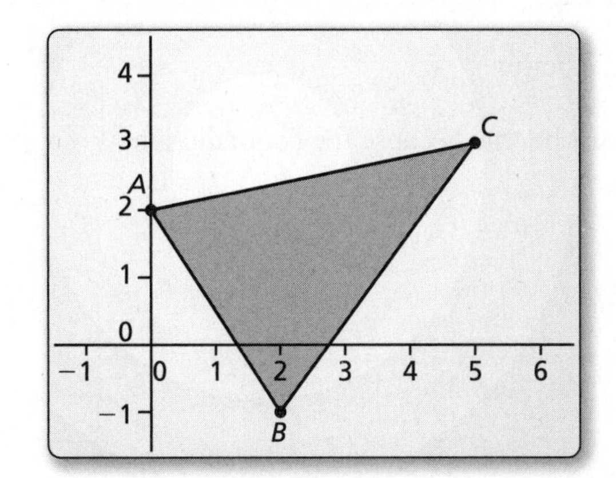

Sample
Points
$A(0, 2)$
$B(2, -1)$
$C(5, 3)$
Segments
$BC = ?$
$AC = ?$
$AB = ?$

c. Drag the vertices of $\triangle ABC$ to form new triangles and repeat parts (a) and (b). Organize your results in a table. Write a conjecture about your findings.

BC	*AC*	*AB*	Comparisons

Communicate Your Answer

3. How are the sides related to the angles of a triangle? How are any two sides of a triangle related to the third side?

4. Is it possible for a triangle to have side lengths of 3, 4, and 10? Explain.

Name ______________________________ Date __________

6.5 Notetaking with Vocabulary
For use after Lesson 6.5

In your own words, write the meaning each vocabulary term.

indirect proof

Core Concepts

How to Write an Indirect Proof (Proof by Contradiction)

Step 1 Identify the statement you want to prove. Assume temporarily that this statement is false by assuming that its opposite is true.

Step 2 Reason logically until you reach a contradiction.

Step 3 Point out that the desired conclusion must be true because the contradiction proves the temporary assumption false.

Notes:

Theorems

Theorem 6.9 Triangle Longer Side Theorem

If one side of a triangle is longer than another side, then the angle opposite the longer side is larger than the angle opposite the shorter side.

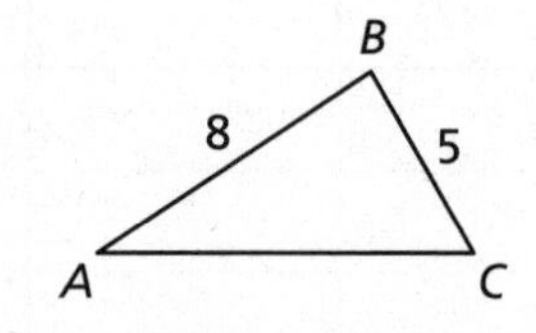

$AB > BC$, so $m\angle C > m\angle A$.

Notes:

Theorem 6.10 Triangle Larger Angle Theorem

If one angle of a triangle is larger than another angle, then the side opposite the larger angle is longer than the side opposite the smaller angle.

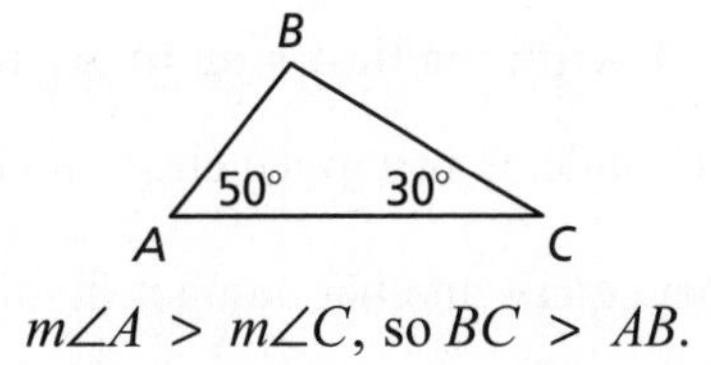

$m\angle A > m\angle C$, so $BC > AB$.

Notes:

Theorem 6.11 Triangle Inequality Theorem

The sum of the lengths of any two sides of a triangle is greater than the length of the third side.

$AB + BC > AC$ $\quad$ $AC + BC > AB$ $\quad$ $AB + AC > BC$

Notes:

Extra Practice

In Exercises 1–3, write the first step in an indirect proof of the statement.

1. Not all the students in a given class can be above average.

2. No number equals another number divided by zero.

3. The square root of 2 is not equal to the quotient of any two integers.

In Exercises 4 and 5, determine which two statements contradict each other. Explain your reasoning.

4. **A** ΔLMN is equilateral.

B $LM \neq MN$

C $\angle L = \angle M$

5. **A** ΔABC is a right triangle.

B $\angle A$ is acute.

C $\angle C$ is obtuse.

In Exercises 6–8, list the angles of the given triangle from smallest to largest.

6.

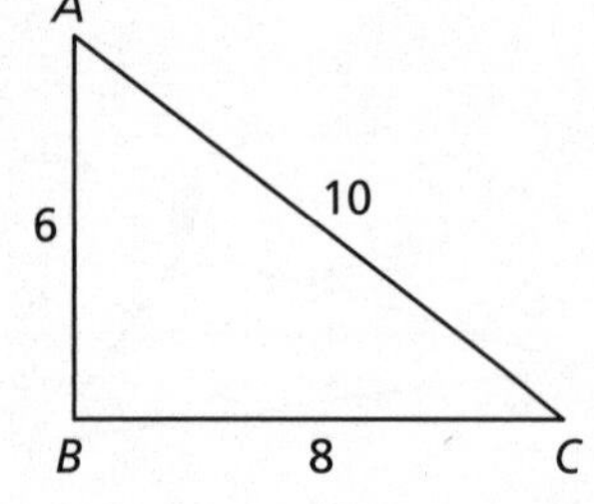

7.

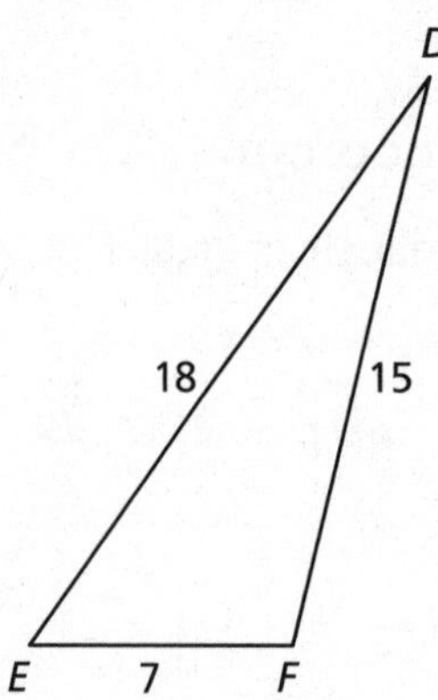

8.

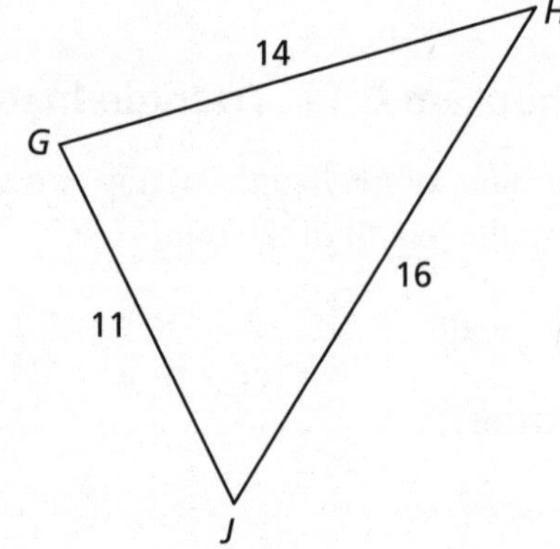

In Exercises 9–12, is it possible to construct a triangle with the given side lengths? If not, explain why not.

9. 3, 12, 17 **10.** 5, 21, 16 **11.** 8, 5, 7 **12.** 10, 3, 11

13. A triangle has two sides with lengths 5 inches and 13 inches. Describe the possible lengths of the third side of the triangle.

Name____ Date__________

6.6 Inequalities in Two Triangles
For use with Exploration 6.6

Essential Question If two sides of one triangle are congruent to two sides of another triangle, what can you say about the third sides of the triangles?

1 EXPLORATION: Comparing Measures in Triangles

Go to *BigIdeasMath.com* for an interactive tool to investigate this exploration.

Work with a partner. Use dynamic geometry software.

a. Draw $\triangle ABC$, as shown below.

b. Draw the circle with center $C(3, 3)$ through the point $A(1, 3)$.

c. Draw $\triangle DBC$ so that D is a point on the circle.

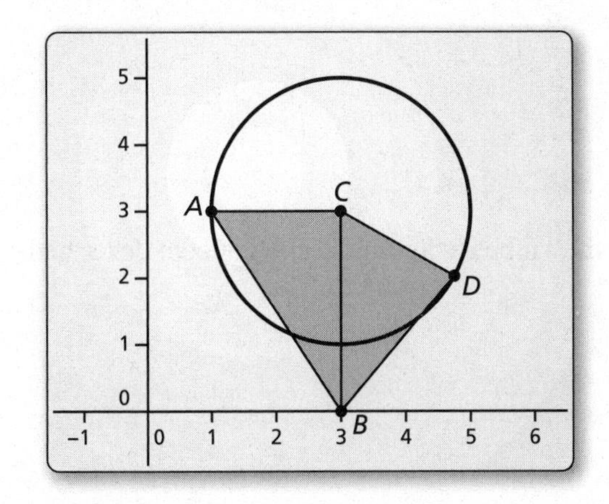

Sample
Points
$A(1, 3)$
$B(3, 0)$
$C(3, 3)$
$D(4.75, 2.03)$
Segments
$BC = 3$
$AC = 2$
$DC = 2$
$AB = 3.61$
$DB = 2.68$

d. Which two sides of $\triangle ABC$ are congruent to two sides of $\triangle DBC$? Justify your answer.

e. Compare the lengths of $\overline{AB}$ and $\overline{DB}$. Then compare the measures of $\angle ACB$ and $\angle DCB$. Are the results what you expected? Explain.

f. Drag point D to several locations on the circle. At each location, repeat part (e). Copy and record your results in the table below.

	D	*AC*	*BC*	*AB*	*BD*	$m\angle ACB$	$m\angle BCD$
1.	(4.75, 2.03)	2	3				
2.		2	3				
3.		2	3				
4.		2	3				
5.		2	3				

1 EXPLORATION: Comparing Measures in Triangles (continued)

g. Look for a pattern of the measures in your table. Then write a conjecture that summarizes your observations.

Communicate Your Answer

2. If two sides of one triangle are congruent to two sides of another triangle, what can you say about the third sides of the triangles?

3. Explain how you can use the hinge shown below to model the concept described in Question 2.

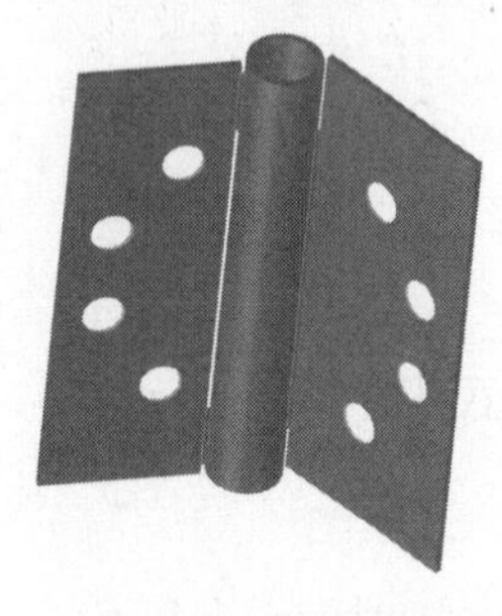

Name____ Date__________

6.6 Notetaking with Vocabulary
For use after Lesson 6.6

In your own words, write the meaning of each vocabulary term.

indirect proof

inequality

Theorems

Theorem 6.12 Hinge Theorem

If two sides of one triangle are congruent to two sides of another triangle, and the included angle of the first is larger than the included angle of the second, then the third side of the first is longer than the third side of the second.

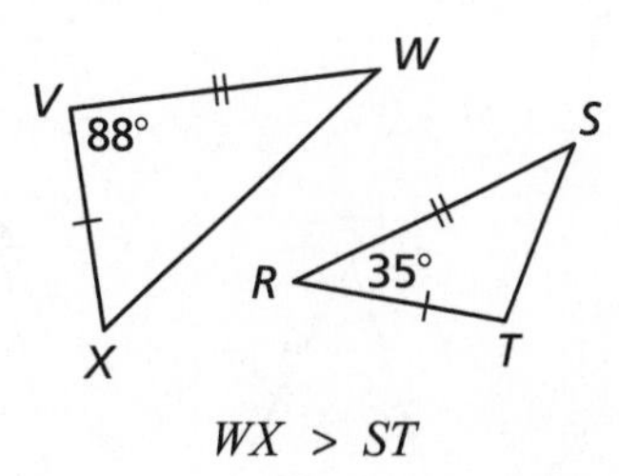

$WX > ST$

Notes:

Theorem 6.13 Converse of the Hinge Theorem

If two sides of one triangle are congruent to two sides of another triangle, and the third side of the first is longer than the third side of the second, then the included angle of the first is larger than the included angle of the second.

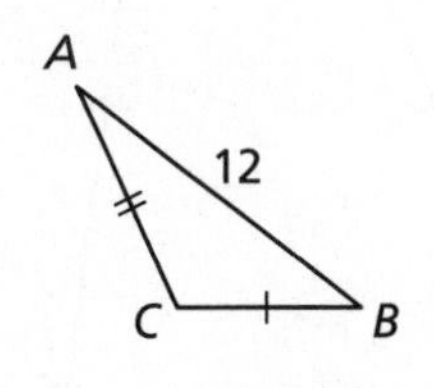

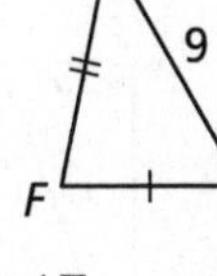

$m\angle C > m\angle F$

Notes:

Name ____ Date __________

6.6 Notetaking with Vocabulary (continued)

Extra Practice

In Exercises 1–9, complete the statement with <, >, or =. Explain your reasoning.

1. BC _____ EF

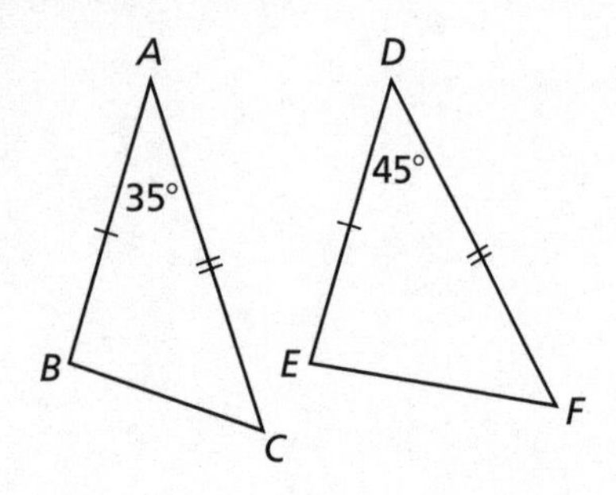

2. BC _____ EF

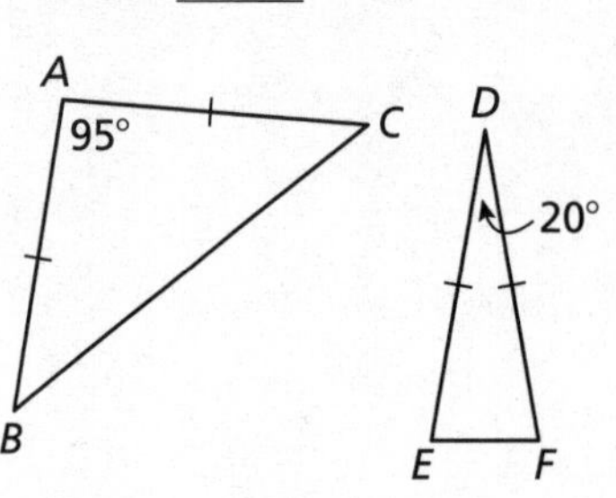

3. BC _____ EF

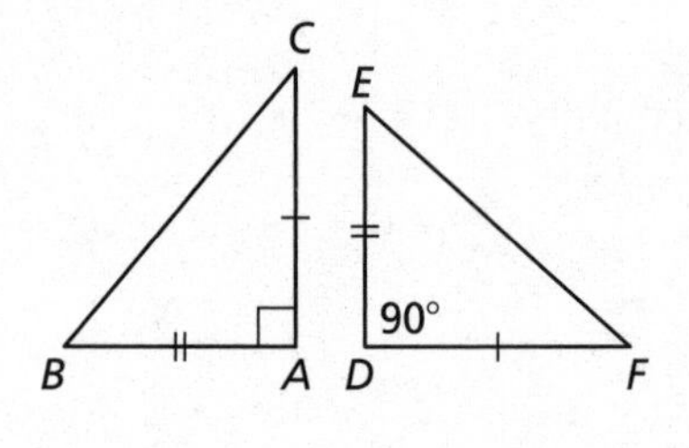

4. $m\angle A$ _____ $m\angle D$

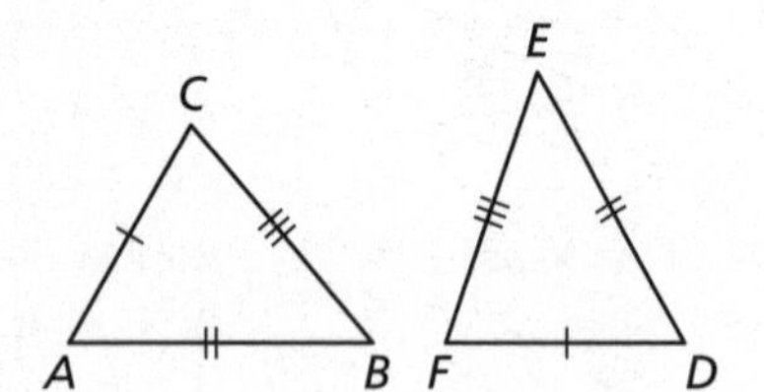

5. $m\angle A$ _____ $m\angle D$

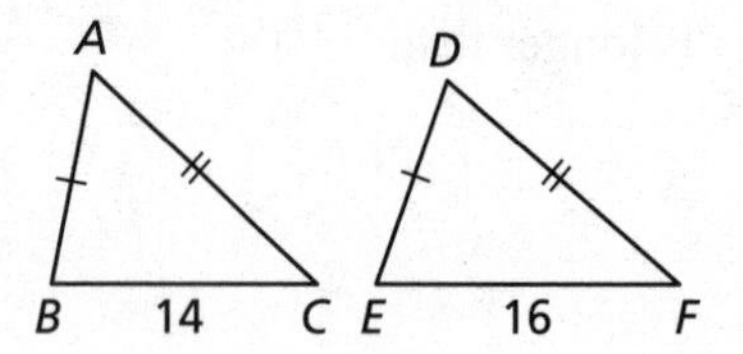

6. $m\angle A$ _____ $m\angle D$

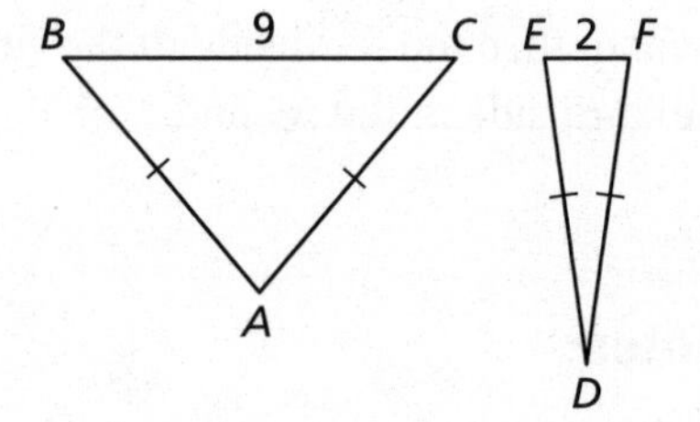

7. AB _____ AC

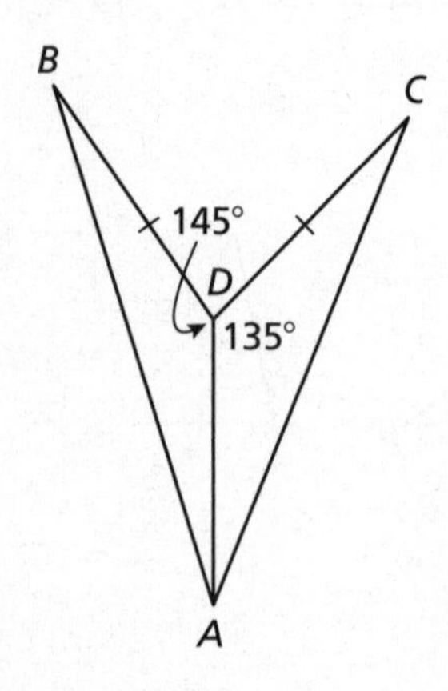

8. AB _____ CD

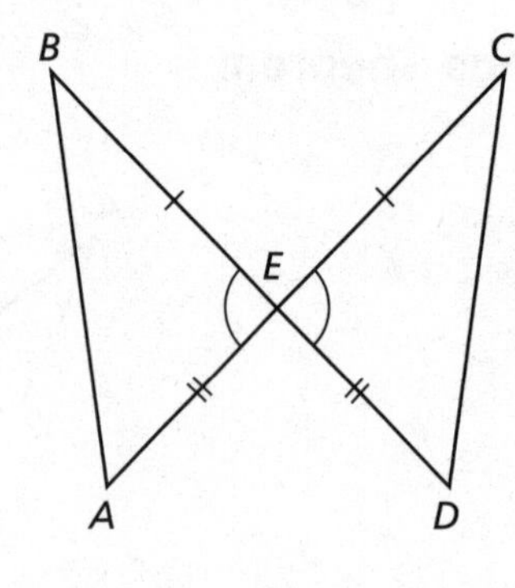

9. $m\angle 1$ _____ $m\angle 2$

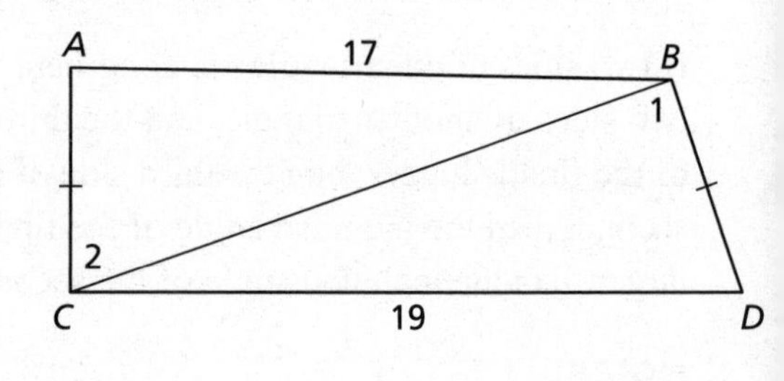

Name____ Date__________

6.6 Notetaking with Vocabulary (continued)

In Exercises 10 and 11, write a proof.

10. **Given** $\overline{XY} \cong \overline{YZ}$, $WX > WZ$

Prove $m\angle WYX > m\angle WYZ$

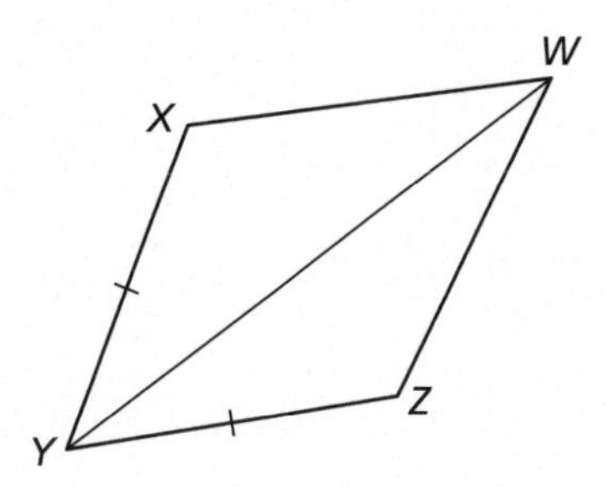

11. **Given** $\overline{AD} \cong \overline{BC}$, $m\angle DAC > m\angle ACB$

Prove $DC > AB$

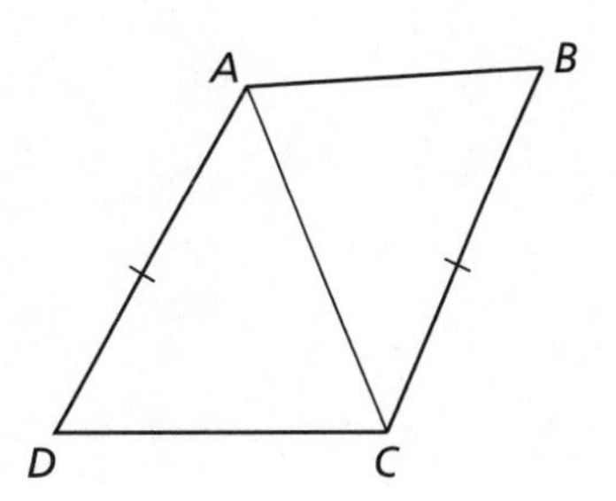

12. Loop a rubber band around the blade ends of a pair of scissors. Describe what happens to the rubber band as you open the scissors. How does that relate to the Hinge Theorem?

13. Starting from a point 10 miles north of Crow Valley, a crow flies northeast for 5 miles. Another crow, starting from a point 10 miles south of Crow Valley, flies due west for 5 miles. Which crow is farther from Crow Valley? Explain.

Name ______________________________ Date __________

Chapter 7 Maintaining Mathematical Proficiency

Solve the equation by interpreting the expression in parentheses as a single quantity.

1. $5(10 - x) = 100$

2. $6(x + 8) - 12 = -48$

3. $3(2 - x) + 4(2 - x) = 56$

Determine which lines are parallel and which are perpendicular.

4.

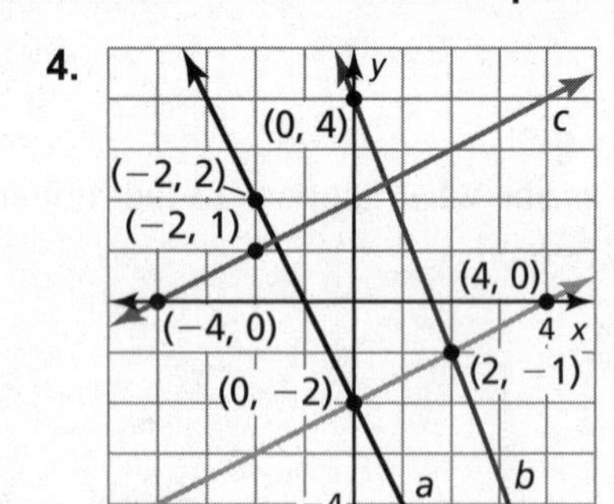

5.

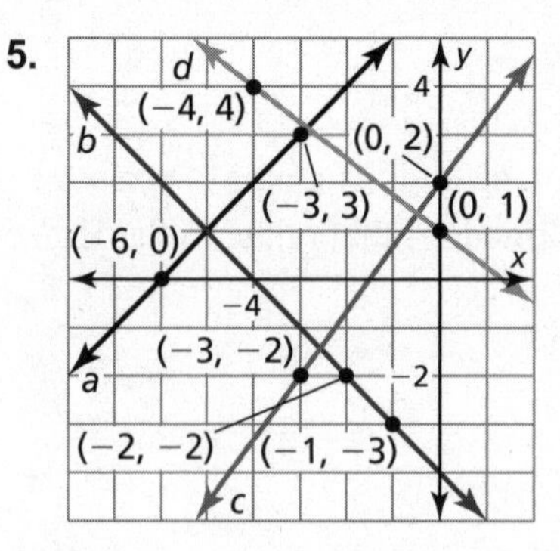

6.

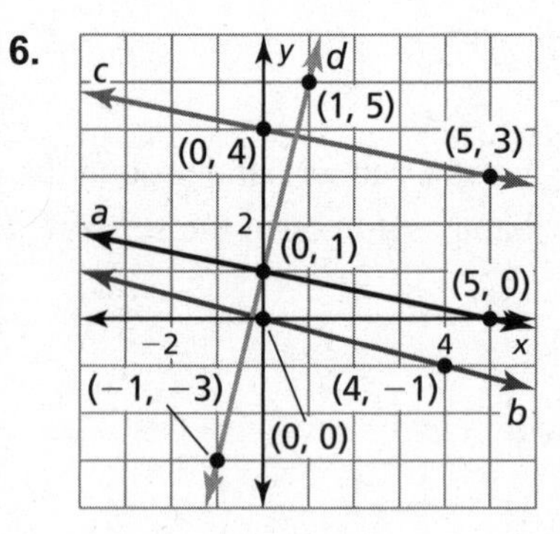

7. Explain why you can rewrite $4(x - 9) + 5(9 - x) = 11$ as $-(x - 9) = 11$? Then solve the equation.

Name____ Date___________

7.1 Angles of Polygons

For use with Exploration 7.1

Essential Question What is the sum of the measures of the interior angles of a polygon?

1 EXPLORATION: The Sum of the Angle Measures of a Polygon

Go to *BigIdeasMath.com* for an interactive tool to investigate this exploration.

Work with a partner. Use dynamic geometry software.

a. Draw a quadrilateral and a pentagon. Find the sum of the measures of the interior angles of each polygon.

Sample

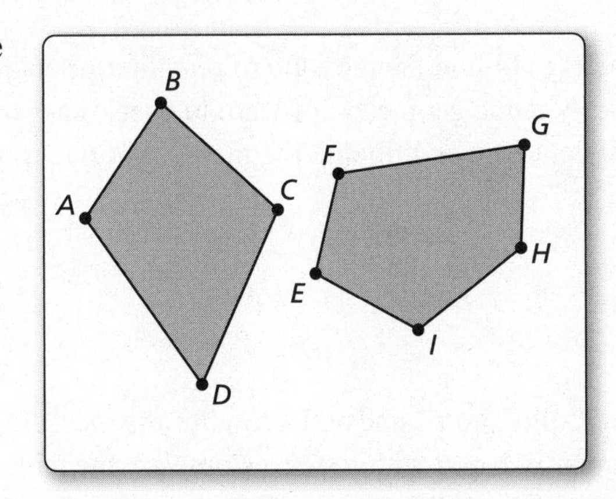

b. Draw other polygons and find the sums of the measures of their interior angles. Record your results in the table below.

Number of sides, *n*	3	4	5	6	7	8	9
Sum of angle measures, *S*							

c. Plot the data from your table in a coordinate plane.

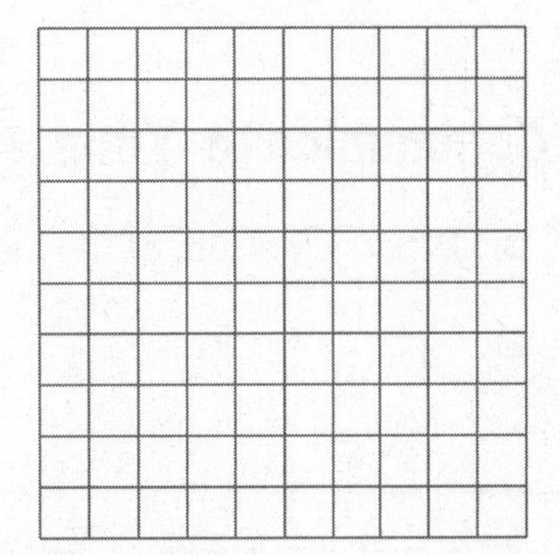

d. Write a function that fits the data. Explain what the function represents.

Name ______________________________ Date __________

2 EXPLORATION: Measure of One Angle in a Regular Polygon

Go to *BigIdeasMath.com* for an interactive tool to investigate this exploration.

Work with a partner.

a. Use the function you found in Exploration 1 to write a new function that gives the measure of one interior angle in a regular polygon with n sides.

b. Use the function in part (a) to find the measure of one interior angle of a regular pentagon. Use dynamic geometry software to check your result by constructing a regular pentagon and finding the measure of one of its interior angles.

c. Copy your table from Exploration 1 and add a row for the measure of one interior angle in a regular polygon with n sides. Complete the table. Use dynamic geometry software to check your results.

Number of sides, n	3	4	5	6	7	8	9
Sum of angle measures, S							
Measure of one interior angle							

Communicate Your Answer

3. What is the sum of the measures of the interior angles of a polygon?

4. Find the measure of one interior angle in a regular dodecagon (a polygon with 12 sides).

Name____ Date__________

7.1 Notetaking with Vocabulary

For use after Lesson 7.1

In your own words, write the meaning of each vocabulary term.

diagonal

equilateral polygon

equiangular polygon

regular polygon

Theorems

Theorem 7.1 Polygon Interior Angles Theorem

The sum of the measures of the interior angles of a convex n-gon is $(n - 2) \bullet 180°$.

$$m\angle 1 + m\angle 2 + \cdots + m\angle n = (n - 2) \bullet 180°$$

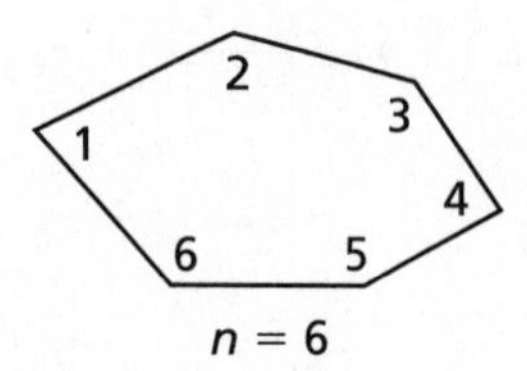

Notes:

Corollary 7.1 Corollary to the Polygon Interior Angles Theorem

The sum of the measures of the interior angles of a quadrilateral is $360°$.

Notes:

Theorem 7.2 Polygon Exterior Angles Theorem

The sum of the measures of the exterior angles of a convex polygon, one angle at each vertex, is $360°$.

$$m\angle 1 + m\angle 2 + \cdots + m\angle n = 360°$$

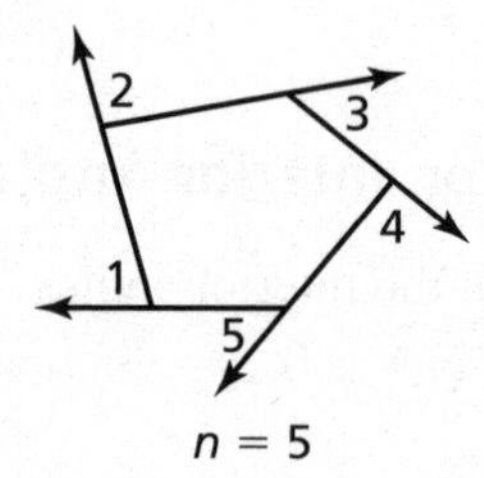

Notes:

Name____ Date__________

7.1 Notetaking with Vocabulary (continued)

Extra Practice

In Exercises 1–3, find the sum of the measures of the interior angles of the indicated convex polygon.

1. octagon **2.** 15-gon **3.** 24-gon

In Exercises 4–6, the sum of the measures of the interior angles of a convex polygon is given. Classify the polygon by the number of sides.

4. 900° **5.** 1620° **6.** 2880°

In Exercises 7–10, find the value of *x*.

7.

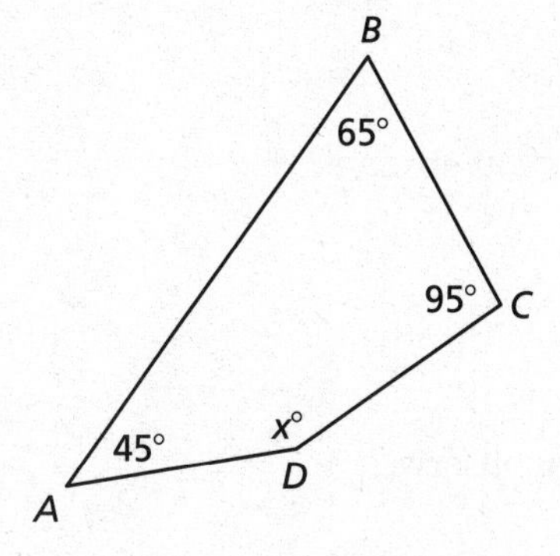

8.

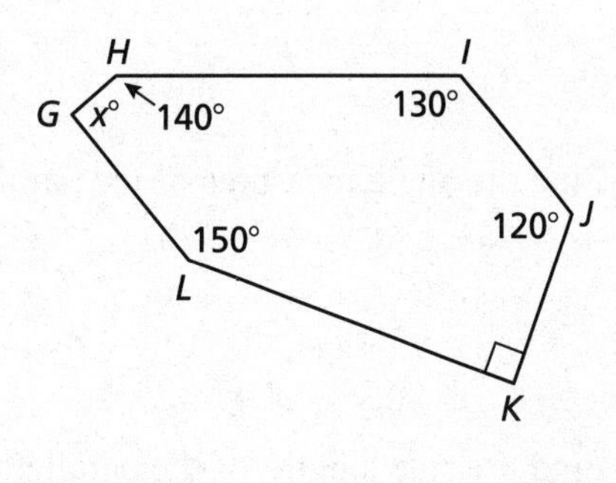

9.

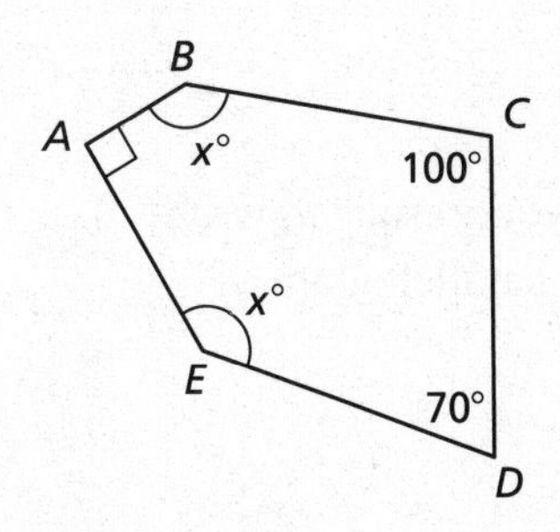

10.

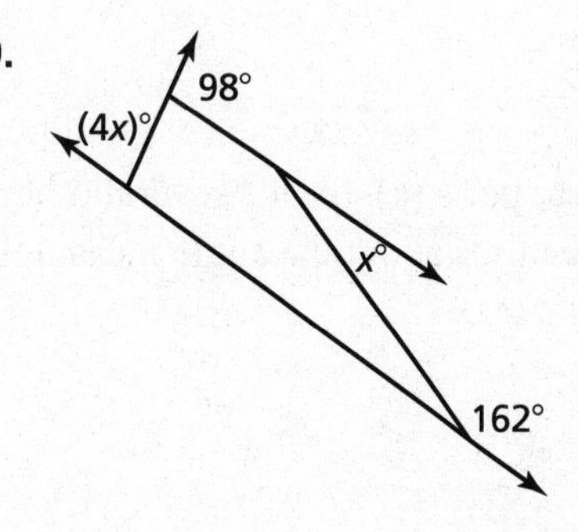

Name ______________________________ Date __________

7.2 Properties of Parallelograms

For use with Exploration 7.2

Essential Question What are the properties of parallelograms?

1 EXPLORATION: Discovering Properties of Parallelograms

Go to *BigIdeasMath.com* for an interactive tool to investigate this exploration.

Work with a partner. Use dynamic geometry software.

a. Construct any parallelogram and label it *ABCD*. Explain your process.

Sample

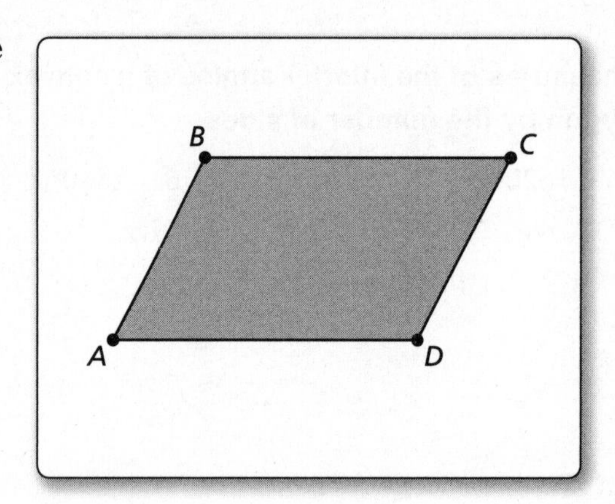

b. Find the angle measures of the parallelogram. What do you observe?

c. Find the side lengths of the parallelogram. What do you observe?

d. Repeat parts (a)–(c) for several other parallelograms. Use your results to write conjectures about the angle measures and side lengths of a parallelogram.

2 EXPLORATION: Discovering a Property of Parallelograms

Go to *BigIdeasMath.com* for an interactive tool to investigate this exploration.

Work with a partner. Use dynamic geometry software.

a. Construct any parallelogram and label it $ABCD$.

b. Draw the two diagonals of the parallelogram. Label the point of intersection E.

Sample

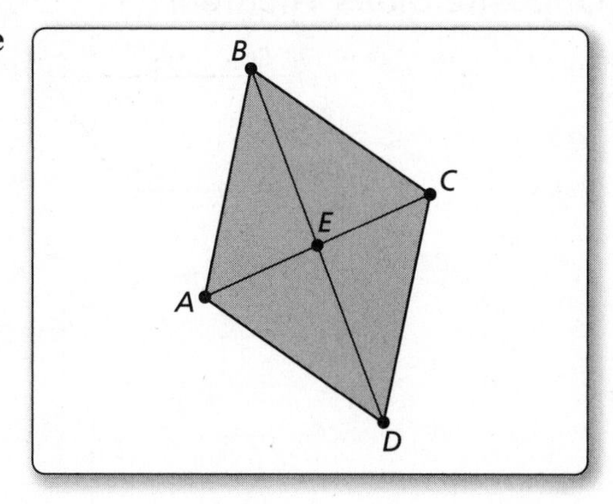

c. Find the segment lengths AE, BE, CE, and DE. What do you observe?

d. Repeat parts (a)–(c) for several other parallelograms. Use your results to write a conjecture about the diagonals of a parallelogram.

Communicate Your Answer

3. What are the properties of parallelograms?

Name ______________________________ Date ________

7.2 Notetaking with Vocabulary

For use after Lesson 7.2

In your own words, write the meaning of each vocabulary term.

parallelogram

Theorems

Theorem 7.3 Parallelogram Opposite Sides Theorem

If a quadrilateral is a parallelogram, then its opposite sides are congruent.

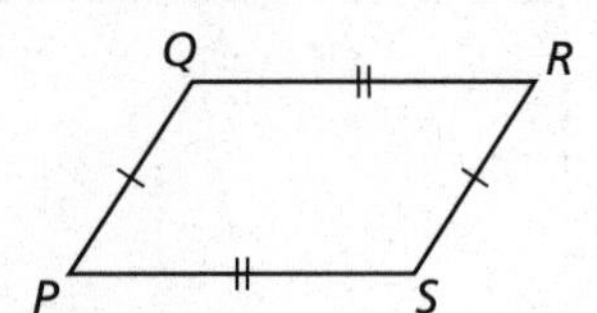

If $PQRS$ is a parallelogram, then $\overline{PQ} \cong \overline{RS}$ and $\overline{QR} \cong \overline{SP}$.

Notes:

Theorem 7.4 Parallelogram Opposite Angles Theorem

If a quadrilateral is a parallelogram, then its opposite angles are congruent.

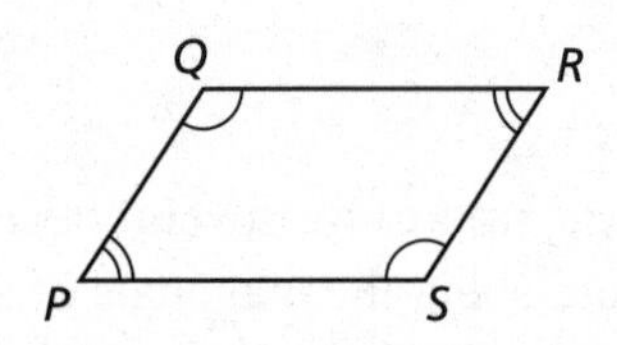

If $PQRS$ is a parallelogram, then $\angle P \cong \angle R$ and $\angle Q \cong \angle S$.

Notes:

Name______________________________ Date__________

7.2 Notetaking with Vocabulary (continued)

Theorem 7.5 Parallelogram Consecutive Angles Theorem

If a quadrilateral is a parallelogram, then its consecutive angles are supplementary.

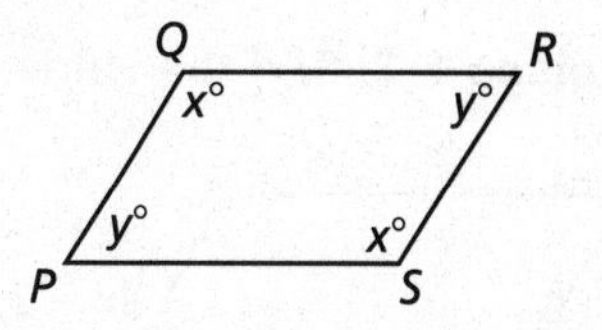

If *PQRS* is a parallelogram, then $x° + y° = 180°$.

Notes:

Theorem 7.6 Parallelogram Diagonals Theorem

If a quadrilateral is a parallelogram, then its diagonals bisect each other.

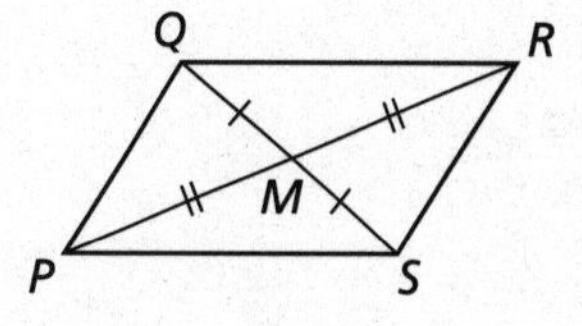

If *PQRS* is a parallelogram, then $\overline{QM} \cong \overline{SM}$ and $\overline{PM} \cong \overline{RM}$.

Notes:

Name ______________________________ Date __________

7.2 Notetaking with Vocabulary (continued)

Extra Practice

In Exercises 1–3, find the value of each variable in the parallelogram.

1.

$4x$

$x + 18$ $3x - 2$

y

2.

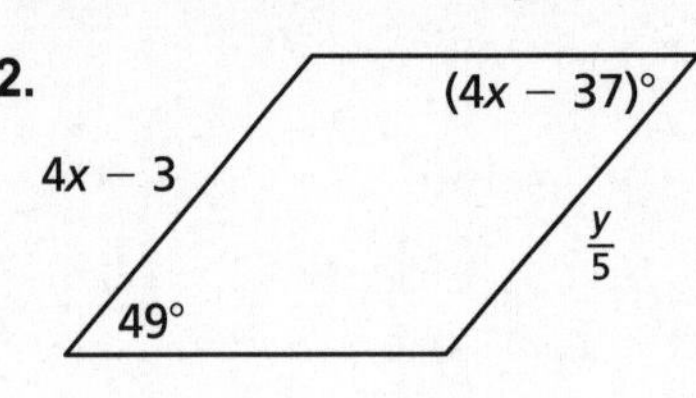

3.

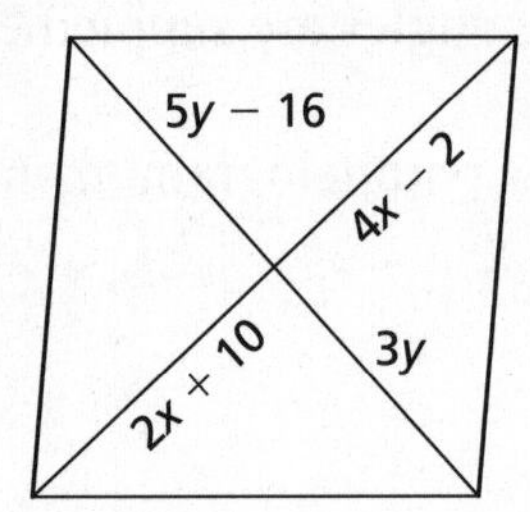

In Exercises 4–11, find the indicated measure in ▱*MNOP*. Explain your reasoning.

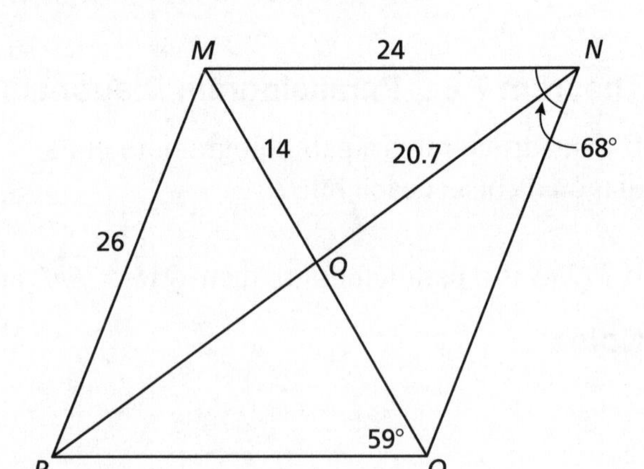

4. PO

5. OQ

6. NO

7. PQ

8. $m\angle PMN$ 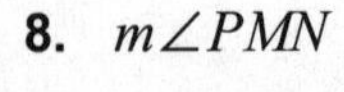

9. $m\angle NOP$

10. $m\angle OPM$

11. $m\angle NMO$

Name____ Date__________

7.3 Proving That a Quadrilateral Is a Parallelogram

For use with Exploration 7.3

Essential Question How can you prove that a quadrilateral is a parallelogram?

1 EXPLORATION: Proving That a Quadrilateral Is a Parallelogram

Go to *BigIdeasMath.com* for an interactive tool to investigate this exploration.

Work with a partner. Use dynamic geometry software.

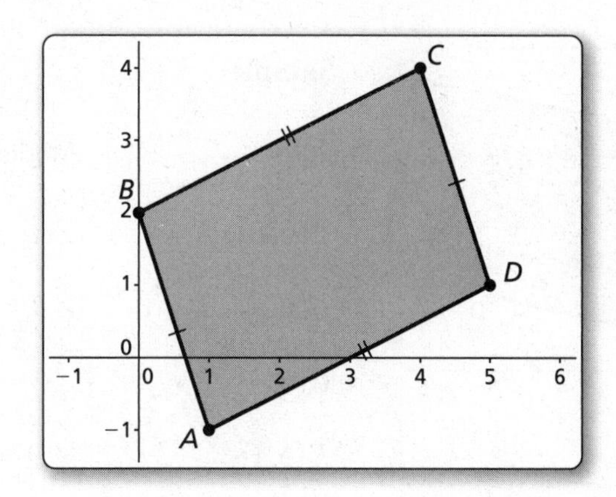

Sample

Points

$A(1, -1)$

$B(0, 2)$

$C(4, 4)$

$D(5, 1)$

Segments

$AB = 3.16$

$BC = 4.47$

$CD = 3.16$

$DA = 4.47$

a. Construct any quadrilateral $ABCD$ whose opposite sides are congruent.

b. Is the quadrilateral a parallelogram? Justify your answer.

c. Repeat parts (a) and (b) for several other quadrilaterals. Then write a conjecture based on your results.

d. Write the converse of your conjecture. Is the converse true? Explain.

2 EXPLORATION: Proving That a Quadrilateral Is a Parallelogram

Go to *BigIdeasMath.com* for an interactive tool to investigate this exploration.

Work with a partner. Use dynamic geometry software.

a. Construct any quadrilateral *ABCD* whose opposite angles are congruent.

b. Is the quadrilateral a parallelogram? Justify your answer.

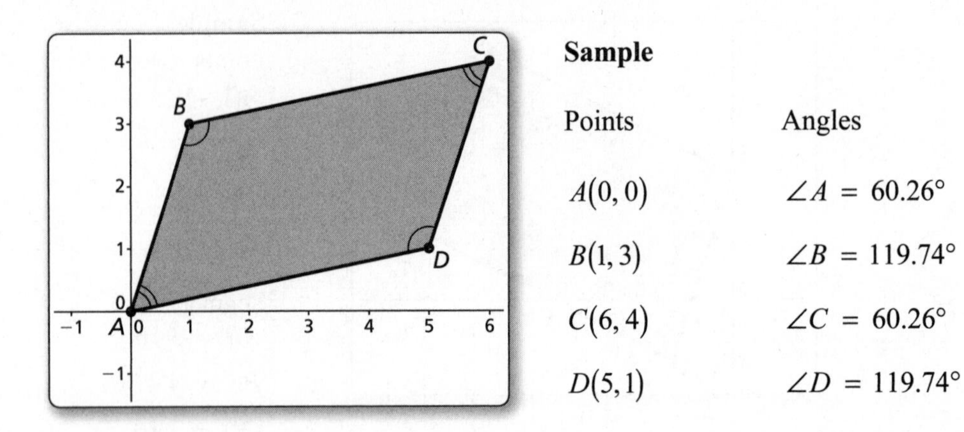

c. Repeat parts (a) and (b) for several other quadrilaterals. Then write a conjecture based on your results.

d. Write the converse of your conjecture. Is the converse true? Explain.

Communicate Your Answer

3. How can you prove that a quadrilateral is a parallelogram?

4. Is the quadrilateral at the right a parallelogram? Explain your reasoning.

B C 53° 53° D A

Name____ Date__________

7.3 Notetaking with Vocabulary

For use after Lesson 7.3

In your own words, write the meaning of each vocabulary term.

diagonal

parallelogram

Theorems

Theorem 7.7 Parallelogram Opposite Sides Converse

If both pairs of opposite sides of a quadrilateral are congruent, then the quadrilateral is a parallelogram.

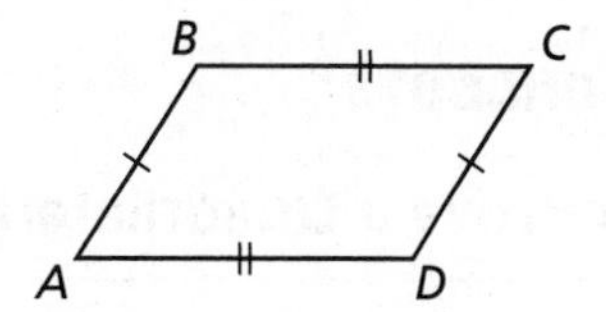

If $\overline{AB} \cong \overline{CD}$ and $\overline{BC} \cong \overline{DA}$, then $ABCD$ is a parallelogram.

Notes:

Theorem 7.8 Parallelogram Opposite Angles Converse

If both pairs of opposite angles of a quadrilateral are congruent, then the quadrilateral is a parallelogram.

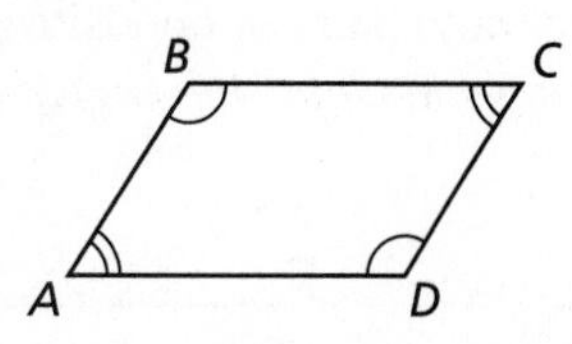

If $\angle A \cong \angle C$ and $\angle B \cong \angle D$, then $ABCD$ is a parallelogram.

Notes:

Theorem 7.9 Opposite Sides Parallel and Congruent Theorem

If one pair of opposite sides of a quadrilateral are congruent and parallel, then the quadrilateral is a parallelogram.

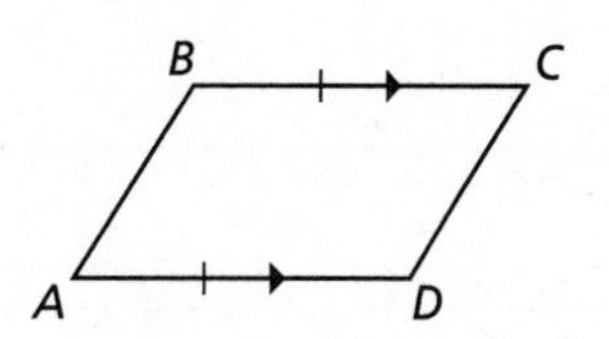

If $\overline{BC} \parallel \overline{AD}$ and $\overline{BC} \cong \overline{AD}$, then $ABCD$ is a parallelogram.

Notes:

Name ______________________________ Date __________

Theorem 7.10 Parallelogram Diagonals Converse

If the diagonals of a quadrilateral bisect each other, then the quadrilateral is a parallelogram.

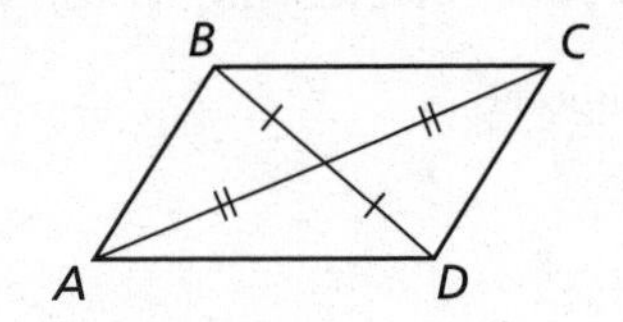

If $\overline{BD}$ and $\overline{AC}$ bisect each other, then $ABCD$ is a parallelogram.

Notes:

Core Concepts

Ways to Prove a Quadrilateral Is a Parallelogram

1. Show that both pairs of opposite sides are parallel. *(Definition)*	
2. Show that both pairs of opposite sides are congruent. *(Parallelogram Opposite Sides Converse)*	
3. Show that both pairs of opposite angles are congruent. *(Parallelogram Opposite Angles Converse)*	
4. Show that one pair of opposite sides are congruent and parallel. *(Opposite Sides Parallel and Congruent Theorem)*	
5. Show that the diagonals bisect each other. *(Parallelogram Diagonals Converse)*	

Extra Practice

In Exercises 1–3, state which theorem you can use to show that the quadrilateral is a parallelogram.

1.

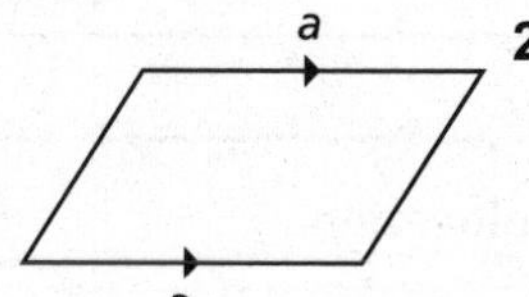

2.

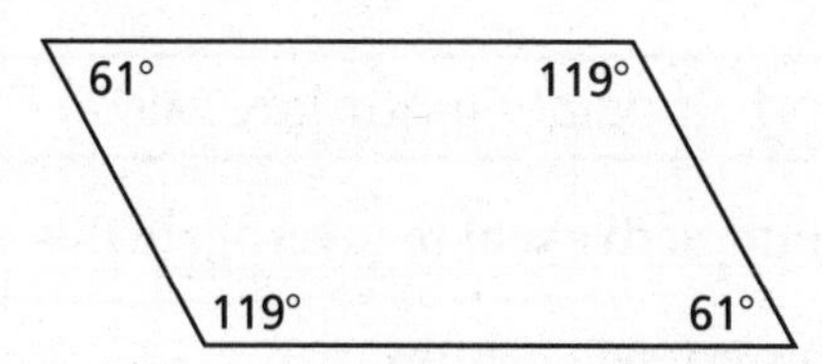

3.

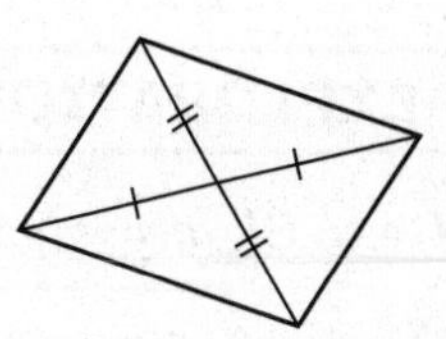

In Exercises 4–7, find the values of *x* and *y* that make the quadrilateral a parallelogram.

4.

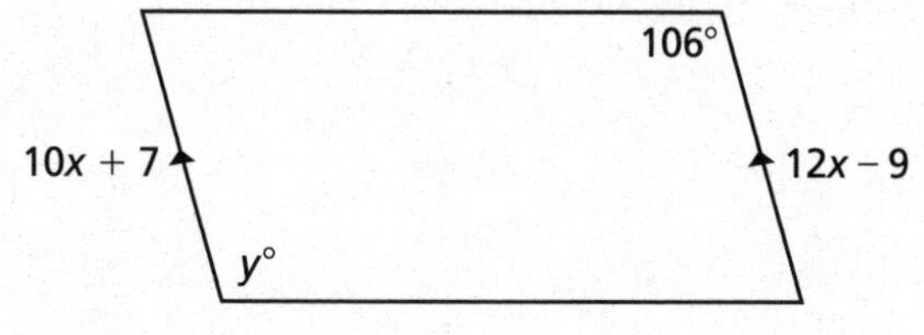

5.

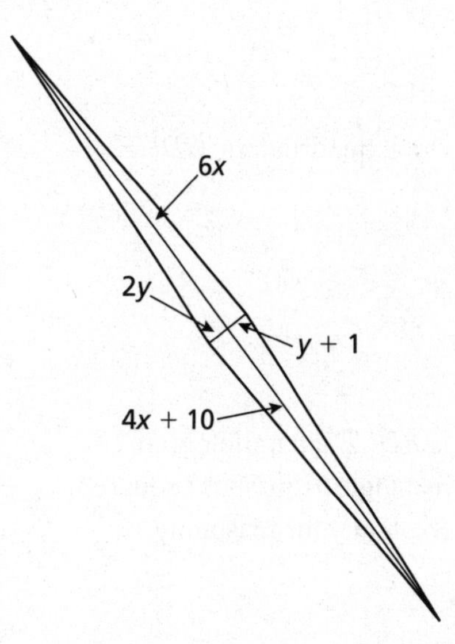

6.

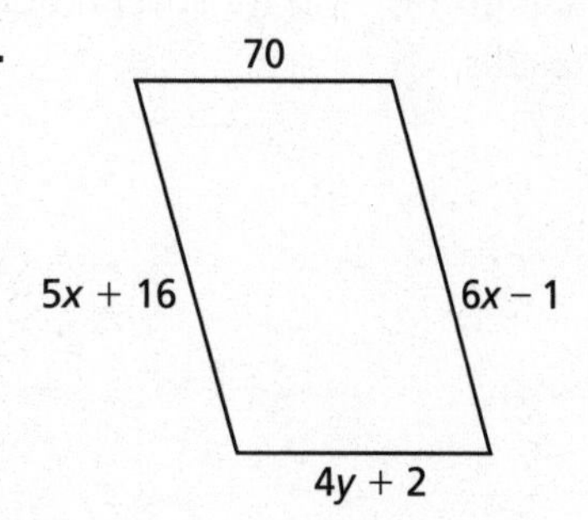

7.

Name ______________________________ Date __________

7.4 Properties of Special Parallelograms

For use with Exploration 7.4

Essential Question What are the properties of the diagonals of rectangles, rhombuses, and squares?

1 EXPLORATION: Identifying Special Quadrilaterals

Go to *BigIdeasMath.com* for an interactive tool to investigate this exploration.

Work with a partner. Use dynamic geometry software.

a. Draw a circle with center *A*.

b. Draw two diameters of the circle. Label the endpoints *B, C, D,* and *E*.

Sample

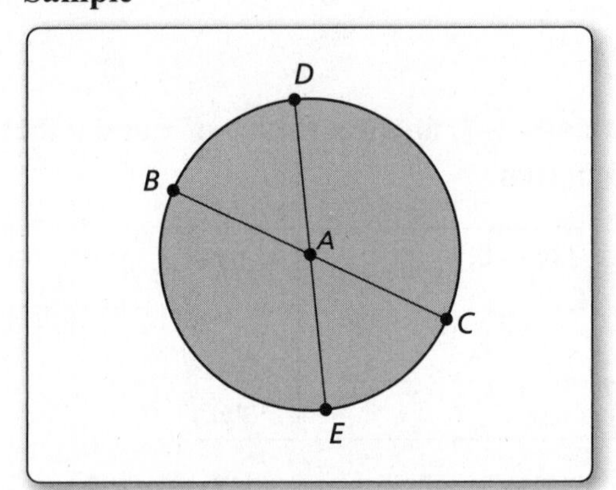

c. Draw quadrilateral *BDCE*.

d. Is *BDCE* a parallelogram? rectangle? rhombus? square? Explain your reasoning.

e. Repeat parts (a) – (d) for several other circles. Write a conjecture based on your results.

2 EXPLORATION: Identifying Special Quadrilaterals

Go to *BigIdeasMath.com* for an interactive tool to investigate this exploration.

Work with a partner. Use dynamic geometry software.

a. Construct two segments that are perpendicular bisectors of each other. Label the endpoints *A, B, D,* and *E.* Label the intersection *C.*

b. Draw quadrilateral *AEBD.*

c. Is *AEBD* a parallelogram? rectangle? rhombus? square? Explain your reasoning.

Sample

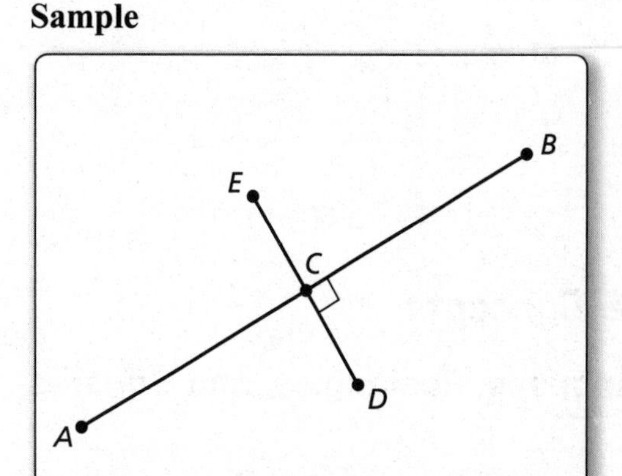

d. Repeat parts (a) – (c) for several other segments. Write a conjecture based on your results.

Communicate Your Answer

3. What are the properties of the diagonals of rectangles, rhombuses, and squares?

4. Is *RSTU* a parallelogram? rectangle? rhombus? square? Explain your reasoning.

5. What type of quadrilateral has congruent diagonals that bisect each other?

Name ________________________________ Date __________

7.4 Notetaking with Vocabulary

For use after Lesson 7.4

In your own words, write the meaning of each vocabulary term.

rhombus

rectangle

square

Core Concepts

Rhombuses, Rectangles, and Squares

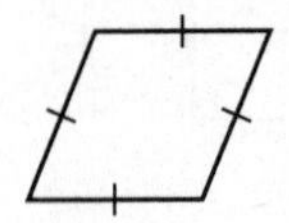

A **rhombus** is a parallelogram with four congruent sides.

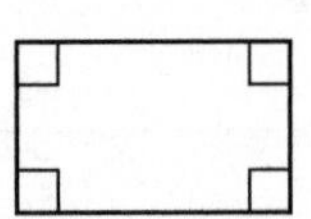

A **rectangle** is a parallelogram with four right angles.

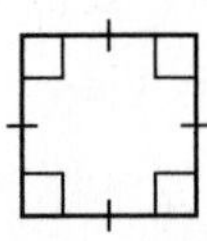

A **square** is a parallelogram with four congruent sides and four right angles.

Notes:

Corollary 7.2 Rhombus Corollary

A quadrilateral is a rhombus if and only if it has four congruent sides.

ABCD is a rhombus if and only if $\overline{AB} \cong \overline{BC} \cong \overline{CD} \cong \overline{AD}$.

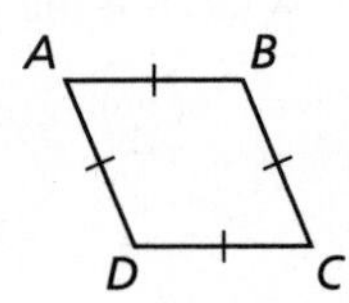

Corollary 7.3 Rectangle Corollary

A quadrilateral is a rectangle if and only if it has four right angles.

ABCD is a rectangle if and only if $\angle A$, $\angle B$, $\angle C$, and $\angle D$ are right angles.

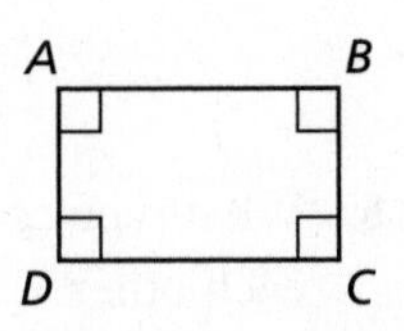

Corollary 7.4 Square Corollary

A quadrilateral is a square if and only if it is a rhombus and a rectangle.

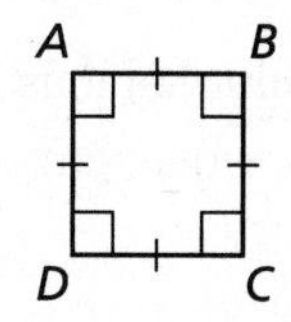

$ABCD$ is a square if and only if $\overline{AB} \cong \overline{BC} \cong \overline{CD} \cong \overline{AD}$ and $\angle A$, $\angle B$, $\angle C$, and $\angle D$ are right angles.

Notes:

Theorem 7.11 Rhombus Diagonals Theorem

A parallelogram is a rhombus if and only if its diagonals are perpendicular.

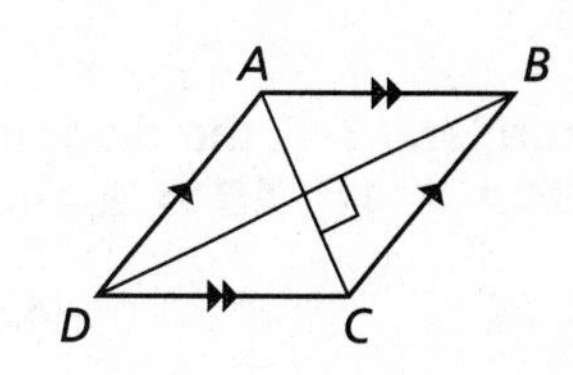

$\square ABCD$ is a rhombus if and only if $\overline{AC} \perp \overline{BD}$.

Notes:

Theorem 7.12 Rhombus Opposite Angles Theorem

A parallelogram is a rhombus if and only if each diagonal bisects a pair of opposite angles.

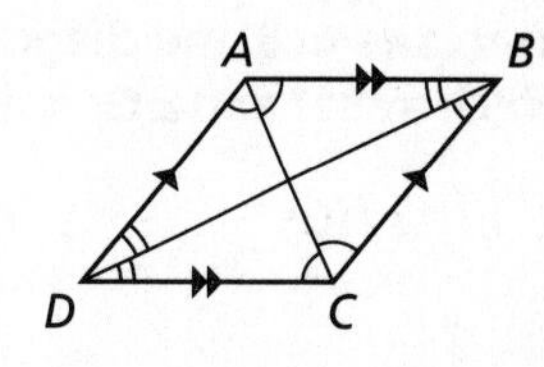

$\square ABCD$ is a rhombus if and only if $\overline{AC}$ bisects $\angle BCD$ and $\angle BAD$, and $\overline{BD}$ bisects $\angle ABC$ and $\angle ADC$.

Notes:

Theorem 7.13 Rectangle Diagonals Theorem

A parallelogram is a rectangle if and only if its diagonals are congruent.

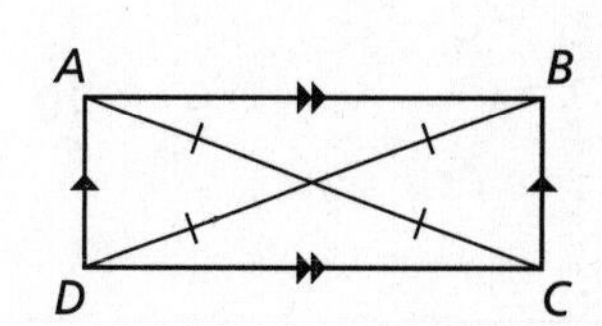

$\square ABCD$ is a rectangle if and only if $\overline{AC} \cong \overline{BD}$.

Notes:

Extra Practice

1. For any rhombus *MNOP*, decide whether the statement $\overline{MO} \cong \overline{NP}$ is *always* or *sometimes* true. Draw a diagram and explain your reasoning.

2. For any rectangle *PQRS*, decide whether the statement $\angle PQS \cong \angle RSQ$ is *always* or *sometimes* true. Draw a diagram and explain your reasoning.

In Exercises 3–5, the diagonals of rhombus *ABCD* intersect at *E*. Given that $m\angle BCA = 44°$, $AB = 9$, and $AE = 7$, find the indicated measure.

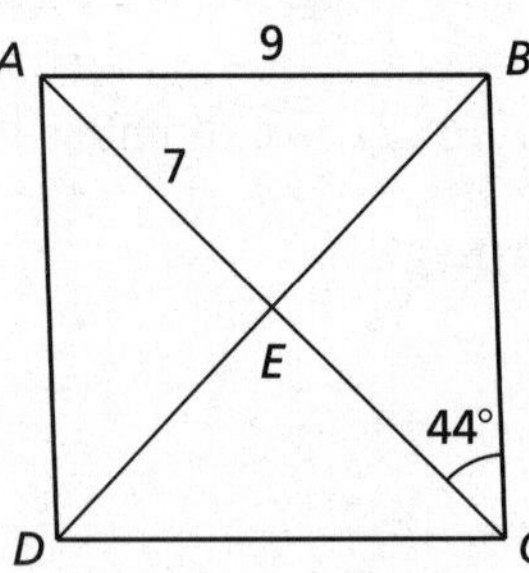

3. BC 4. AC 5. $m\angle ADC$

In Exercises 6–8, the diagonals of rectangle *EFGH* intersect at *I*. Given that $m\angle HFG = 31°$ and $EG = 17$, find the indicated measure.

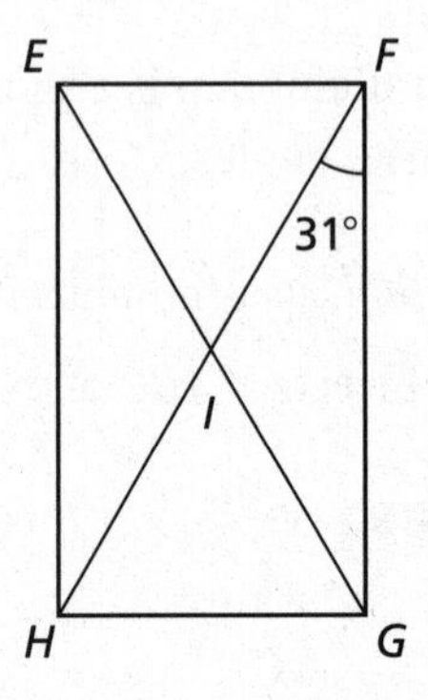

6. $m\angle FHG$ 7. HF 8. $m\angle EFH$

In Exercises 9–11, the diagonals of square *LMNP* intersect at *K*. Given that $MK = \frac{1}{2}$, find the indicated measure.

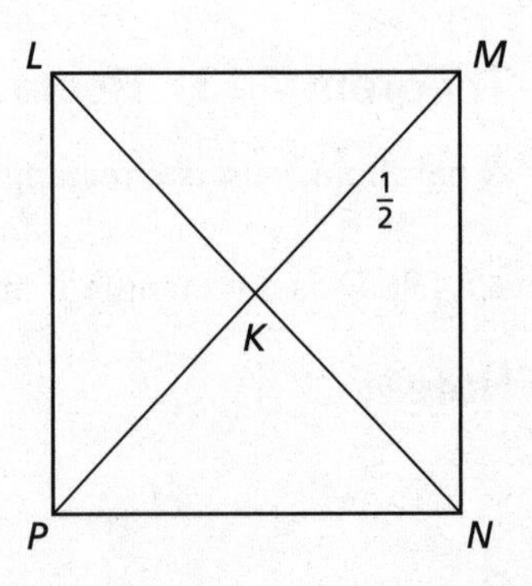

9. PK 10. $m\angle PKN$ 11. $m\angle MNK$

Name____ Date__________

7.5 Properties of Trapezoids and Kites

For use with Exploration 7.5

Essential Question What are some properties of trapezoids and kites?

1 EXPLORATION: Making a Conjecture about Trapezoids

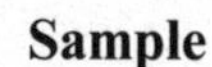

Go to *BigIdeasMath.com* for an interactive tool to investigate this exploration.

Work with a partner. Use dynamic geometry software.

a. Construct a trapezoid whose base angles are congruent. Explain your process.

Sample

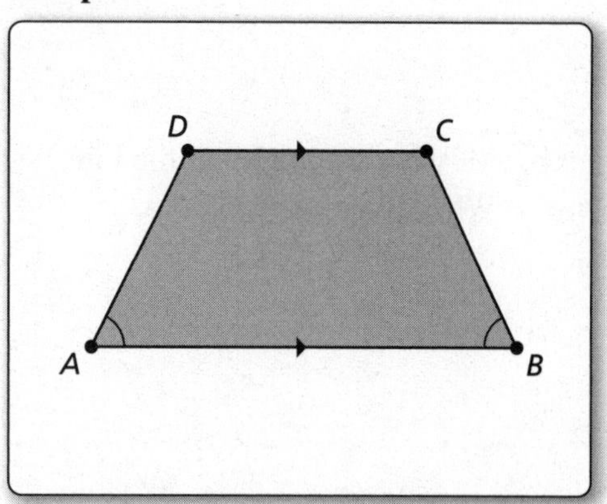

b. Is the trapezoid isosceles? Justify your answer.

c. Repeat parts (a) and (b) for several other trapezoids. Write a conjecture based on your results.

2 EXPLORATION: Discovering a Property of Kites

Go to *BigIdeasMath.com* for an interactive tool to investigate this exploration.

Work with a partner. Use dynamic geometry software.

a. Construct a kite. Explain your process.

Sample

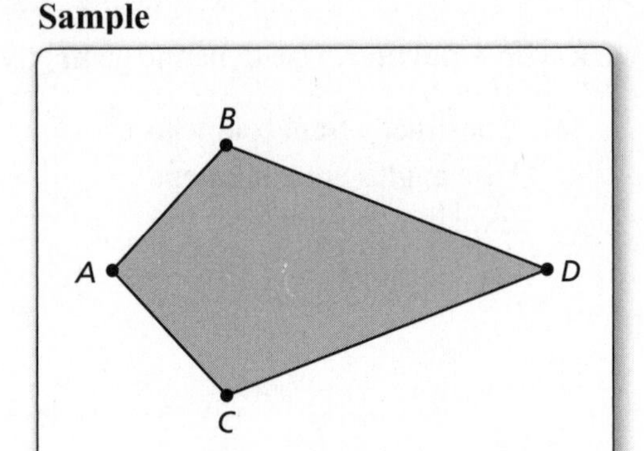

b. Measure the angles of the kite. What do you observe?

c. Repeat parts (a) and (b) for several other kites. Write a conjecture based on your results.

Communicate Your Answer

3. What are some properties of trapezoids and kites?

4. Is the trapezoid at the right isosceles? Explain.

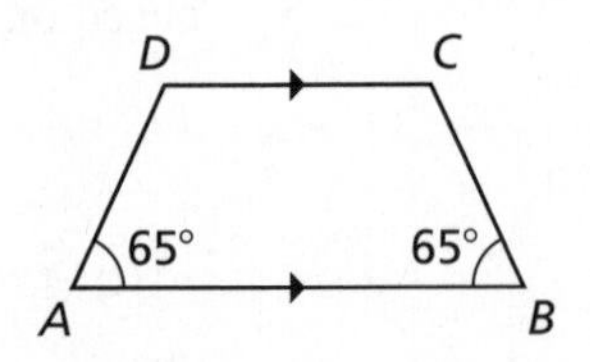

5. A quadrilateral has angle measures of 70°, 70°, 110°, and 110°. Is the quadrilateral a kite? Explain.

Name____ Date__________

7.5 Notetaking with Vocabulary
For use after Lesson 7.5

In your own words, write the meaning of each vocabulary term.

trapezoid

bases

base angles

legs

isosceles trapezoid

midsegment of a trapezoid

kite

Theorems

Theorem 7.14 Isosceles Trapezoid Base Angles Theorem

If a trapezoid is isosceles, then each pair of base angles is congruent.

If trapezoid $ABCD$ is isosceles, then $\angle A \cong \angle D$ and $\angle B \cong \angle C$.

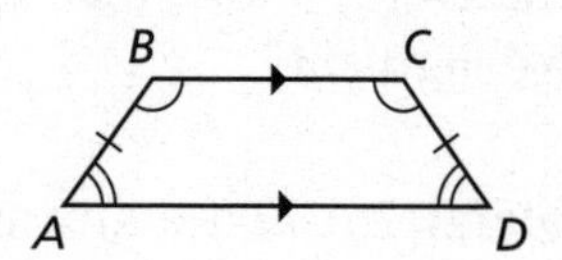

Theorem 7.15 Isosceles Trapezoid Base Angles Converse

If a trapezoid has a pair of congruent base angles, then it is an isosceles trapezoid.

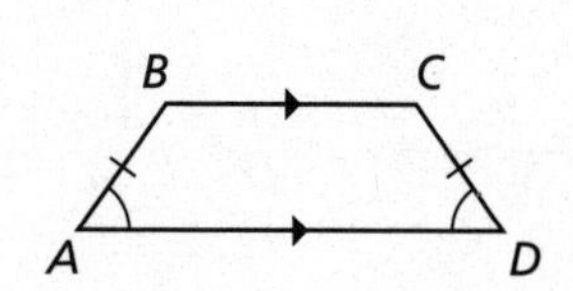

If $\angle A \cong \angle D$ (or if $\angle B \cong \angle C$), then trapezoid $ABCD$ is isosceles.

Theorem 7.16 Isosceles Trapezoid Diagonals Theorem

A trapezoid is isosceles if and only if its diagonals are congruent.

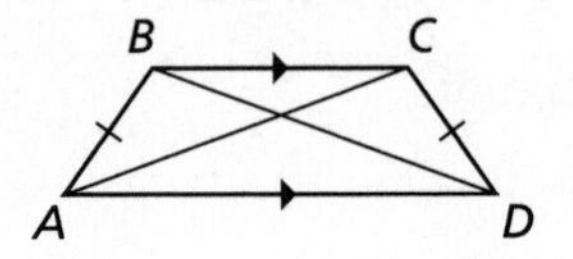

Trapezoid $ABCD$ is isosceles if and only if $\overline{AC} \cong \overline{BD}$.

Theorem 7.17 Trapezoid Midsegment Theorem

The midsegment of a trapezoid is parallel to each base, and its length is one-half the sum of the lengths of the bases.

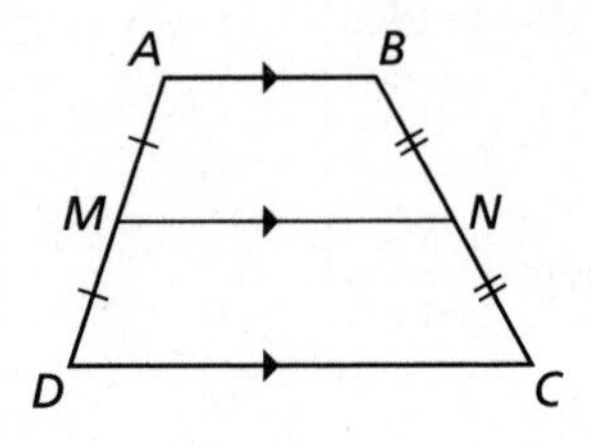

If $\overline{MN}$ is the midsegment of trapezoid $ABCD$, then $\overline{MN} \parallel \overline{AB}$, $\overline{MN} \parallel \overline{DC}$, and $MN = \frac{1}{2}(AB + CD)$.

Theorem 7.18 Kite Diagonals Theorem

If a quadrilateral is a kite, then its diagonals are perpendicular.

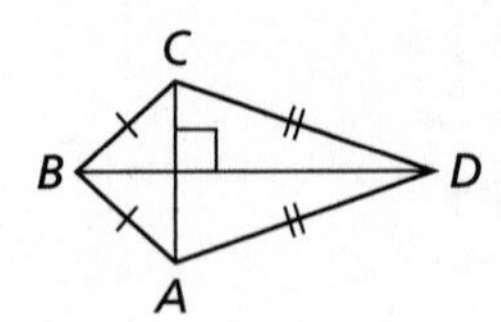

If quadrilateral $ABCD$ is a kite, then $\overline{AC} \perp \overline{BD}$.

Theorem 7.19 Kite Opposite Angles Theorem

If a quadrilateral is a kite, then exactly one pair of opposite angles are congruent.

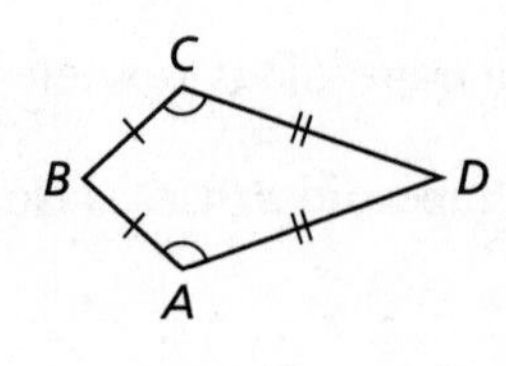

If quadrilateral $ABCD$ is a kite and $\overline{BC} \cong \overline{BA}$, then $\angle A \cong \angle C$ and $\angle B \not\cong \angle D$.

Notes:

Name______________________________ Date__________

Extra Practice

1. Show that the quadrilateral with vertices at $Q(0, 3), R(0, 6), S(-6, 0)$, and $T(-3, 0)$ is a trapezoid. Decide whether the trapezoid is isosceles. Then find the length of the midsegment of the trapezoid.

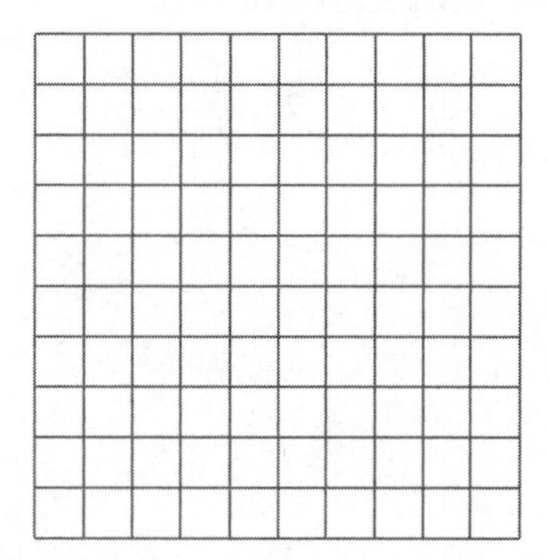

In Exercises 2 and 3, find $m\angle K$ and $m\angle L$.

2.

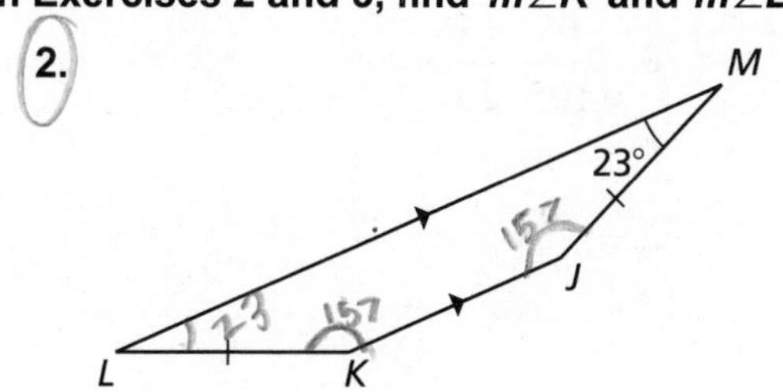

3.

In Exercises 4 and 5, find *CD*.

4.

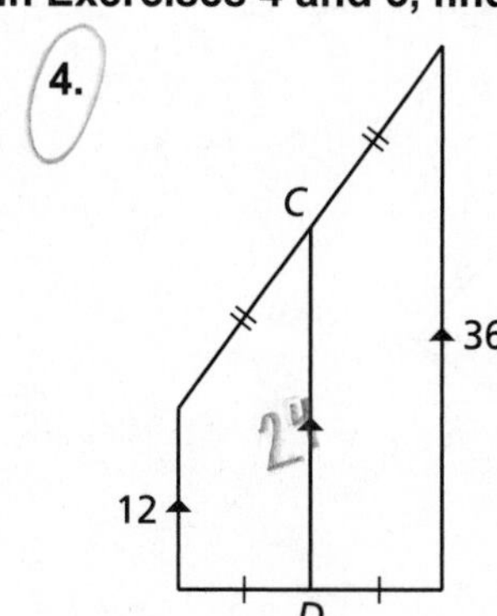

5.

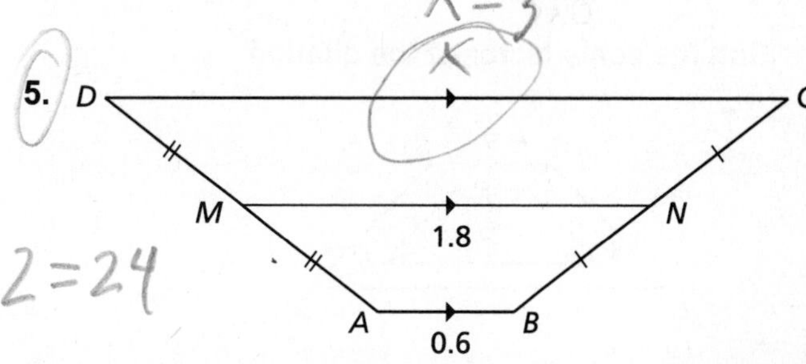

In Exercises 6 and 7, find the value of *x*.

6.

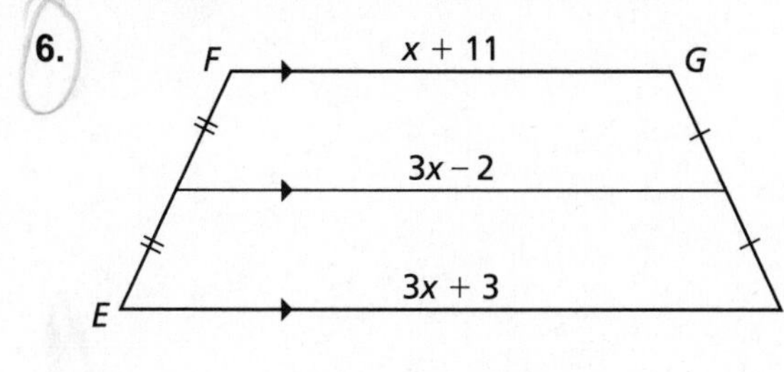

7.

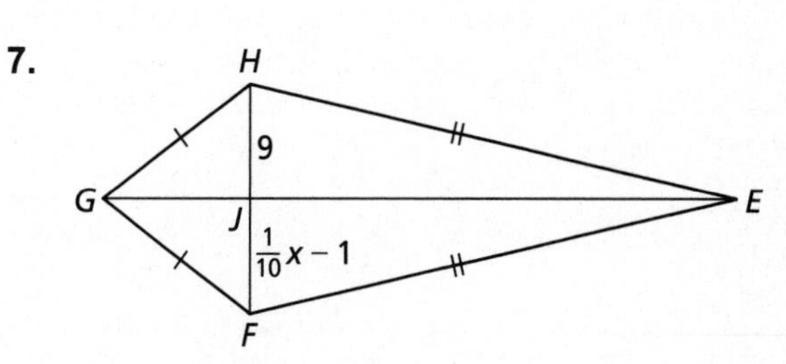

Name ____________________ Date __________

Chapter 8 Maintaining Mathematical Proficiency

Tell whether the ratios form a proportion.

1. $\frac{3}{4}, \frac{16}{12}$

2. $\frac{35}{63}, \frac{45}{81}$

3. $\frac{12}{96}, \frac{16}{100}$

4. $\frac{15}{24}, \frac{75}{100}$

5. $\frac{17}{68}, \frac{32}{128}$

6. $\frac{65}{105}, \frac{156}{252}$

Find the scale factor of the dilation.

7.

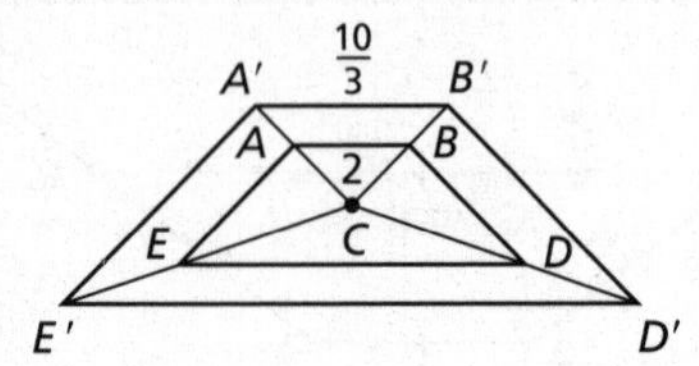

8.

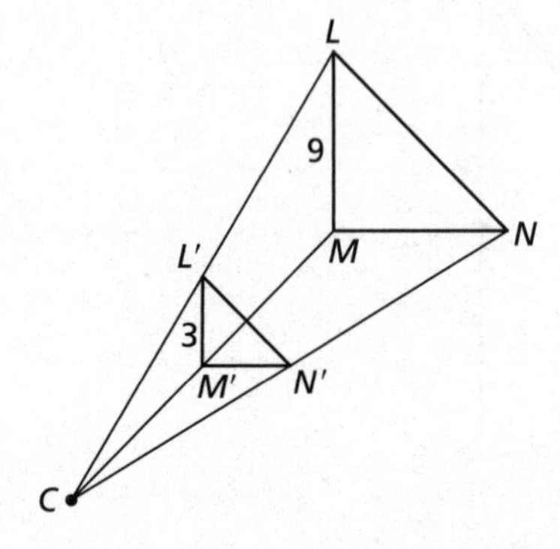

Name____ Date__________

8.1 Similar Polygons

For use with Exploration 8.1

Essential Question How are similar polygons related?

1 EXPLORATION: Comparing Triangles after a Dilation

Go to *BigIdeasMath.com* for an interactive tool to investigate this exploration.

Work with a partner. Use dynamic geometry software to draw any $\triangle ABC$. Dilate $\triangle ABC$ to form a similar $\triangle A'B'C'$ using any scale factor k and any center of dilation.

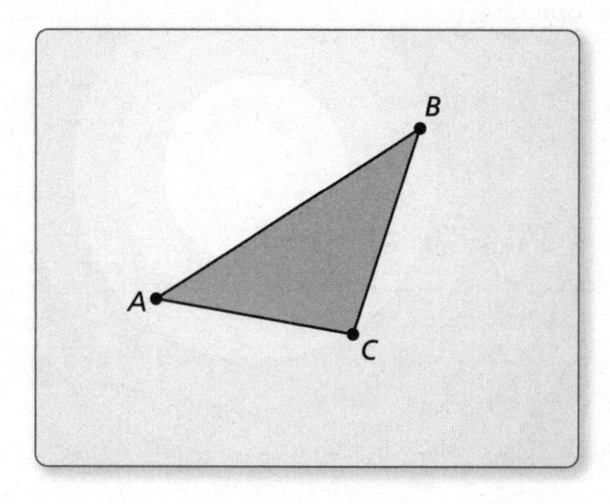

a. Compare the corresponding angles of $\triangle A'B'C'$ and $\triangle ABC$.

b. Find the ratios of the lengths of the sides of $\triangle A'B'C'$ to the lengths of the corresponding sides of $\triangle ABC$. What do you observe?

c. Repeat parts (a) and (b) for several other triangles, scale factors, and centers of dilation. Do you obtain similar results?

2 EXPLORATION: Comparing Triangles after a Dilation

Go to *BigIdeasMath.com* for an interactive tool to investigate this exploration.

Work with a partner. Use dynamic geometry software to draw any $\triangle ABC$. Dilate $\triangle ABC$ to form a similar $\triangle A'B'C'$ using any scale factor k and any center of dilation.

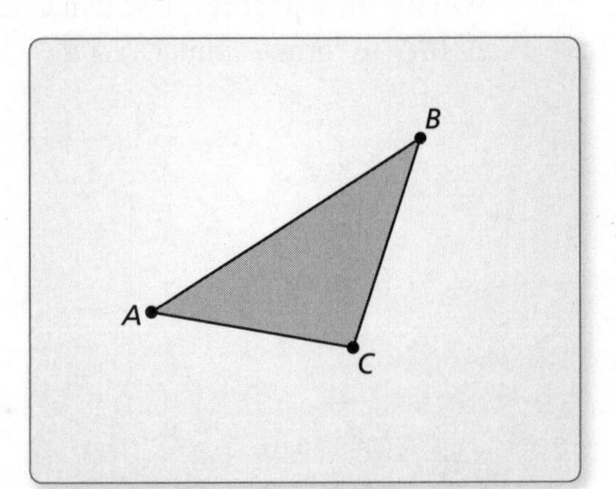

a. Compare the perimeters of $\triangle A'B'C'$ and $\triangle ABC$. What do you observe?

b. Compare the areas of $\triangle A'B'C'$ and $\triangle ABC$. What do you observe?

c. Repeat parts (a) and (b) for several other triangles, scale factors, and centers of dilation. Do you obtain similar results?

Communicate Your Answer

3. How are similar polygons related?

4. A $\triangle RST$ is dilated by a scale factor of 3 to form $\triangle R'S'T'$. The area of $\triangle RST$ is 1 square inch. What is the area of $\triangle R'S'T'$?

Name____ Date__________

8.1 Notetaking with Vocabulary

For use after Lesson 8.1

In your own words, write the meaning of each vocabulary term.

similar figures

similarity transformation

corresponding parts

Core Concepts

Corresponding Parts of Similar Polygons

In the diagram below, $\triangle ABC$ is similar to $\triangle DEF$. You can write "$\triangle ABC$ is similar to $\triangle DEF$" as $\triangle ABC \sim \triangle DEF$. A similarity transformation preserves angle measure. So, corresponding angles are congruent. A similarity transformation also enlarges or reduces side lengths by a scale factor k. So, corresponding side lengths are proportional.

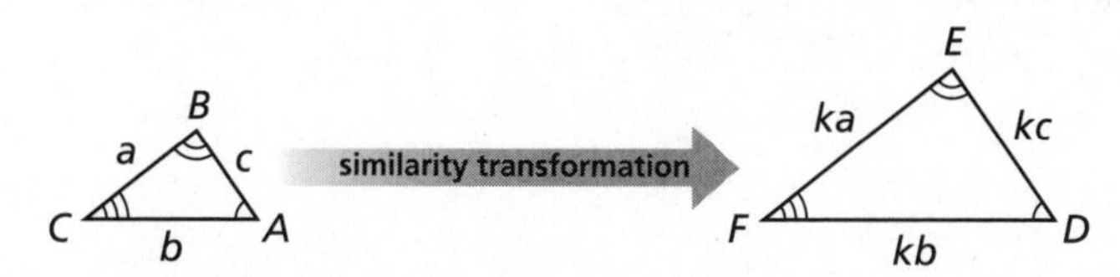

Corresponding angles

$\angle A \cong \angle D, \angle B \cong \angle E, \angle C \cong \angle F$

Ratios of corresponding side lengths

$$\frac{DE}{AB} = \frac{EF}{BC} = \frac{FD}{CA} = k$$

Notes:

Name ______________________________ Date __________

Corresponding Lengths in Similar Polygons

If two polygons are similar, then the ratio of any two corresponding lengths in the polygons is equal to the scale factor of the similar polygons.

Notes:

Theorems

Theorem 8.1 Perimeters of Similar Polygons

If two polygons are similar, then the ratio of their perimeters is equal to the ratios of their corresponding side lengths.

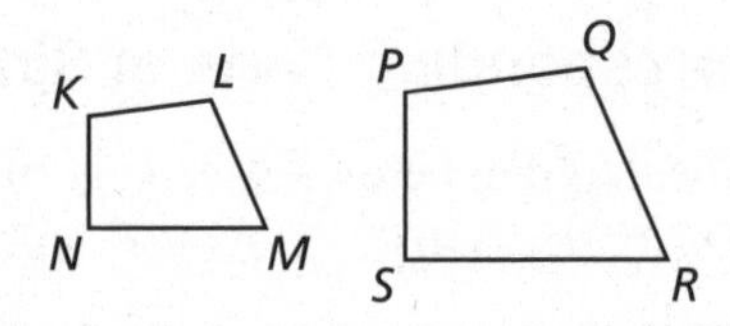

If $KLMN \sim PQRS$, then

$$\frac{PQ + QR + RS + SP}{KL + LM + MN + NK} = \frac{PQ}{KL} = \frac{QR}{LM} = \frac{RS}{MN} = \frac{SP}{NK}.$$

Notes:

Theorem 8.2 Areas of Similar Polygons

If two polygons are similar, then the ratio of their areas is equal to the squares of the ratios of their corresponding side lengths.

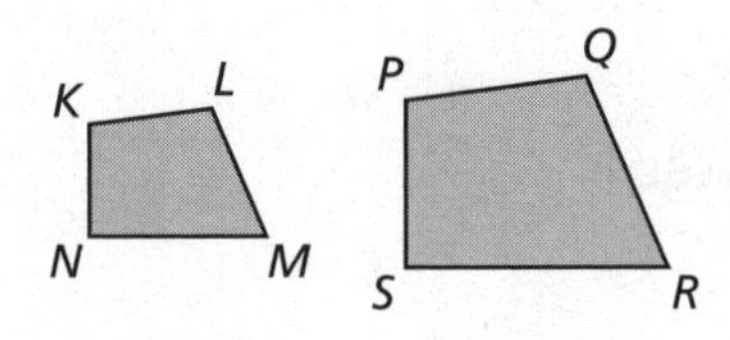

If $KLMN \sim PQRS$, then

$$\frac{\text{Area of } PQRS}{\text{Area of } KLMN} = \left(\frac{PQ}{KL}\right)^2 = \left(\frac{QR}{LM}\right)^2 = \left(\frac{RS}{MN}\right)^2 = \left(\frac{SP}{NK}\right)^2.$$

Notes:

8.1 Notetaking with Vocabulary (continued)

Extra Practice

In Exercises 1 and 2, the polygons are similar. Find the value of *x*.

1.

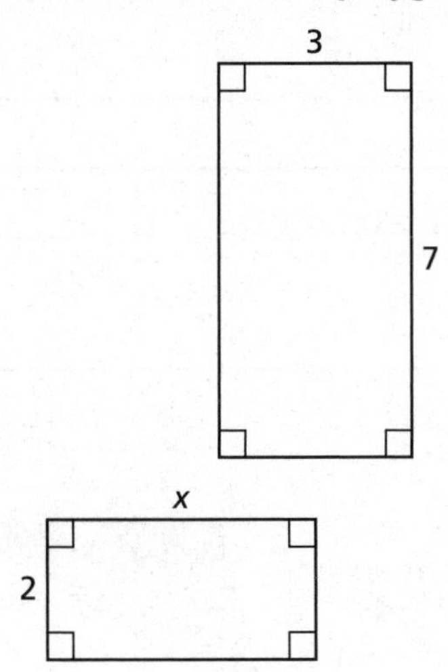

2.

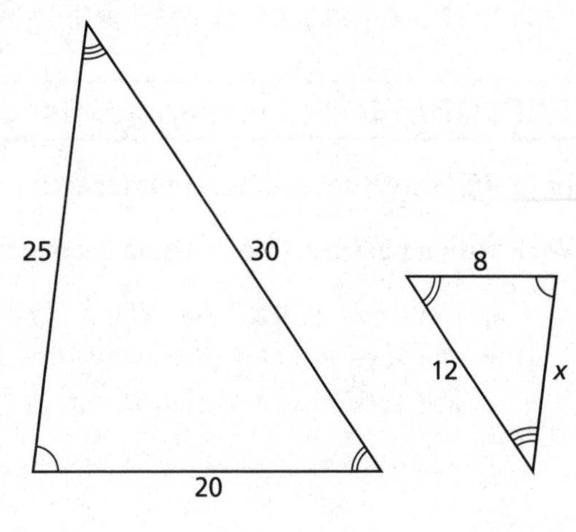

In Exercises 3–8, *ABCDE* ~ *KLMNP*.

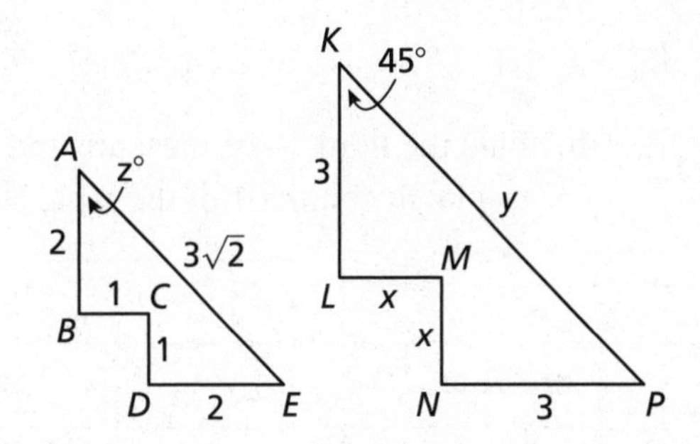

3. Find the scale factor from *ABCDE* to *KLMNP*.

4. Find the scale factor from *KLMNP* to *ABCDE*.

5. Find the values of x, y, and z.

6. Find the perimeter of each polygon.

7. Find the ratio of the perimeters of *ABCDE* to *KLMNP*.

8. Find the ratio of the areas of *ABCDE* to *KLMNP*.

Name ______________________________ Date __________

8.2 Proving Triangle Similarity by AA

For use with Exploration 8.2

Essential Question What can you conclude about two triangles when you know that two pairs of corresponding angles are congruent?

1 EXPLORATION: Comparing Triangles

Go to *BigIdeasMath.com* for an interactive tool to investigate this exploration.

Work with a partner. Use dynamic geometry software.

a. Construct $\triangle ABC$ and $\triangle DEF$ so that $m\angle A = m\angle D = 106°$, $m\angle B = m\angle E = 31°$, and $\triangle DEF$ is not congruent to $\triangle ABC$.

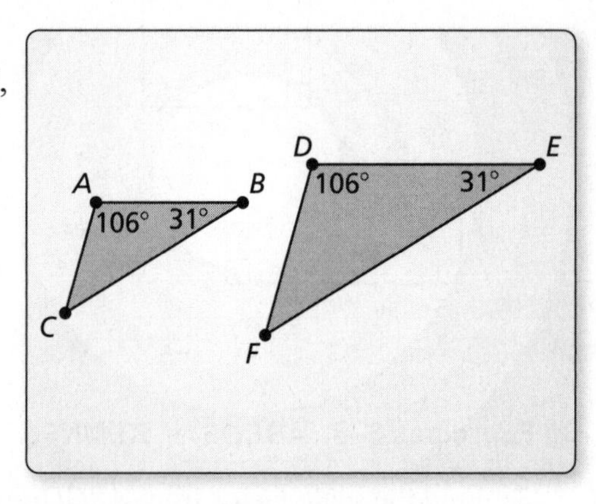

b. Find the third angle measure and the side lengths of each triangle. Record your results in column 1 of the table below.

	1.	2.	3.	4.	5.	6.
$m\angle A, m\angle D$	106°	88°	40°			
$m\angle B, m\angle E$	31°	42°	65°			
$m\angle C$						
$m\angle F$						
AB						
DE						
BC						
EF						
AC						
DF						

Name______________________________ Date__________

1 EXPLORATION: Comparing Triangles (continued)

c. Are the two triangles similar? Explain.

d. Repeat parts (a)–(c) to complete columns 2 and 3 of the table for the given angle measures.

e. Complete each remaining column of the table using your own choice of two pairs of equal corresponding angle measures. Can you construct two triangles in this way that are *not* similar?

f. Make a conjecture about any two triangles with two pairs of congruent corresponding angles.

Communicate Your Answer

2. What can you conclude about two triangles when you know that two pairs of corresponding angles are congruent?

3. Find RS in the figure at the right.

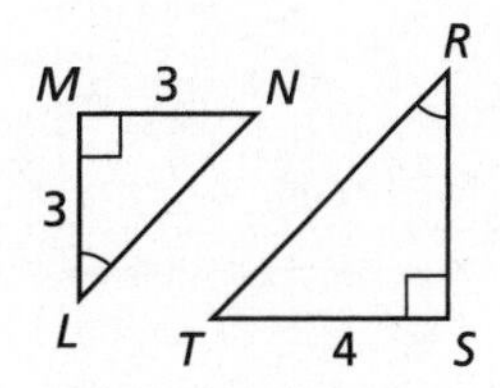

Name ______________________________ Date __________

8.2 Notetaking with Vocabulary
For use after Lesson 8.2

In your own words, write the meaning of each vocabulary term.

similar figures

similarity transformation

Theorems

Theorem 8.3 Angle-Angle (AA) Similarity Theorem

If two angles of one triangle are congruent to two angles of another triangle, then the two triangles are similar.

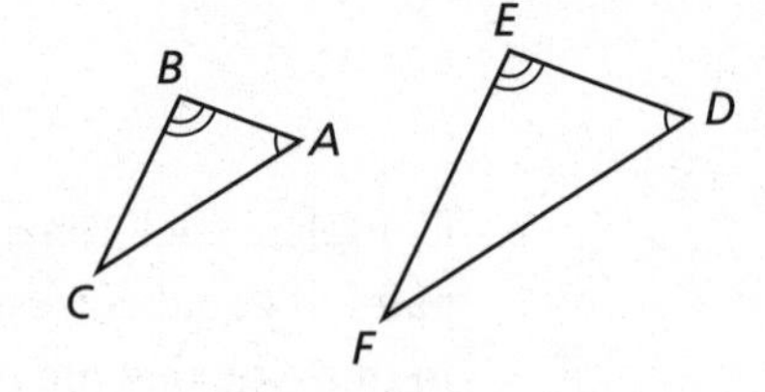

If $\angle A \cong \angle D$ and $\angle B \cong \angle E$, then $\triangle ABC \sim \triangle DEF$.

Notes:

Extra Practice

In Exercises 1 and 2, determine whether the triangles are similar. If they are, write a similarity statement. Explain your reasoning.

1.

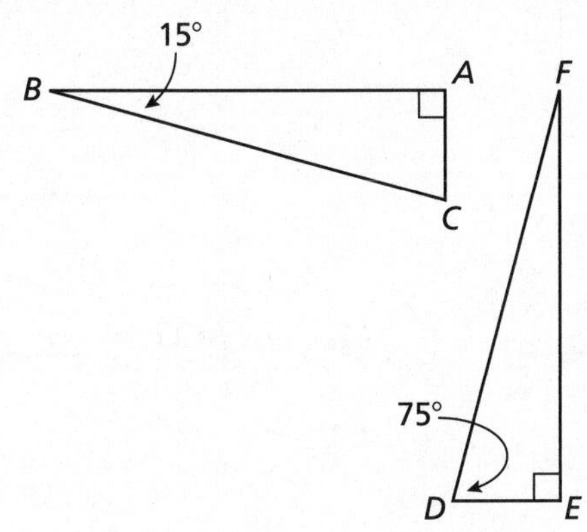

2.

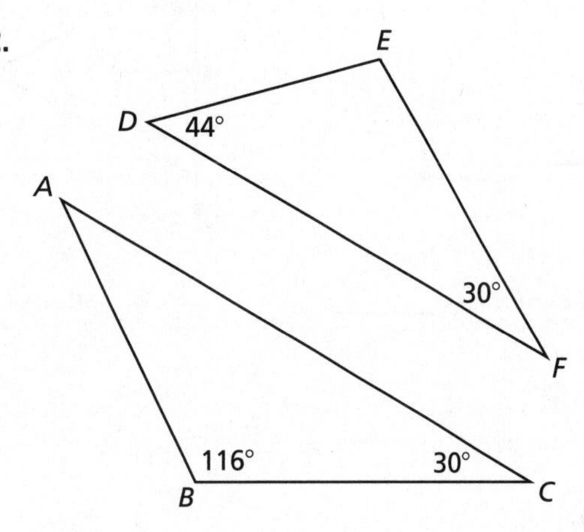

In Exercises 3 and 4, show that the two triangles are similar.

3.

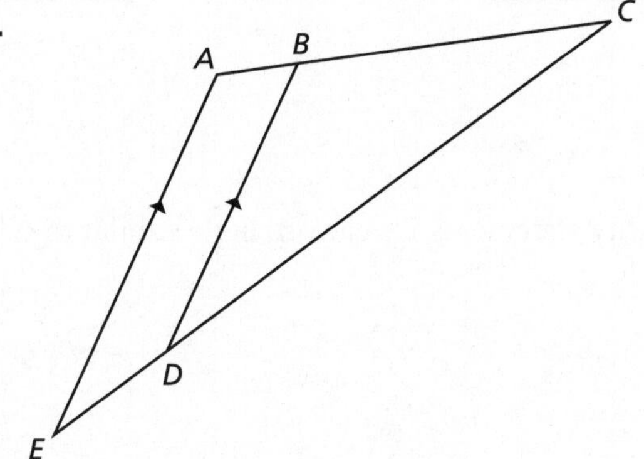

4.

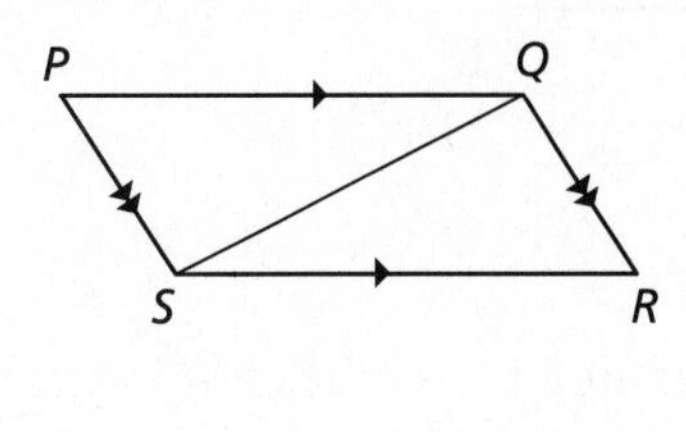

Name ______________________________ Date __________

8.2 Notetaking with Vocabulary (continued)

In Exercises 5–13, use the diagram to complete the statement.

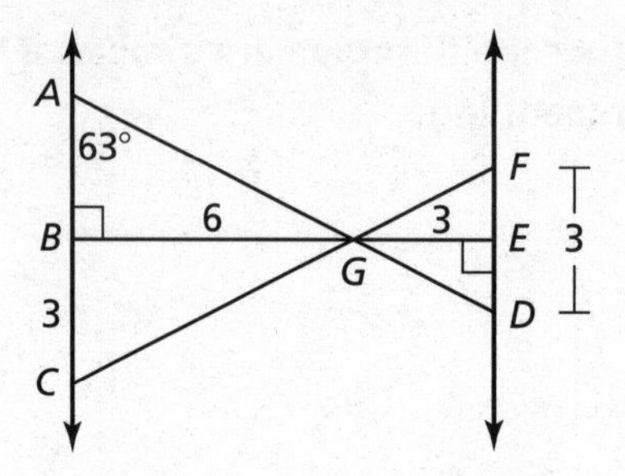

5. $m\angle AGB$ = ______ **6.** $m\angle EGD$ = ______ **7.** $m\angle BCG$ = ______

8. AG = ______ **9.** AB = ______ **10.** FE = ______

11. ED = ______ **12.** GF = ______ **13.** $\triangle AGC \sim$ ______

14. Using the diagram for Exercises 5–13, write similarity statements for each triangle similar to $\triangle EFG$.

15. Determine if it is possible for $\triangle HJK$ and $\triangle PQR$ to be similar. Explain your reasoning.

$$m\angle H = 100°, m\angle K = 46°, m\angle P = 44°, \text{and } m\angle Q = 46°$$

Name____ Date__________

8.3 Proving Triangle Similarity by SSS and SAS

For use with Exploration 8.3

Essential Question What are two ways to use corresponding sides of two triangles to determine that the triangles are similar?

1 EXPLORATION: Deciding Whether Triangles Are Similar

Go to *BigIdeasMath.com* for an interactive tool to investigate this exploration.

Work with a partner. Use dynamic geometry software.

a. Construct $\triangle ABC$ and $\triangle DEF$ with the side lengths given in column 1 of the table below.

	1.	2.	3.	4.	5.	6.	7.
AB	5	5	6	15	9	24	
BC	8	8	8	20	12	18	
AC	10	10	10	10	8	16	
DE	10	15	9	12	12	8	
EF	16	24	12	16	15	6	
DF	20	30	15	8	10	8	
$m\angle A$							
$m\angle B$							
$m\angle C$							
$m\angle D$							
$m\angle E$							
$m\angle F$							

b. Complete column 1 in the table above.

c. Are the triangles similar? Explain your reasoning.

d. Repeat parts (a)–(c) for columns 2–6 in the table.

e. How are the corresponding side lengths related in each pair of triangles that are similar? Is this true for each pair of triangles that are not similar?

1 EXPLORATION: Deciding Whether Triangles Are Similar (continued)

f. Make a conjecture about the similarity of two triangles based on their corresponding side lengths.

g. Use your conjecture to write another set of side lengths of two similar triangles. Use the side lengths to complete column 7 of the table.

2 EXPLORATION: Deciding Whether Triangles Are Similar

Go to *BigIdeasMath.com* for an interactive tool to investigate this exploration.

Work with a partner. Use dynamic geometry software. Construct any $\triangle ABC$.

a. Find AB, AC, and $m\angle A$. Choose any positive rational number k and construct $\triangle DEF$ so that $DE = k \bullet AB$, $DF = k \bullet AC$, and $m\angle D = m\angle A$.

b. Is $\triangle DEF$ similar to $\triangle ABC$? Explain your reasoning.

c. Repeat parts (a) and (b) several times by changing $\triangle ABC$ and k. Describe your results.

Communicate Your Answer

3. What are two ways to use corresponding sides of two triangles to determine that the triangles are similar?

Name____ Date__________

8.3 Notetaking with Vocabulary

For use after Lesson 8.3

In your own words, write the meaning of each vocabulary term.

similar figures

corresponding parts

slope

parallel lines

perpendicular lines

Theorems

Theorem 8.4 Side-Side-Side (SSS) Similarity Theorem

If the corresponding side lengths of two triangles are proportional, then the triangles are similar.

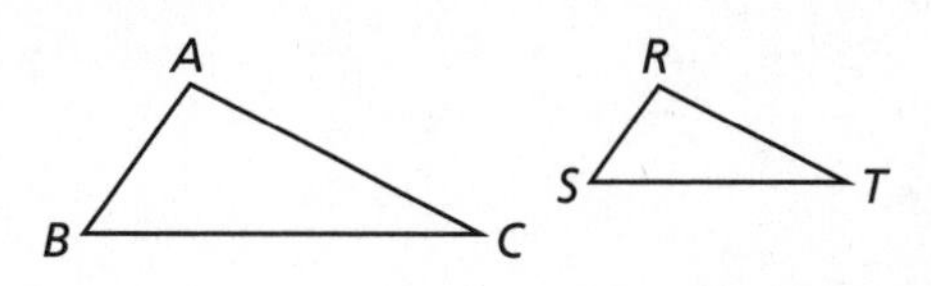

If $\frac{AB}{RS} = \frac{BC}{ST} = \frac{CA}{TR}$, then $\triangle ABC \sim \triangle RST$.

Notes:

Theorem 8.5 Side-Angle-Side (SAS) Similarity Theorem

If an angle of one triangle is congruent to an angle of a second triangle and the lengths of the sides including these angles are proportional, then the triangles are similar.

If $\angle X \cong \angle M$ and $\dfrac{ZX}{PM} = \dfrac{XY}{MN}$, then $\triangle XYZ \sim \triangle MNP$.

Notes:

Extra Practice

In Exercises 1 and 2, determine whether $\triangle RST$ is similar to $\triangle ABC$.

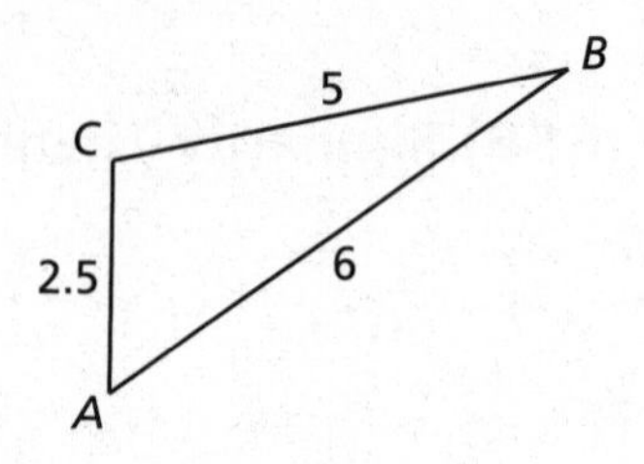

1.

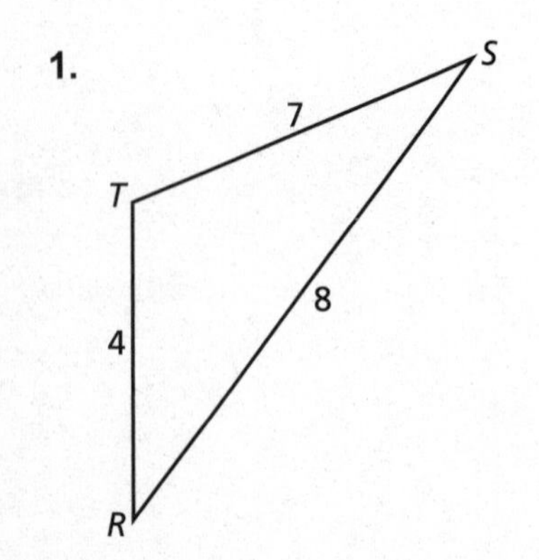

2.

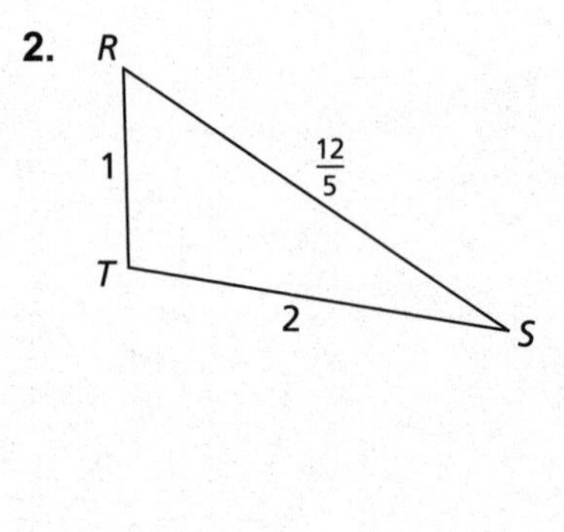

3. Find the value of x that makes $\triangle RST \sim \triangle HGK$.

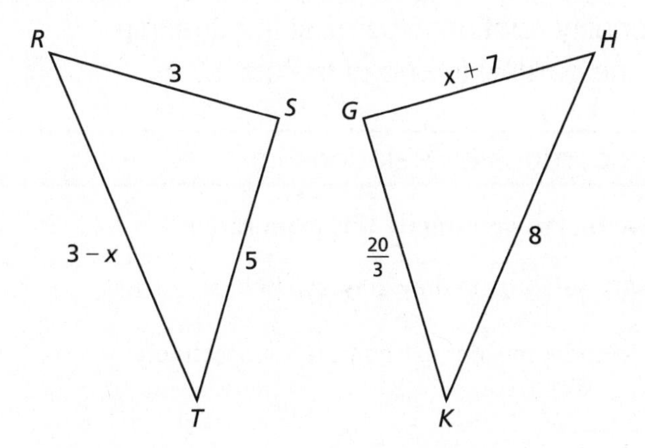

4. Verify that $\triangle RST \sim \triangle XYZ$. Find the scale factor of $\triangle RST$ to $\triangle XYZ$.

$$\triangle RST : RS = 12, ST = 15, TR = 24$$
$$\triangle XYZ : XY = 28, YZ = 35, ZX = 56$$

In Exercises 5 and 6, use $\triangle ABC$.

5. The shortest side of a triangle similar to $\triangle ABC$ is 15 units long. Find the other side lengths of the triangle.

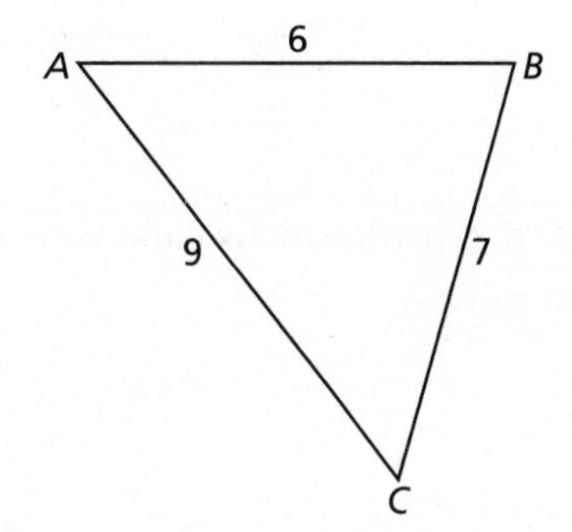

6. The longest side of a triangle similar to $\triangle ABC$ is 6 units long. Find the other side lengths of the triangle.

Name ______________________________ Date __________

8.4 Proportionality Theorems

For use with Exploration 8.4

Essential Question What proportionality relationships exist in a triangle intersected by an angle bisector or by a line parallel to one of the sides?

1 EXPLORATION: Discovering a Proportionality Relationship

Go to *BigIdeasMath.com* for an interactive tool to investigate this exploration.

Work with a partner. Use dynamic geometry software to draw any $\triangle ABC$.

a. Construct $\overline{DE}$ parallel to $\overline{BC}$ with endpoints on $\overline{AB}$ and $\overline{AC}$, respectively.

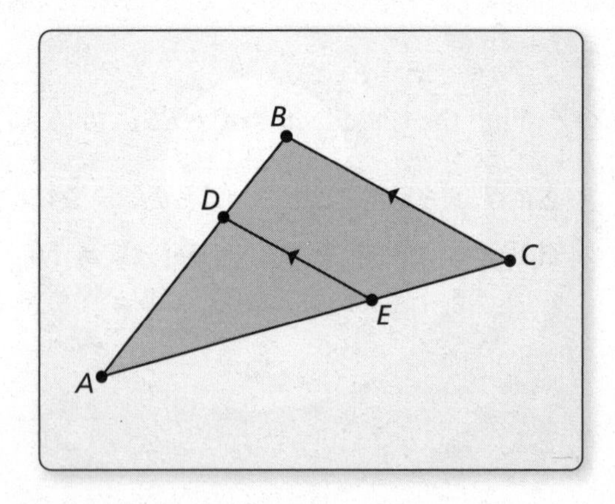

b. Compare the ratios of AD to BD and AE to CE.

c. Move $\overline{DE}$ to other locations parallel to $\overline{BC}$ with endpoints on $\overline{AB}$ and $\overline{AC}$, and repeat part (b).

d. Change $\triangle ABC$ and repeat parts (a)–(c) several times. Write a conjecture that summarizes your results.

2 EXPLORATION: Discovering a Proportionality Relationship

Go to *BigIdeasMath.com* for an interactive tool to investigate this exploration.

Work with a partner. Use dynamic geometry software to draw any $\triangle ABC$.

a. Bisect $\angle B$ and plot point D at the intersection of the angle bisector and $\overline{AC}$.

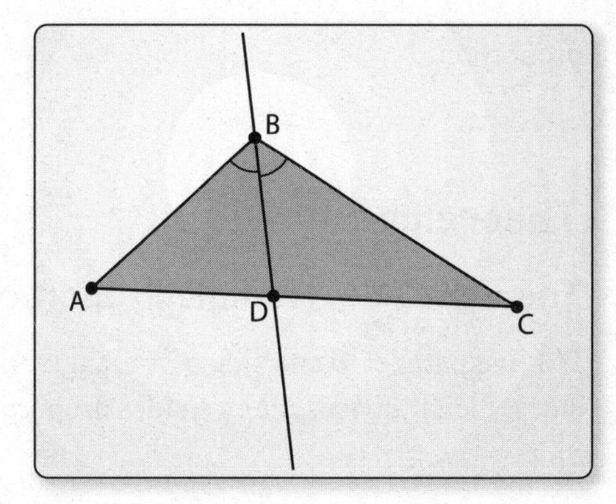

b. Compare the ratios of AD to DC and BA to BC.

c. Change $\triangle ABC$ and repeat parts (a) and (b) several times. Write a conjecture that summarizes your results.

Communicate Your Answer

3. What proportionality relationships exist in a triangle intersected by an angle bisector or by a line parallel to one of the sides?

4. Use the figure at the right to write a proportion.

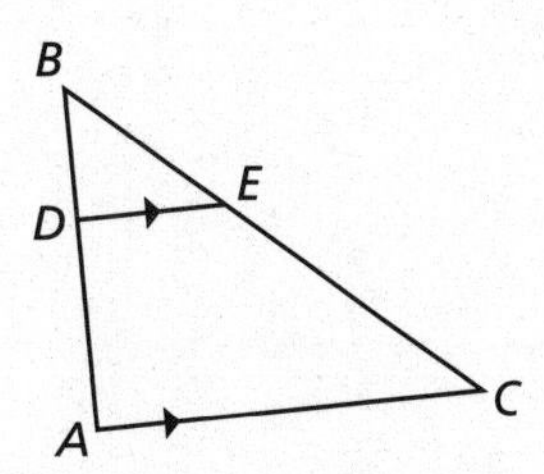

Name ____ Date __________

8.4 Notetaking with Vocabulary

For use after Lesson 8.4

In your own words, write the meaning of each vocabulary term.

corresponding angles

ratio

proportion

Theorems

Theorem 8.6 Triangle Proportionality Theorem

If a line parallel to one side of a triangle intersects the other two sides, then it divides the two sides proportionally.

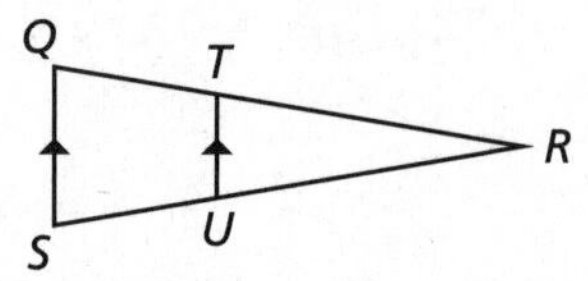

Notes:

If $\overline{TU} \parallel \overline{QS}$, then $\frac{RT}{TQ} = \frac{RU}{US}$.

Theorem 8.7 Converse of the Triangle Proportionality Theorem

If a line divides two sides of a triangle proportionally, then it is parallel to the third side.

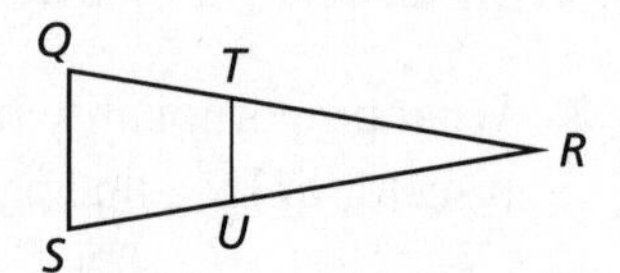

Notes:

If $\frac{RT}{TQ} = \frac{RU}{US}$, then $\overline{TU} \parallel \overline{QS}$.

Theorem 8.8 Three Parallel Lines Theorem

If three parallel lines intersect two transversals, then they divide the transversals proportionally.

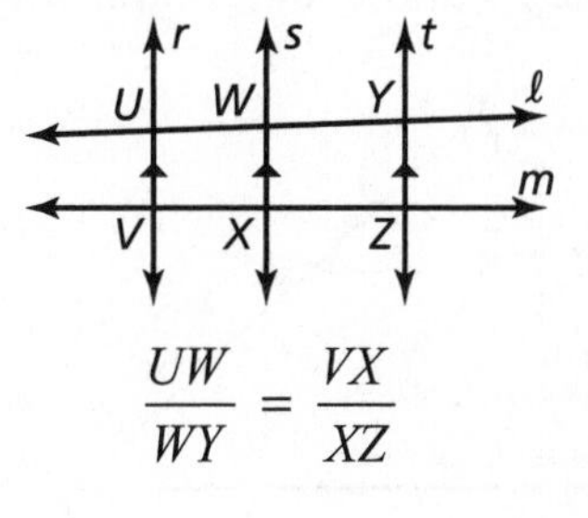

$$\frac{UW}{WY} = \frac{VX}{XZ}$$

Notes:

Theorem 8.9 Triangle Angle Bisector Theorem

If a ray bisects an angle of a triangle, then it divides the opposite side into segments whose lengths are proportional to the lengths of the other two sides.

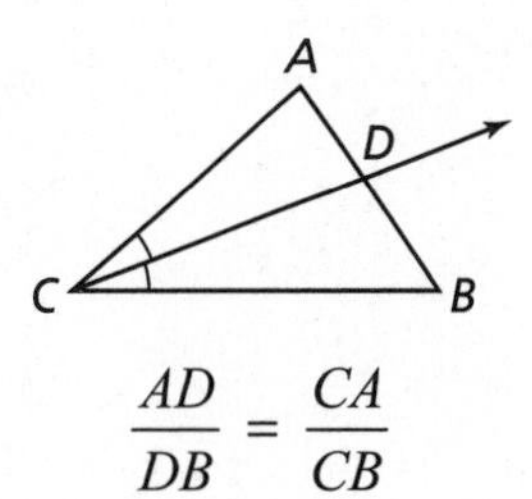

$$\frac{AD}{DB} = \frac{CA}{CB}$$

Notes:

Extra Practice

In Exercises 1 and 2, find the length of $\overline{AB}$.

1.

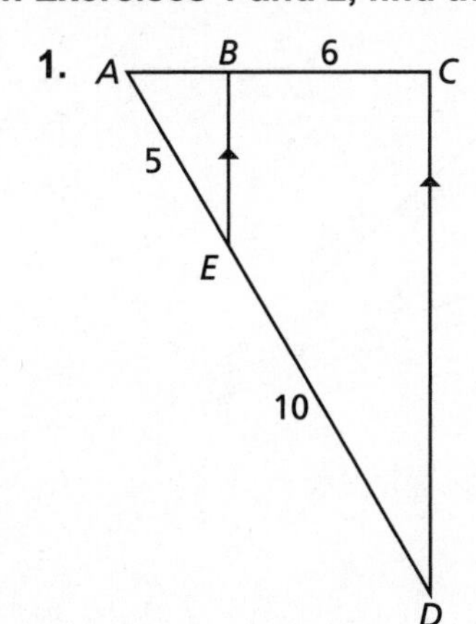

2.

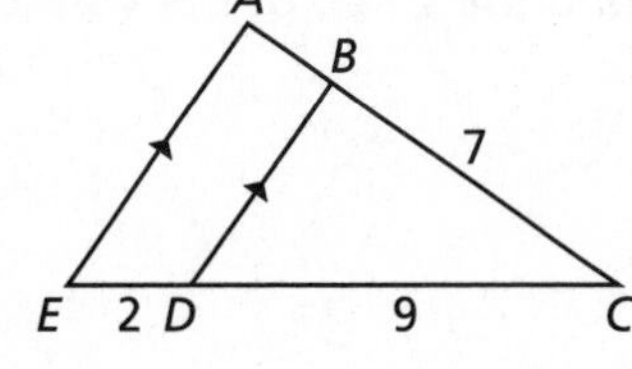

Name ______________________________ Date __________

8.4 Notetaking with Vocabulary (continued)

In Exercises 3 and 4, determine whether $\overline{AB} \parallel \overline{XY}$.

3.

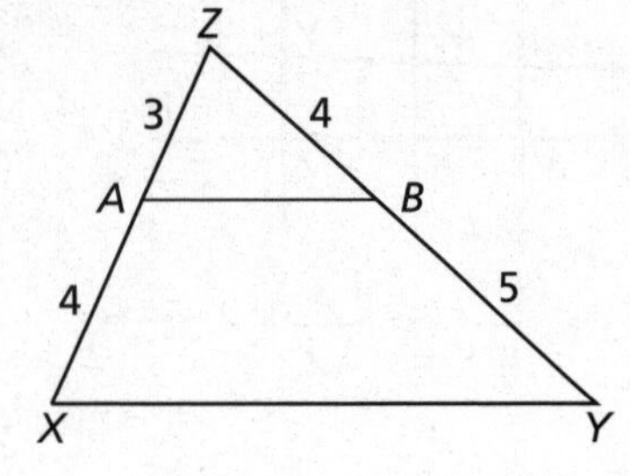

4.

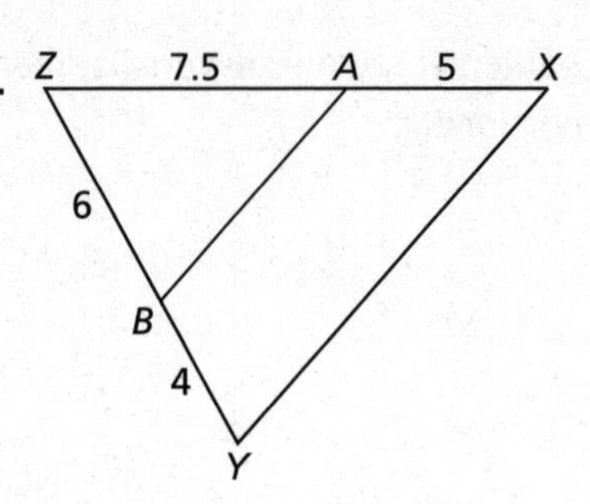

In Exercises 5–7, use the diagram to complete the proportion.

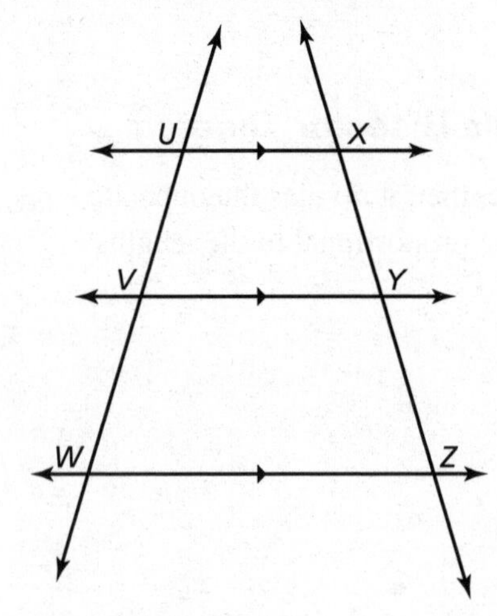

5. $\frac{UV}{UW} = \frac{XY}{\square}$

6. $\frac{XY}{YZ} = \frac{\square}{VW}$

7. $\frac{\square}{ZY} = \frac{WU}{WV}$

In Exercises 8 and 9, find the value of the variable.

8.

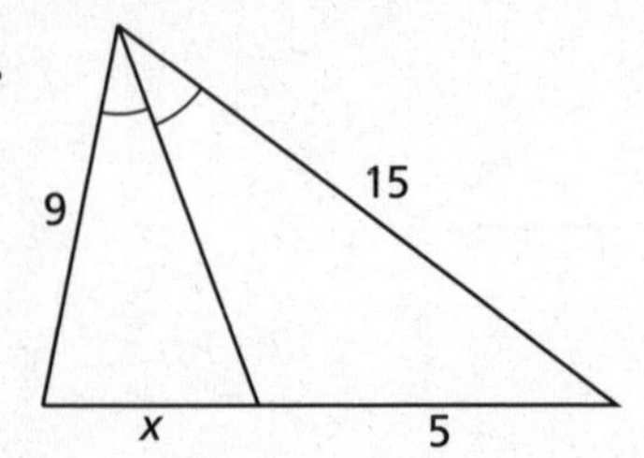

9.

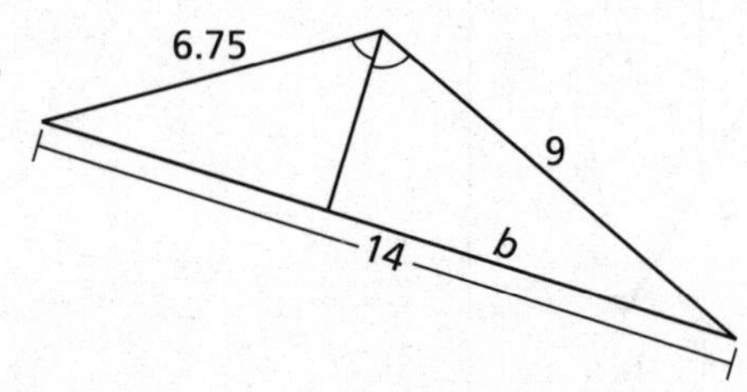

Name____ Date__________

Chapter 9 Maintaining Mathematical Proficiency

Simplify the expression.

1. $\sqrt{500}$

2. $\sqrt{189}$

3. $\sqrt{252}$

4. $\frac{4}{\sqrt{3}}$

5. $\frac{11}{\sqrt{5}}$

6. $\frac{8}{\sqrt{2}}$

Solve the proportion.

7. $\frac{x}{21} = \frac{2}{7}$

8. $\frac{x}{5} = \frac{9}{4}$

9. $\frac{3}{x} = \frac{14}{42}$

10. $\frac{20}{27} = \frac{6}{x}$

11. $\frac{x - 4}{5} = \frac{10}{9}$

12. $\frac{15}{5x + 25} = \frac{3}{9}$

13. The Pythagorean Theorem states that $a^2 + b^2 = c^2$, where a and b are legs of a right triangle and c is the hypotenuse. Are you able to simplify the Pythagorean Theorem further to say that $a + b = c$? Explain.

Name ______________________________ Date __________

9.1 The Pythagorean Theorem

For use with Exploration 9.1

Essential Question How can you prove the Pythagorean Theorem?

1 EXPLORATION: Proving the Pythagorean Theorem without Words

Work with a partner.

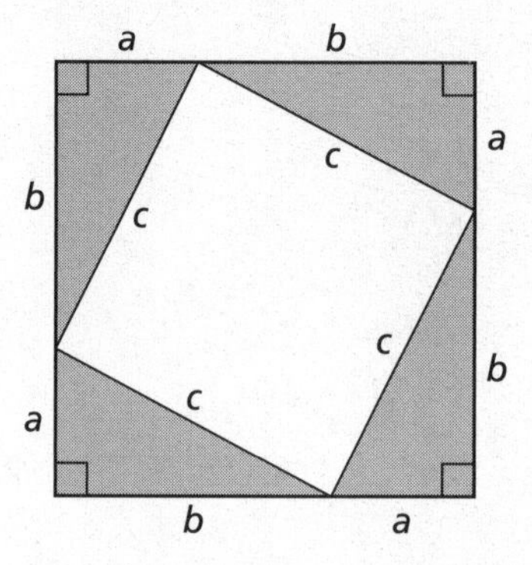

a. Draw and cut out a right triangle with legs a and b, and hypotenuse c.

b. Make three copies of your right triangle. Arrange all four triangles to form a large square as shown.

c. Find the area of the large square in terms of a, b, and c by summing the areas of the triangles and the small square.

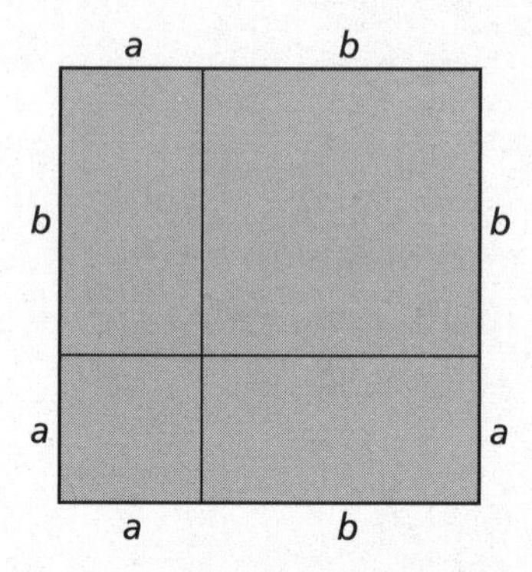

d. Copy the large square. Divide it into two smaller squares and two equally-sized rectangles, as shown.

e. Find the area of the large square in terms of a and b by summing the areas of the rectangles and the smaller squares.

f. Compare your answers to parts (c) and (e). Explain how this proves the Pythagorean Theorem.

2 EXPLORATION: Proving the Pythagorean Theorem

Work with a partner.

a. Consider the triangle shown.

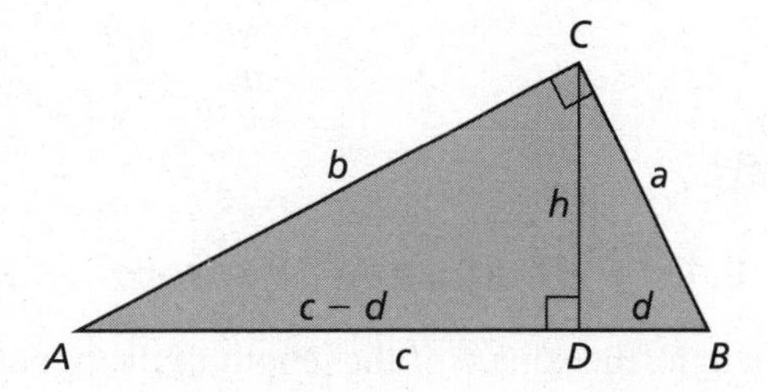

b. Explain why $\triangle ABC$, $\triangle ACD$, and $\triangle CBD$ are similar.

c. Write a two-column proof using the similar triangles in part (b) to prove that $a^2 + b^2 = c^2$.

Communicate Your Answer

3. How can you prove the Pythagorean Theorem?

4. Use the Internet or some other resource to find a way to prove the Pythagorean Theorem that is different from Explorations 1 and 2.

Name ______________________________ Date ________

9.1 Notetaking with Vocabulary

For use after Lesson 9.1

In your own words, write the meaning of each vocabulary term.

Pythagorean triple

Theorems

Theorem 9.1 Pythagorean Theorem

In a right triangle, the square of the length of the hypotenuse is equal to the sum of the squares of the lengths of the legs.

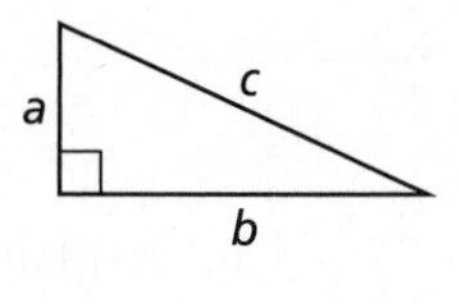

$c^2 = a^2 + b^2$

Notes:

Core Concepts

Common Pythagorean Triples and Some of Their Multiples

3, 4, 5	**5, 12, 13**	**8, 15, 17**	**7, 24, 25**
6, 8, 10	10, 24, 26	16, 30, 34	14, 48, 50
9, 12, 15	15, 36, 39	24, 45, 51	21, 72, 75
$3x$, $4x$, $5x$	$5x$, $12x$, $13x$	$8x$, $15x$, $17x$	$7x$, $24x$, $25x$

The most common Pythagorean triples are in bold. The other triples are the result of multiplying each integer in a bold-faced triple by the same factor.

Notes:

9.1 Notetaking with Vocabulary (continued)

Theorems

Theorem 9.2 Converse of the Pythagorean Theorem

If the square of the length of the longest side of a triangle is equal to the sum of the squares of the lengths of the other two sides, then the triangle is a right triangle.

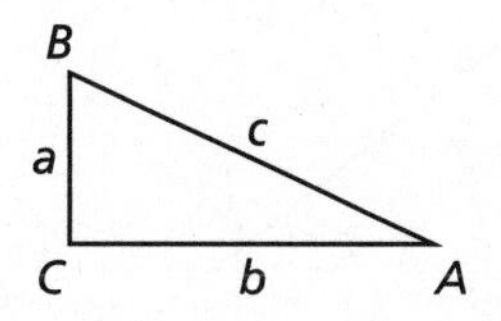

If $c^2 = a^2 + b^2$, then $\triangle ABC$ is a right triangle.

Notes:

Theorem 9.3 Pythagorean Inequalities Theorem

For any $\triangle ABC$, where c is the length of the longest side, the following statements are true.

If $c^2 < a^2 + b^2$, then $\triangle ABC$ is acute.

If $c^2 > a^2 + b^2$, then $\triangle ABC$ is obtuse.

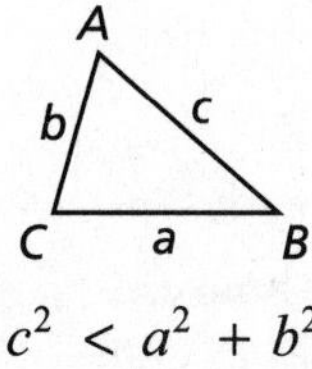

$c^2 < a^2 + b^2$

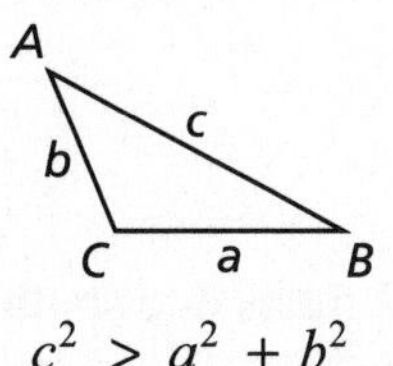

$c^2 > a^2 + b^2$

Notes:

Name ____ Date __________

Extra Practice

In Exercises 1–6, find the value of *x*. Then tell whether the side lengths form a Pythagorean triple.

1.

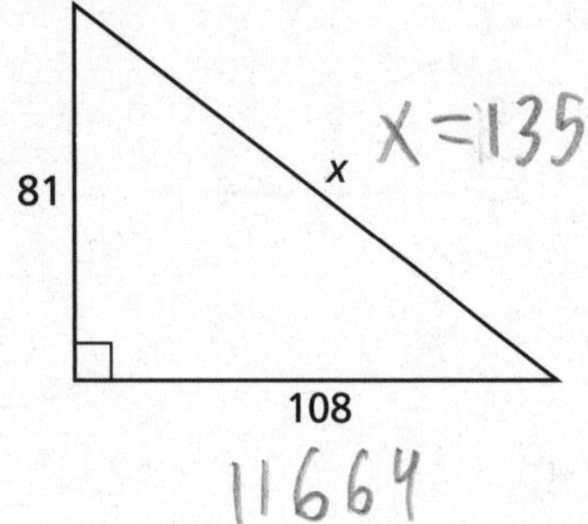

2.

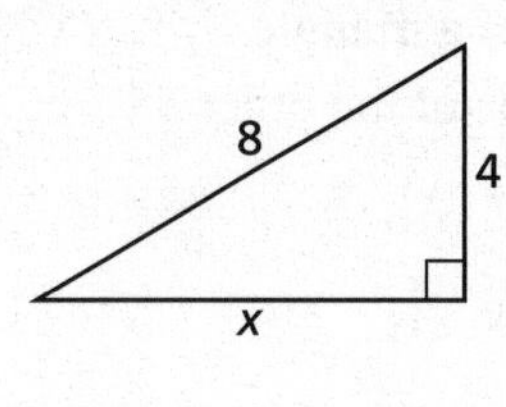

3.

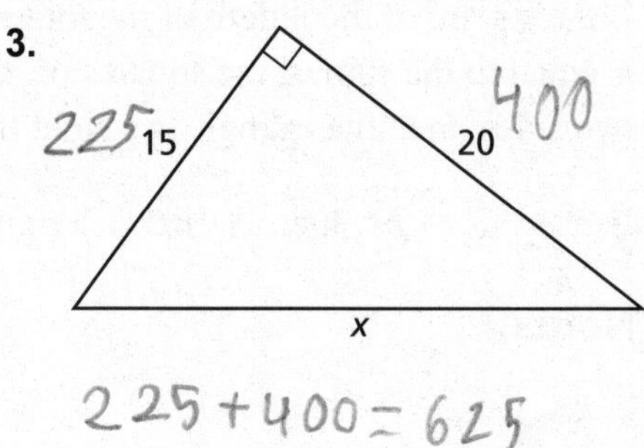

4.

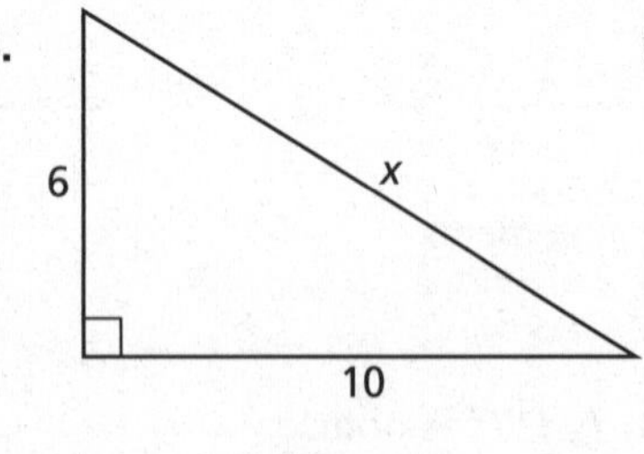

5.

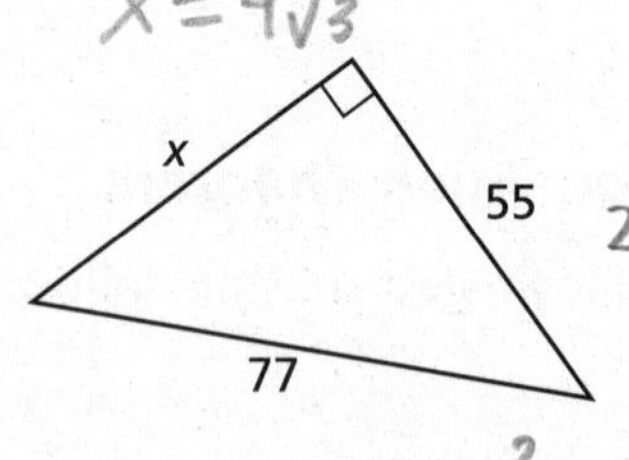

6.

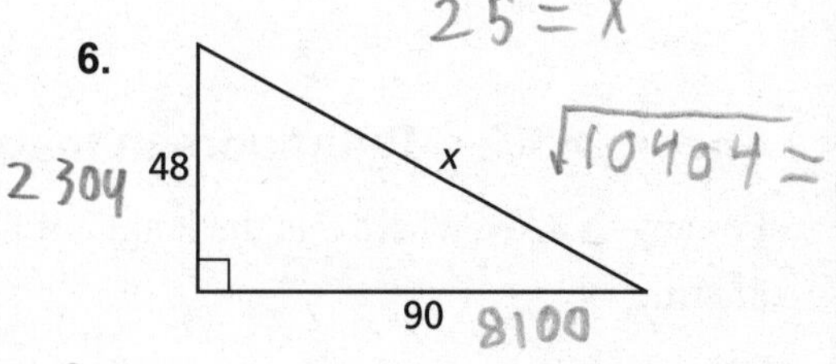

7. From school, you biked 1.2 miles due south and then 0.5 mile due east to your house. If you had biked home on the street that runs directly diagonal from your school to your house, how many fewer miles would you have biked?

In Exercises 8 and 9, verify that the segment lengths form a triangle. Is the triangle *acute*, *right*, or *obtuse*?

8. 90, 216, and 234

9. 1, 1, and $\sqrt{3}$

Name____ Date__________

9.2 Special Right Triangles

For use with Exploration 9.2

Essential Question What is the relationship among the side lengths of 45°-45°-90° triangles? 30°-60°-90° triangles?

1 EXPLORATION: Side Ratios of an Isosceles Right Triangle

Go to *BigIdeasMath.com* for an interactive tool to investigate this exploration.

Work with a partner.

a. Use dynamic geometry software to construct an isosceles right triangle with a leg length of 4 units.

b. Find the acute angle measures. Explain why this triangle is called a 45°-45°-90° triangle.

c. Find the exact ratios of the side lengths (using square roots).

$\frac{AB}{AC}$ = ______

$\frac{AB}{BC}$ = ______

$\frac{AC}{BC}$ = ______

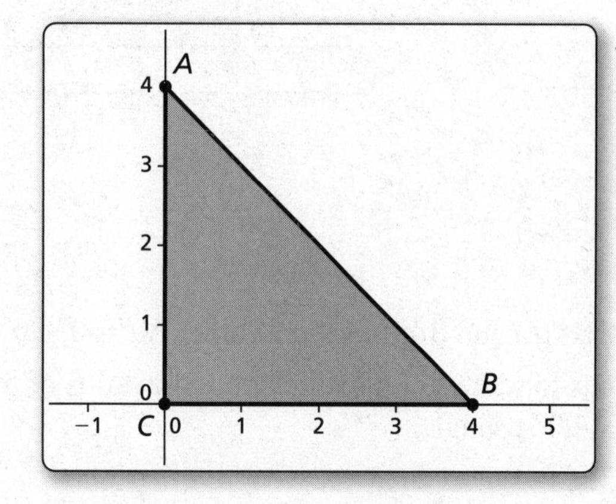

Sample

Points
$A(0, 4)$
$B(4, 0)$
$C(0, 0)$

Segments
$AB = 5.66$
$BC = 4$
$AC = 4$

Angles
$m\angle A = 45°$
$m\angle B = 45°$

d. Repeat parts (a) and (c) for several other isosceles right triangles. Use your results to write a conjecture about the ratios of the side lengths of an isosceles right triangle.

Name ______________________________ Date __________

2 EXPLORATION: Side Ratios of a 30°-60°-90° Triangle

Go to *BigIdeasMath.com* for an interactive tool to investigate this exploration.

Work with a partner.

a. Use dynamic geometry software to construct a right triangle with acute angle measures of 30° and 60° (a 30°-60°-90° triangle), where the shorter leg length is 3 units.

b. Find the exact ratios of the side lengths (using square roots).

$\frac{AB}{AC} =$ ______

$\frac{AB}{BC} =$ ______

$\frac{AC}{BC} =$ ______

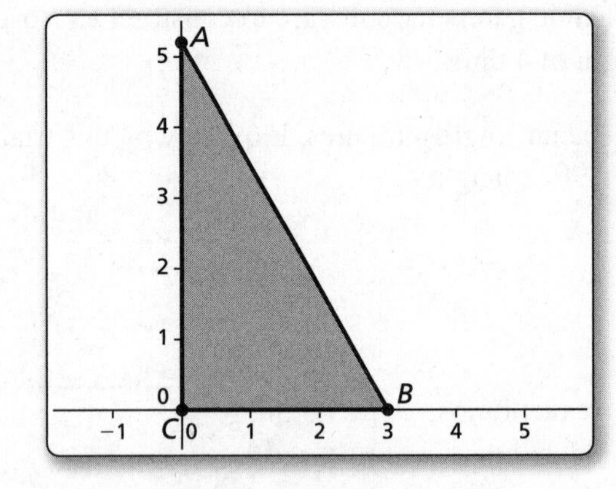

Sample

Points

$A(0, 5.20)$

$B(3, 0)$

$C(0, 0)$

Segments

$AB = 6$

$BC = 3$

$AC = 5.20$

Angles

$m\angle A = 30°$

$m\angle B = 60°$

c. Repeat parts (a) and (b) for several other 30°-60°-90° triangles. Use your results to write a conjecture about the ratios of the side lengths of a 30°-60°-90° triangle.

Communicate Your Answer

3. What is the relationship among the side lengths of 45°-45°-90° triangles? 30°-60°-90° triangles?

Name____ Date___________

9.2 Notetaking with Vocabulary

For use after Lesson 9.2

In your own words, write the meaning of each vocabulary term.

isosceles triangle

Theorems

Theorem 9.4 45°-45°-90° Triangle Theorem

In a 45°-45°-90° triangle, the hypotenuse is $\sqrt{2}$ times as long as each leg.

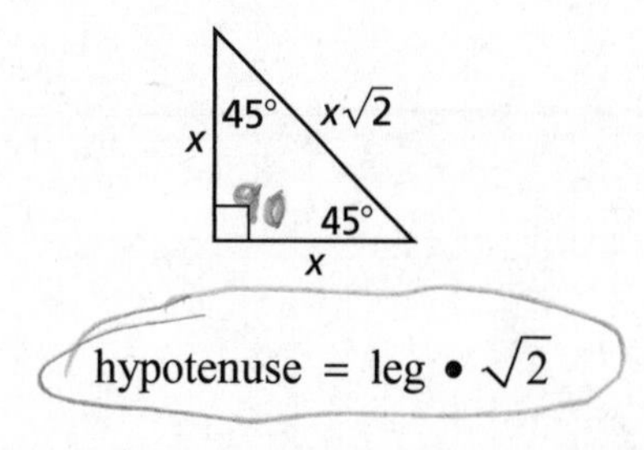

Notes:

Theorem 9.5 30°-60°-90° Triangle Theorem

In a 30°-60°-90° triangle, the hypotenuse is twice as long as the shorter leg, and the longer leg is $\sqrt{3}$ times as long as the shorter leg.

Notes:

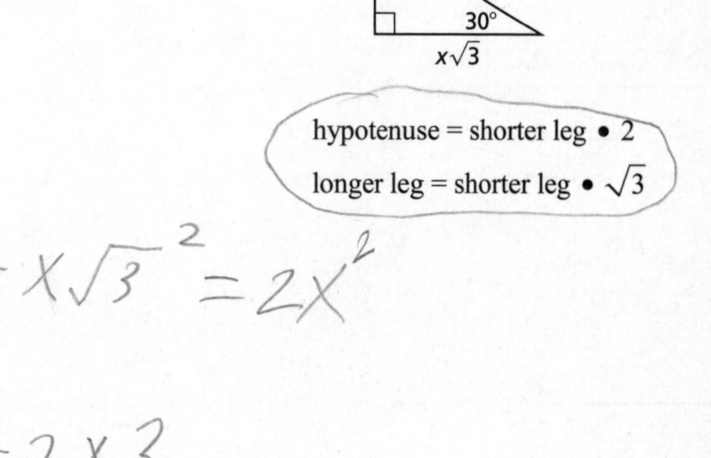

9.2 Notetaking with Vocabulary (continued)

Extra Practice

In Exercises 1–4, find the value of x. Write your answer in simplest form.

1. 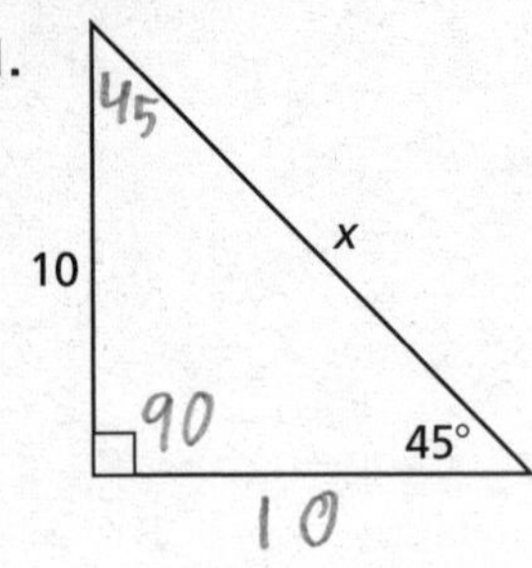

2. 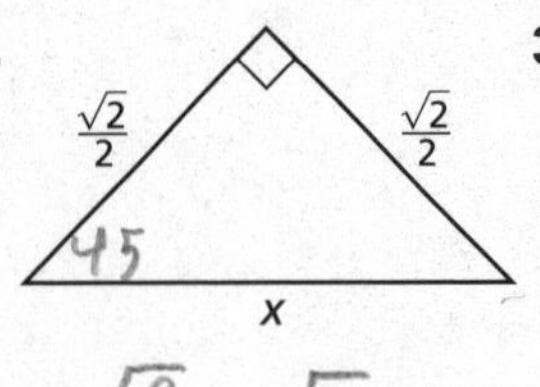

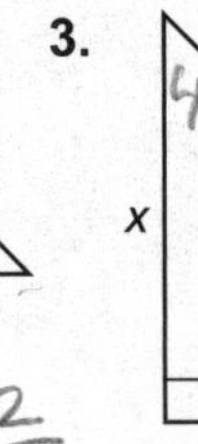

$\frac{\sqrt{2}}{2} \cdot \sqrt{2} = \frac{2}{2}$

$\frac{2}{2} = 1$

3. 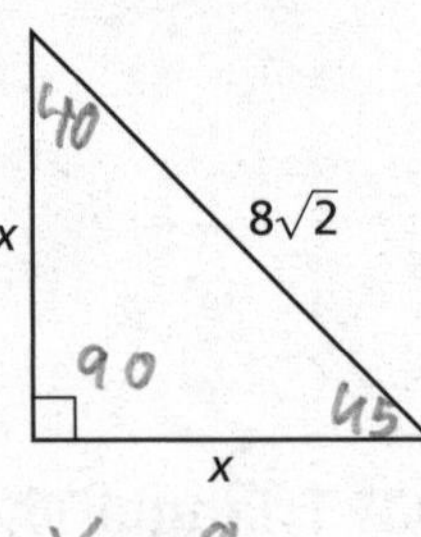

$x = 8$

4. 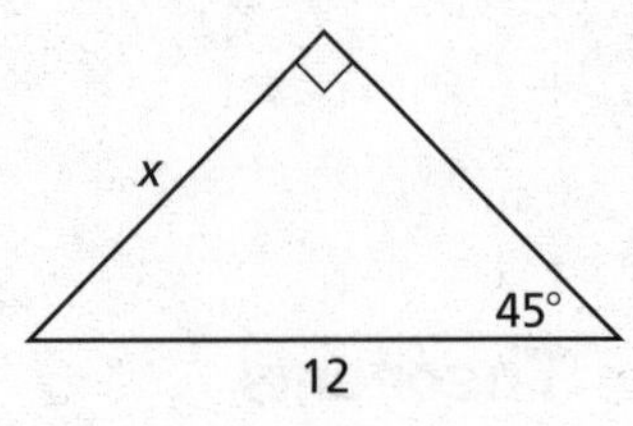

$12^2 = x^2 + x^2$

In Exercises 5–7, find the values of x and y. Write your answers in simplest form.

5. 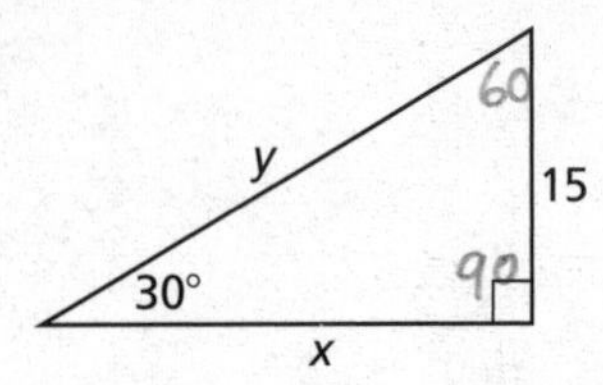

$15^2 + x^2 = y^2$

$30 + x^2 = y^2$

6. 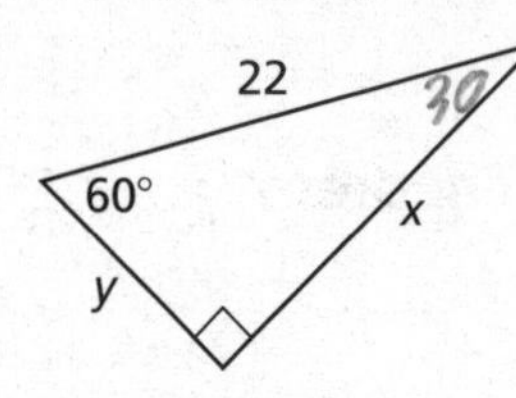

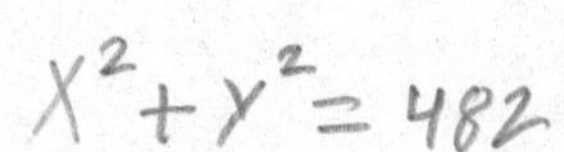

$x^2 + y^2 = 22^2$

$x^2 + y^2 = 482$

7. 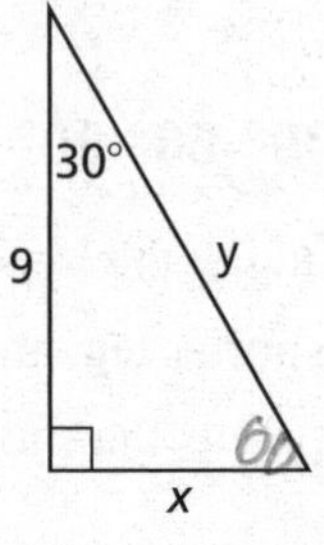

$9^2 + x^2 = y^2$

$81 + x^2 = y^2$

In Exercises 8 and 9, sketch the figure that is described. Find the indicated length. Round decimal answers to the nearest tenth.

8. The length of a diagonal in a square is 32 inches. Find the perimeter of the square.

9. An isosceles triangle with $30°$ base angles has an altitude of $\sqrt{3}$ meters. Find the length of the base of the isosceles triangle.

10. Find the area of $\triangle DEF$. Round decimal answers to the nearest tenth.

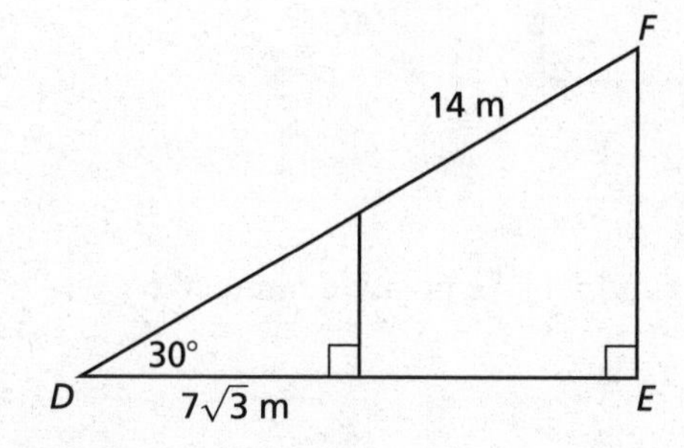

Name ______________________________ Date __________

9.3 Similar Right Triangles
For use with Exploration 9.3

Essential Question How are altitudes and geometric means of right triangles related?

1 EXPLORATION: Writing a Conjecture

Go to *BigIdeasMath.com* for an interactive tool to investigate this exploration.

Work with a partner.

a. Use dynamic geometry software to construct right $\triangle ABC$, as shown. Draw $\overline{CD}$ so that it is an altitude from the right angle to the hypotenuse of $\triangle ABC$.

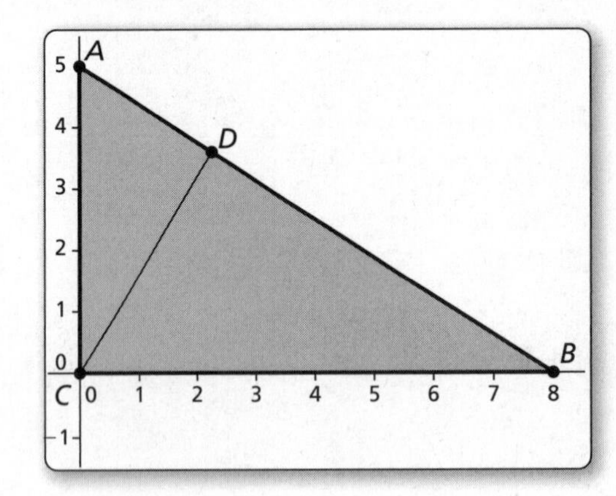

Points
$A(0, 5)$
$B(8, 0)$
$C(0, 0)$
$D(2.25, 3.6)$

Segments
$AB = 9.43$
$BC = 8$
$AC = 5$

b. The **geometric mean** of two positive numbers a and b is the positive number x that satisfies

$$\frac{a}{x} = \frac{x}{b}.$$

x is the geometric mean of a and b.

Write a proportion involving the side lengths of $\triangle CBD$ and $\triangle ACD$ so that CD is the geometric mean of two of the other side lengths. Use similar triangles to justify your steps.

Name____ Date__________

1 EXPLORATION: Writing a Conjecture (continued)

c. Use the proportion you wrote in part (b) to find *CD*.

d. Generalize the proportion you wrote in part (b). Then write a conjecture about how the geometric mean is related to the altitude from the right angle to the hypotenuse of a right triangle.

2 EXPLORATION: Comparing Geometric and Arithmetic Means

Go to *BigIdeasMath.com* for an interactive tool to investigate this exploration.

Work with a partner. Use a spreadsheet to find the arithmetic mean and the geometric mean of several pairs of positive numbers. Compare the two means. What do you notice?

	A	B	C	D
1	a	b	Arithmetic Mean	Geometric Mean
2	3	4	3.5	3.464
3	4	5		
4	6	7		
5	0.5	0.5		
6	0.4	0.8		
7	2	5		
8	1	4		
9	9	16		
10	10	100		
11				

Communicate Your Answer

3. How are altitudes and geometric means of right triangles related?

Name ____ Date __________

9.3 Notetaking with Vocabulary

For use after Lesson 9.3

In your own words, write the meaning of each vocabulary term.

geometric mean

Theorems

Theorem 9.6 Right Triangle Similarity Theorem

If the altitude is drawn to the hypotenuse of a right triangle, then the two triangles formed are similar to the original triangle and to each other.

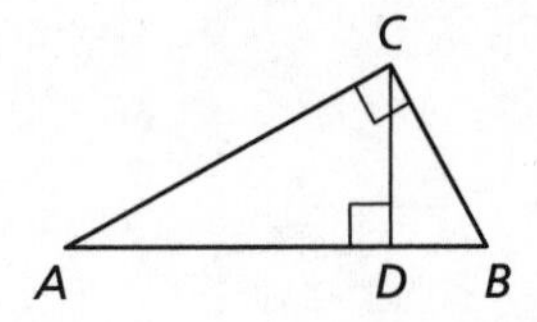

$\triangle CBD \sim \triangle ABC$, $\triangle ACD \sim \triangle ABC$, and $\triangle CBD \sim \triangle ACD$.

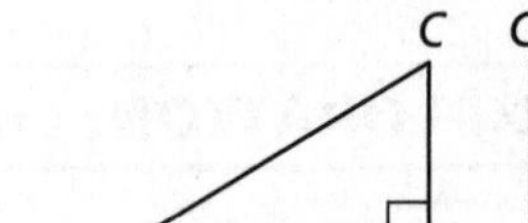

Notes:

Core Concepts

Geometric Mean

The **geometric mean** of two positive numbers a and b is the positive number x that satisfies $\frac{a}{x} = \frac{x}{b}$. So, $x^2 = ab$ and $x = \sqrt{ab}$.

Notes:

Name____ Date__________

Theorems

Theorem 9.7 Geometric Mean (Altitude) Theorem

In a right triangle, the altitude from the right angle to the hypotenuse divides the hypotenuse into two segments.

The length of the altitude is the geometric mean of the lengths of the two segments of the hypotenuse.

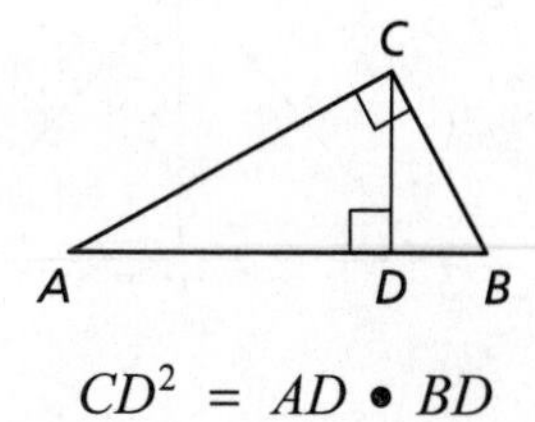

$CD^2 = AD \bullet BD$

Notes:

Theorem 9.8 Geometric Mean (Leg) Theorem

In a right triangle, the altitude from the right angle to the hypotenuse divides the hypotenuse into two segments.

The length of each leg of the right triangle is the geometric mean of the lengths of the hypotenuse and the segment of the hypotenuse that is adjacent to the leg.

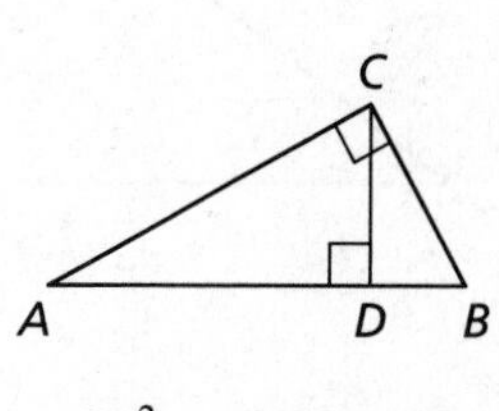

$CB^2 = DB \bullet AB$

$AC^2 = AD \bullet AB$

Notes:

Name ______________________ Date ________

Extra Practice

In Exercises 1 and 2, identify the similar triangles.

1.

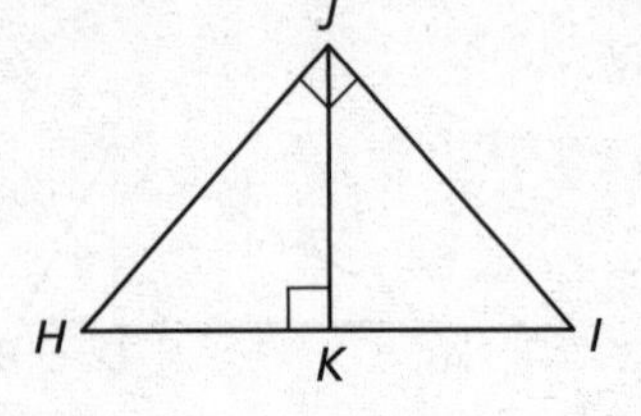

2.

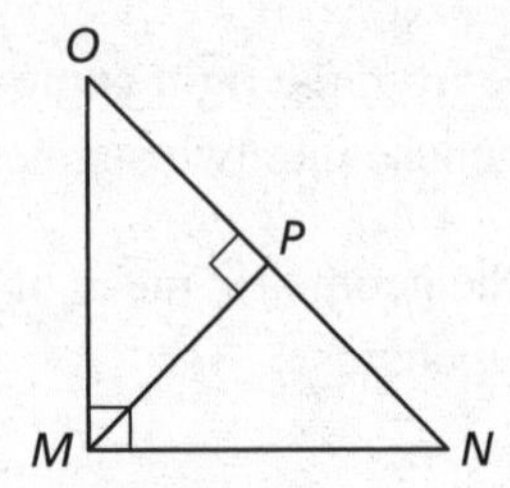

In Exercises 3 and 4, find the geometric mean of the two numbers.

3. 2 and 6

4. 5 and 45

In Exercises 5–8, find the value of the variable.

5.

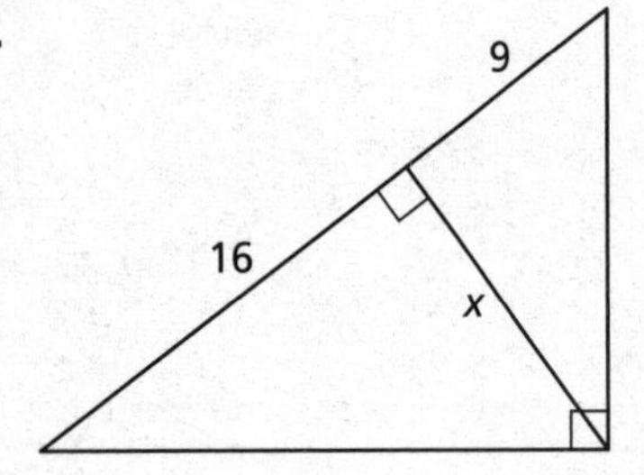

6.

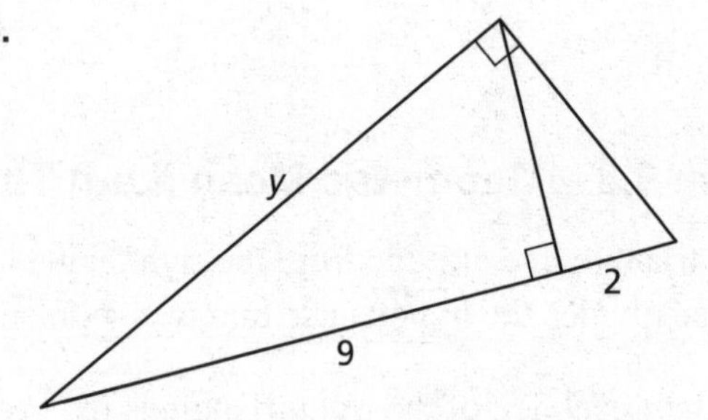

7.

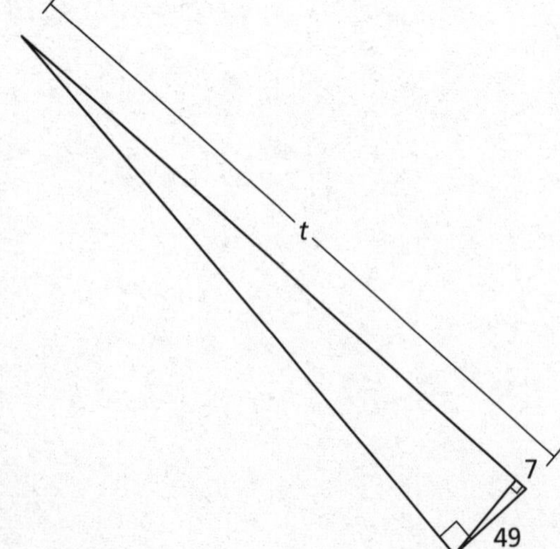

8.

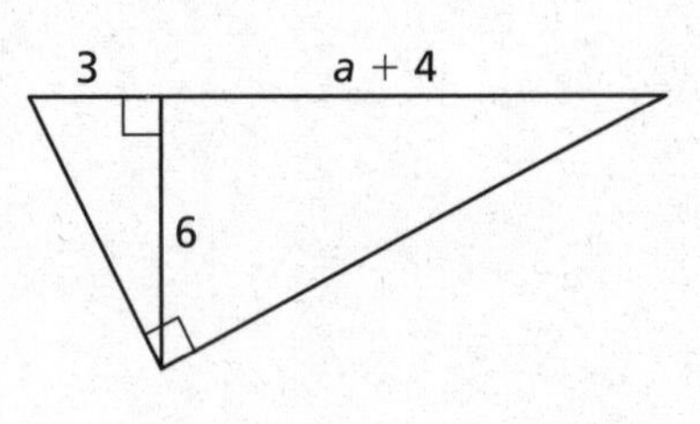

Name______________________________ Date__________

9.4 The Tangent Ratio

For use with Exploration 9.4

Essential Question How is a right triangle used to find the tangent of an acute angle? Is there a unique right triangle that must be used?

Let $\triangle ABC$ be a right triangle with acute $\angle A$. The *tangent* of $\angle A$ (written as $\tan A$) is defined as follows.

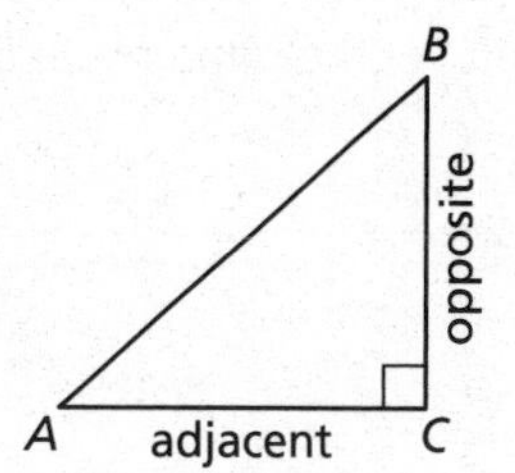

$$\tan A = \frac{\text{length of leg opposite } \angle A}{\text{length of leg adjacent to } \angle A} = \frac{BC}{AC}$$

1 EXPLORATION: Calculating a Tangent Ratio

Go to *BigIdeasMath.com* for an interactive tool to investigate this exploration.

Work with a partner. Use dynamic geometry software.

a. Construct $\triangle ABC$, as shown. Construct segments perpendicular to $\overline{AC}$ to form right triangles that share vertex A and are similar to $\triangle ABC$ with vertices, as shown.

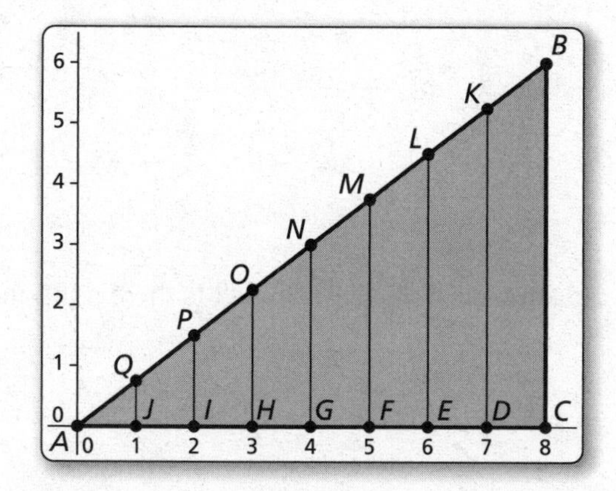

Sample
Points
$A(0, 0)$
$B(8, 6)$
$C(8, 0)$
Angle
$m\angle BAC = 36.87°$

b. Calculate each given ratio to complete the table for the decimal value of $\tan A$ for each right triangle. What can you conclude?

Ratio	$\frac{BC}{AC}$	$\frac{KD}{AD}$	$\frac{LE}{AE}$	$\frac{MF}{AF}$	$\frac{NG}{AG}$	$\frac{OH}{AH}$	$\frac{PI}{AI}$	$\frac{QJ}{AJ}$
tan A								

2 EXPLORATION: Using a Calculator

Work with a partner. Use a calculator that has a tangent key to calculate the tangent of 36.87°. Do you get the same result as in Exploration 1? Explain.

Communicate Your Answer

3. Repeat Exploration 1 for $\triangle ABC$ with vertices $A(0, 0)$, $B(8, 5)$, and $C(8, 0)$. Construct the seven perpendicular segments so that not all of them intersect $\overline{AC}$ at integer values of x. Discuss your results.

4. How is a right triangle used to find the tangent of an acute angle? Is there a unique right triangle that must be used?

Name____ Date__________

9.4 Notetaking with Vocabulary
For use after Lesson 9.4

In your own words, write the meaning of each vocabulary term.

trigonometric ratio

tangent

angle of elevation

Core Concepts

Tangent Ratio

Let $\triangle ABC$ be a right triangle with acute $\angle A$.

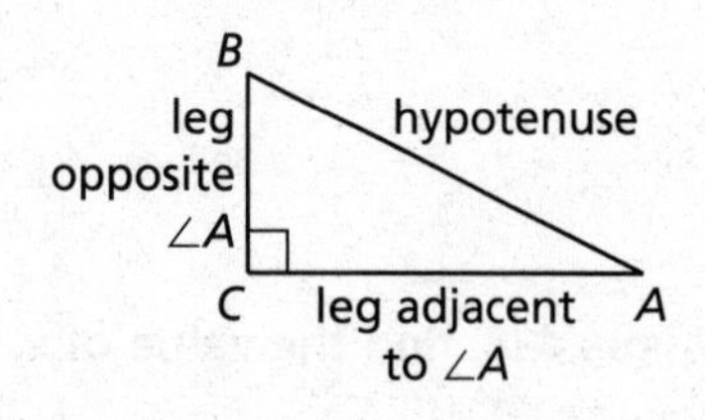

The tangent of $\angle A$ (written as tan A) is defined as follows.

$$\tan A = \frac{\text{length of leg opposite } \angle A}{\text{length of leg adjacent to } \angle A} = \frac{BC}{AC}$$

Notes:

Name ______________________________ Date __________

Extra Practice

In Exercises 1–3, find the tangents of the acute angles in the right triangle. Write each answer as a fraction and as a decimal rounded to four decimal places.

1.

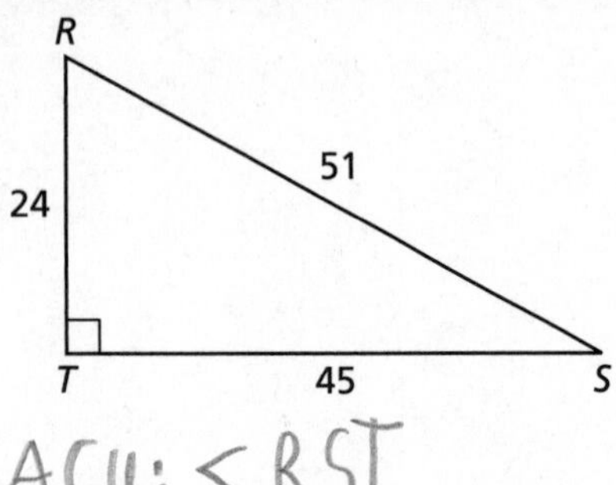

ACU: <RST
ACU: <TRS

$\frac{45}{24} = 1.875$

$\frac{24}{45} = 0.5\overline{3}$

2.

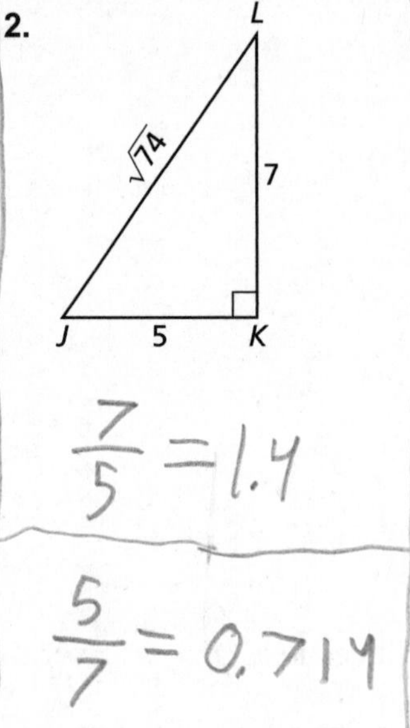

$\frac{7}{5} = 1.4$

$\frac{5}{7} = 0.714$

3.

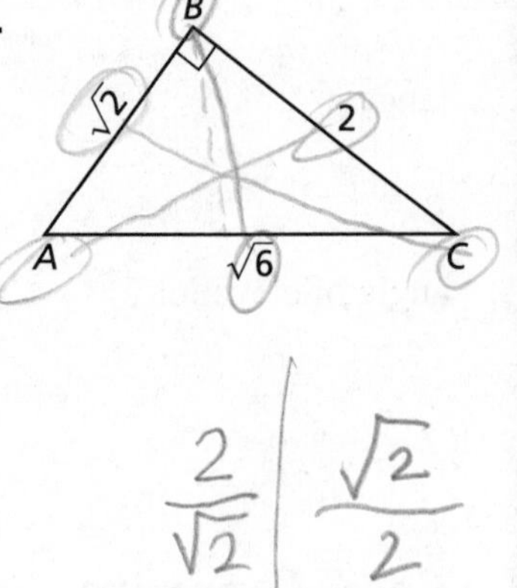

$\frac{2}{\sqrt{2}}$ | $\frac{\sqrt{2}}{2}$

$\frac{2}{\sqrt{2}} = \sqrt{2}$ | $\frac{\sqrt{2}}{2} =$

In Exercises 4–6, find the value of *x*. Round your answer to the nearest tenth.

4.

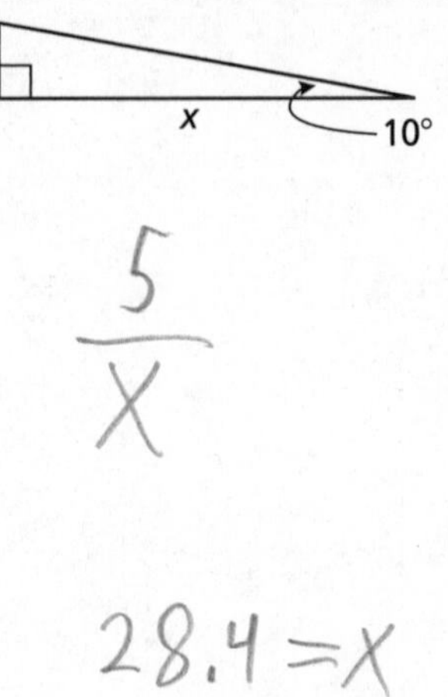

$\frac{5}{X}$

$28.4 = X$

5.

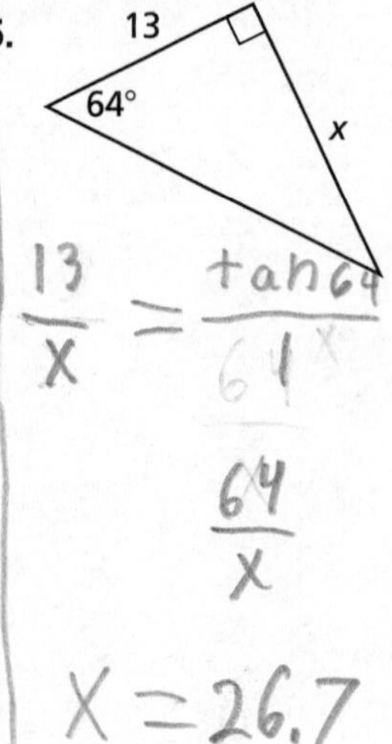

$\frac{13}{X} = \frac{\tan 64}{1}$

$\frac{64}{X}$

$X = 26.7$

6.

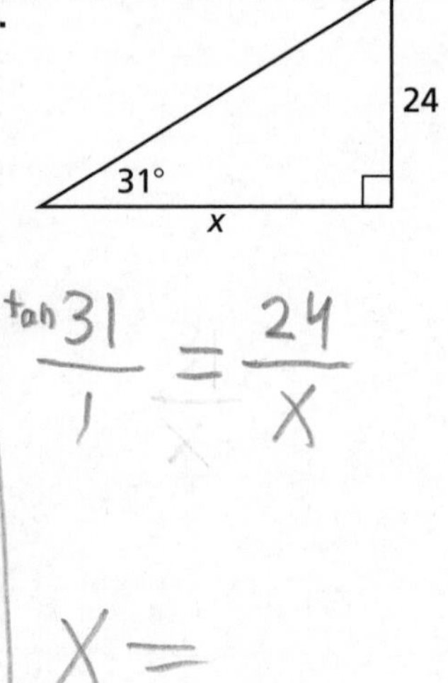

$\frac{\tan 31}{1} = \frac{24}{X}$

$X =$

7. In $\triangle CDE$, $\angle E = 90°$ and $\tan C = \frac{4}{3}$. Find tan *D*? Write your answer as a fraction.

$\tan C = \frac{120}{90} = \frac{4}{3}$ $\quad \frac{\tan \frac{4}{3}}{1} = \frac{X}{90}$

$<E = 90$

8. An environmentalist wants to measure the width of a river to monitor its erosion. From point A, she walks downstream 100 feet and measures the angle from this point to point C to be $40°$.

a. How wide is the river? Round to the nearest tenth.

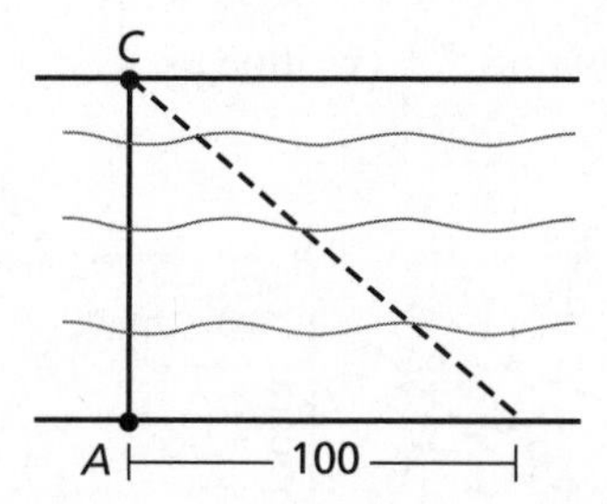

b. One year later, the environmentalist returns to measure the same river. From point A, she again walks downstream 100 feet and measures the angle from this point to point C to be now $51°$. By how many feet has the width of the river increased?

9. A boy flies a kite at an angle of elevation of $18°$. The kite reaches its maximum height 300 feet away from the boy. What is the maximum height of the kite? Round to the nearest tenth.

10. Find the perimeter of the figure.

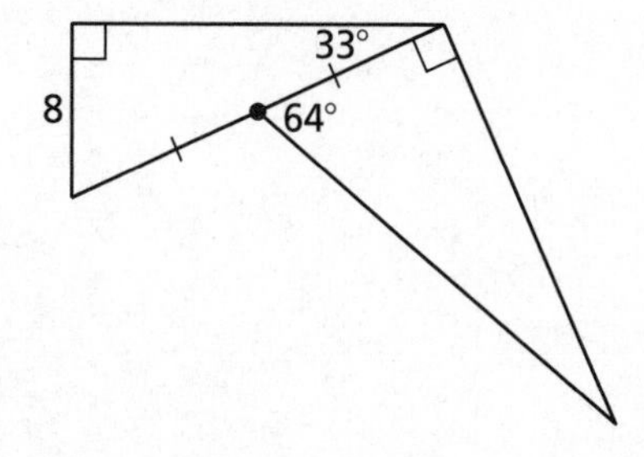

Name ____ Date __________

9.5 The Sine and Cosine Ratios

For use with Exploration 9.5

Essential Question How is a right triangle used to find the sine and cosine of an acute angle? Is there a unique right triangle that must be used?

Let $\triangle ABC$ be a right triangle with acute $\angle A$. The *sine* of $\angle A$ and *cosine* of $\angle A$ (written as $\sin A$ and $\cos A$, respectively) are defined as follows.

$$\sin A = \frac{\text{length of leg opposite } \angle A}{\text{length of hypotenuse}} = \frac{BC}{AB}$$

$$\cos A = \frac{\text{length of leg adjacent to } \angle A}{\text{length of hypotenuse}} = \frac{AC}{AB}$$

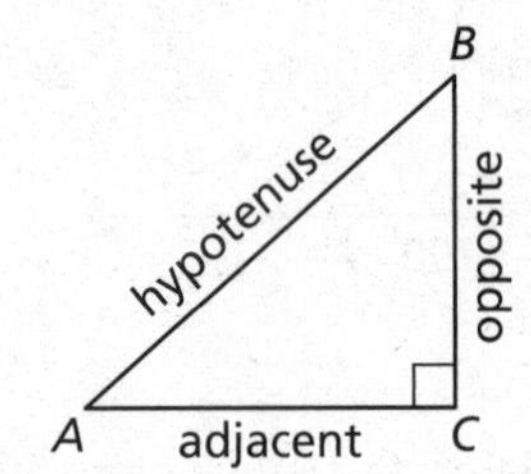

1 EXPLORATION: Calculating Sine and Cosine Ratios

Go to *BigIdeasMath.com* for an interactive tool to investigate this exploration.

Work with a partner. Use dynamic geometry software.

a. Construct $\triangle ABC$, as shown. Construct segments perpendicular to $\overline{AC}$ to form right triangles that share vertex A and are similar to $\triangle ABC$ with vertices, as shown.

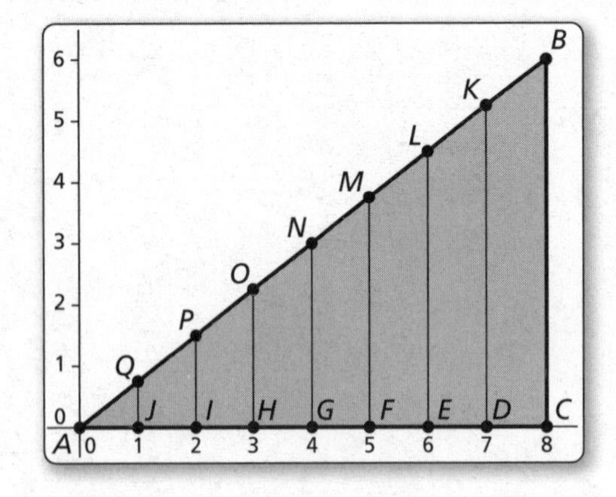

Sample

Points

$A(0, 0)$

$B(8, 6)$

$C(8, 0)$

Angle

$m\angle BAC = 36.87°$

1 EXPLORATION: Calculating Sine and Cosine Ratios (continued)

b. Calculate each given ratio to complete the table for the decimal values of sin A and cos A for each right triangle. What can you conclude?

Sine ratio	$\frac{BC}{AB}$	$\frac{KD}{AK}$	$\frac{LE}{AL}$	$\frac{MF}{AM}$	$\frac{NG}{AN}$	$\frac{OH}{AO}$	$\frac{PI}{AP}$	$\frac{QJ}{AQ}$
sin *A*								
Cosine ratio	$\frac{AC}{AB}$	$\frac{AD}{AK}$	$\frac{AE}{AL}$	$\frac{AF}{AM}$	$\frac{AG}{AN}$	$\frac{AH}{AO}$	$\frac{AI}{AP}$	$\frac{AJ}{AQ}$
cos *A*								

Communicate Your Answer

2. How is a right triangle used to find the sine and cosine of an acute angle? Is there a unique right triangle that must be used?

3. In Exploration 1, what is the relationship between $\angle A$ and $\angle B$ in terms of their measures? Find sin B and cos B. How are these two values related to sin A and cos A? Explain why these relationships exist.

Name ______________________________ Date __________

9.5 Notetaking with Vocabulary
For use after Lesson 9.5

In your own words, write the meaning of each vocabulary term.

sine

cosine

angle of depression

Core Concepts

Sine and Cosine Ratios

Let $\triangle ABC$ be a right triangle with acute $\angle A$. The sine of $\angle A$ and cosine of $\angle A$ (written as $\sin A$ and $\cos A$) are defined as follows.

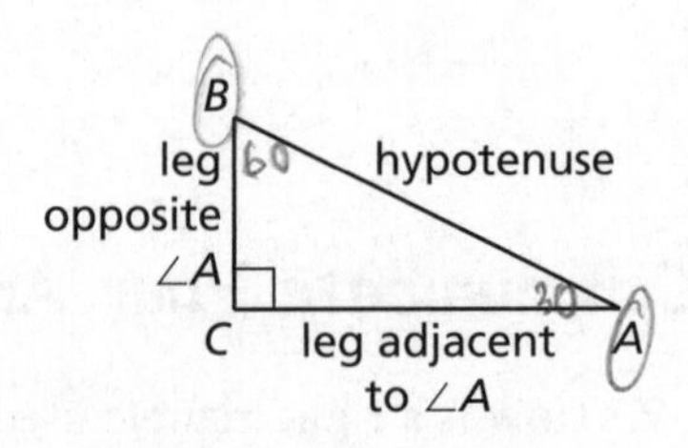

$$\sin A = \frac{\text{length of leg opposite } \angle A}{\text{length of hypotenuse}} = \frac{BC}{AB}$$

$$\cos A = \frac{\text{length of leg adjacent to } \angle A}{\text{length of hypotenuse}} = \frac{AC}{AB}$$

Notes:

9.5 Notetaking with Vocabulary (continued)

Sine and Cosine of Complementary Angles

The sine of an acute angle is equal to the cosine of its complement. The cosine of an acute angle is equal to the sine of its complement.

Let A and B be complementary angles. Then the following statements are true.

$$\sin A = \cos(90° - A) = \cos B \qquad \sin B = \cos(90° - B) = \cos A$$

$$\cos A = \sin(90° - A) = \sin B \qquad \cos B = \sin(90° - B) = \sin A$$

Notes:

Extra Practice

In Exercises 1–3, find sin *F*, sin *G*, cos *F*, and cos *G*. Write each answer as a fraction and as a decimal rounded to four places.

1.

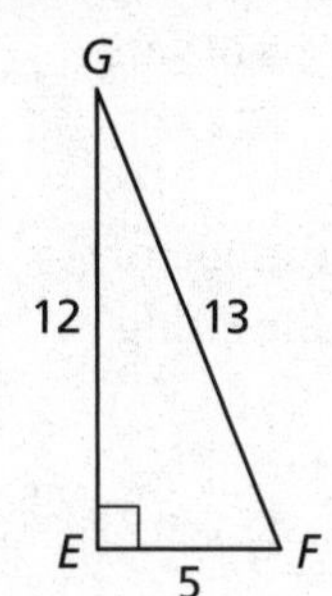

2.

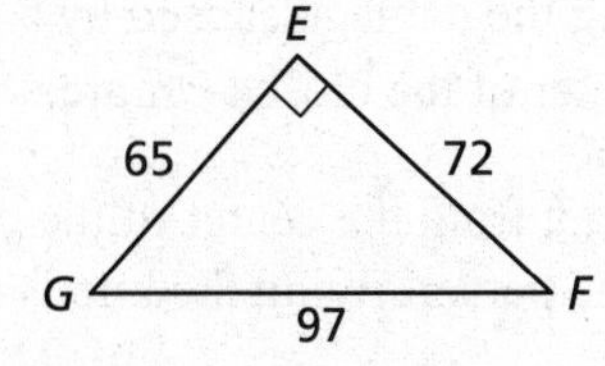

3.

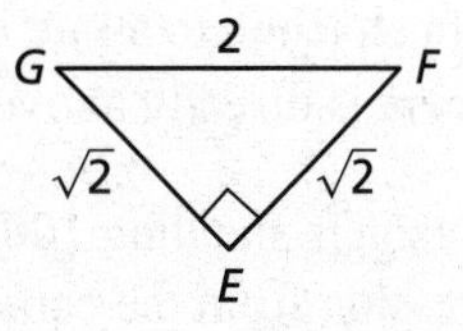

In Exercises 4–6, write the expression in terms of cosine.

4. $\sin 9°$

5. $\sin 30°$

6. $\sin 77°$

Name ____ Date __________

9.5 Notetaking with Vocabulary (continued)

In Exercises 7–9, write the expression in terms of sine.

7. $\cos 15°$ **8.** $\cos 83°$ **9.** $\cos 45°$

In Exercises 10–13, find the value of each variable using sine and cosine. Round your answers to the nearest tenth.

10.

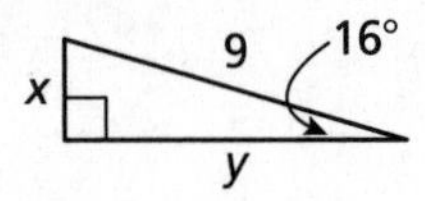

11.

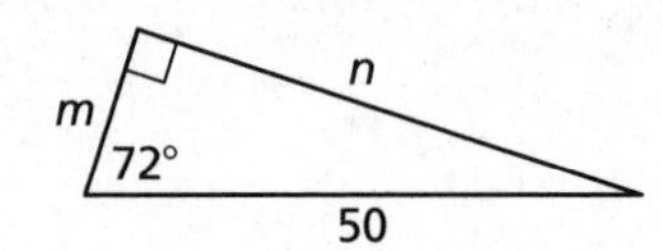

12.

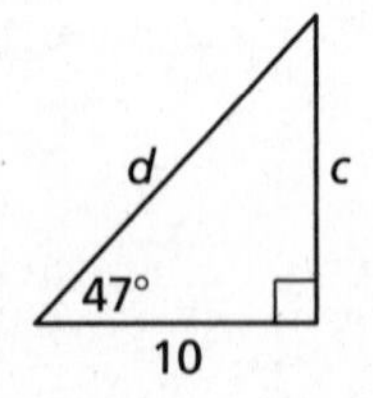

13.

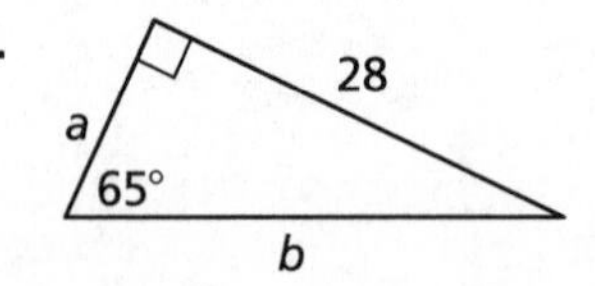

14. A camera attached to a kite is filming the damage caused by a brush fire in a closed-off area. The camera is directly above the center of the closed-off area.

a. A person is standing 100 feet away from the center of the closed-off area. The angle of depression from the camera to the person flying the kite is 25°. How long is the string on the kite?

b. If the string on the kite is 200 feet long, how far away must the person flying the kite stand from the center of the closed-off area, assuming the same angle of depression of 25°, to film the damage?

Name____ Date__________

9.6 Solving Right Triangles

For use with Exploration 9.6

Essential Question When you know the lengths of the sides of a right triangle, how can you find the measures of the two acute angles?

1 EXPLORATION: Solving Special Right Triangles

Go to *BigIdeasMath.com* for an interactive tool to investigate this exploration.

Work with a partner. Use the figures to find the values of the sine and cosine of $\angle A$ and $\angle B$. Use these values to find the measures of $\angle A$ and $\angle B$. Use dynamic geometry software to verify your answers.

a.

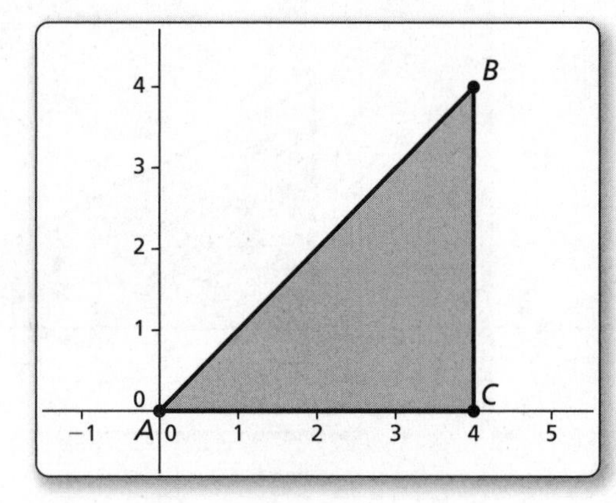

b.

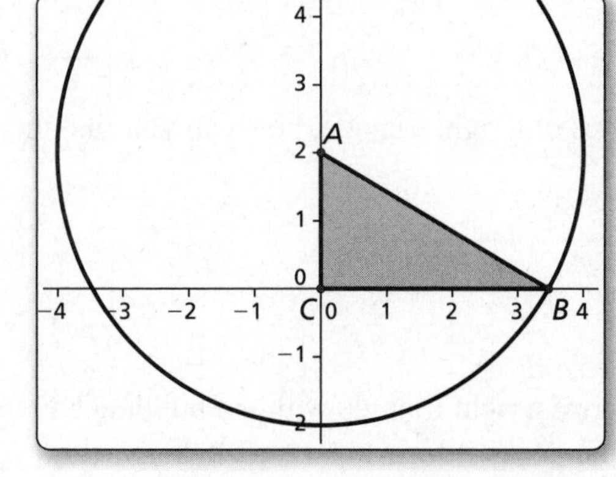

Name ______________________________________ Date __________

2 EXPLORATION: Solving Right Triangles

Go to *BigIdeasMath.com* for an interactive tool to investigate this exploration.

Work with a partner. You can use a calculator to find the measure of an angle when you know the value of the sine, cosine, or tangent of the angle. Use the inverse sine, inverse cosine, or inverse tangent feature of your calculator to approximate the measures of $\angle A$ and $\angle B$ to the nearest tenth of a degree. Then use dynamic geometry software to verify your answers.

a.

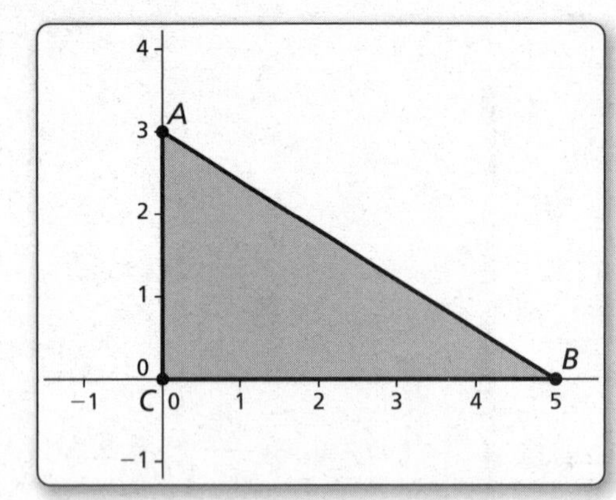

b.

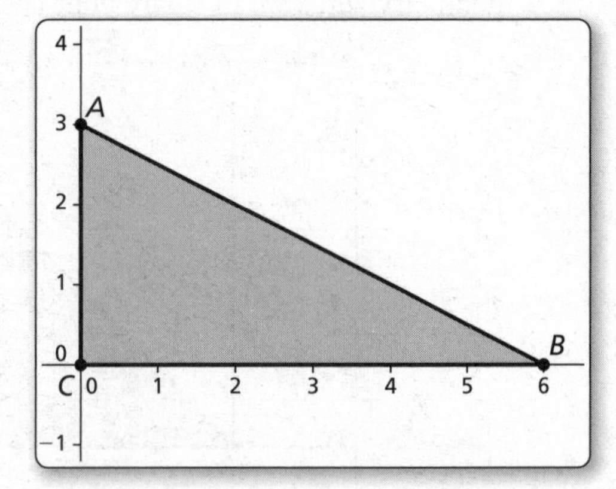

Communicate Your Answer

3. When you know the lengths of the sides of a right triangle, how can you find the measures of the two acute angles?

4. A ladder leaning against a building forms a right triangle with the building and the ground. The legs of the right triangle (in meters) form a 5-12-13 Pythagorean triple. Find the measures of the two acute angles to the nearest tenth of a degree.

Name____ Date__________

9.6 Notetaking with Vocabulary
For use after Lesson 9.6

In your own words, write the meaning of each vocabulary term.

inverse tangent

inverse sine

inverse cosine

solve a right triangle

Core Concepts

Inverse Trigonometric Ratios

Let $\angle A$ be an acute angle.

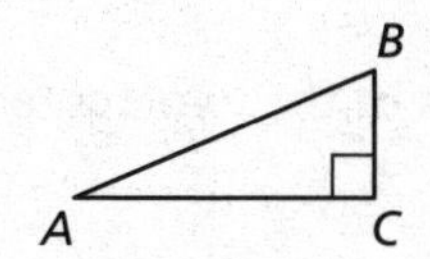

Inverse Tangent If $\tan A = x$, then $\tan^{-1} x = m\angle A$. $\quad \tan^{-1}\dfrac{BC}{AC} = m\angle A$

Inverse Sine If $\sin A = y$, then $\sin^{-1} y = m\angle A$. $\quad \sin^{-1}\dfrac{BC}{AB} = m\angle A$

Inverse Cosine If $\cos A = z$, then $\cos^{-1} z = m\angle A$. $\quad \cos^{-1}\dfrac{AC}{AB} = m\angle A$

Notes:

Solving a Right Triangle

To **solve a right triangle** means to find all unknown side lengths and angle measures. You can solve a right triangle when you know either of the following.

- two side lengths
- one side length and the measure of one acute angle

Notes:

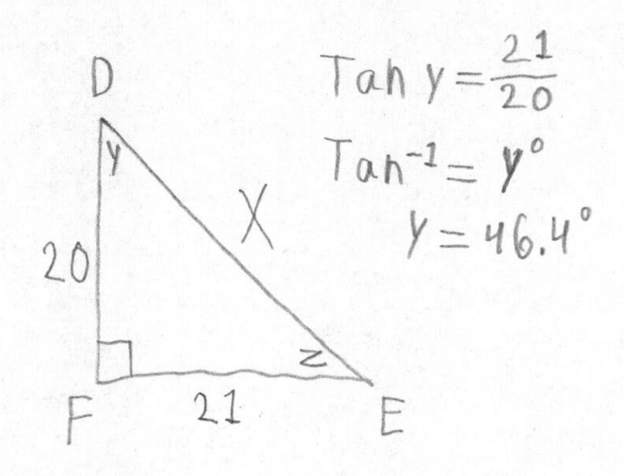

90 + 46.4 = 136.4

180 − 136.4 = 43.6

Z = 43.6

Extra Practice

In Exercises 1 and 2, determine which of the two acute angles has the given trigonometric ratio.

1. The cosine of the angle is $\frac{24}{25}$.

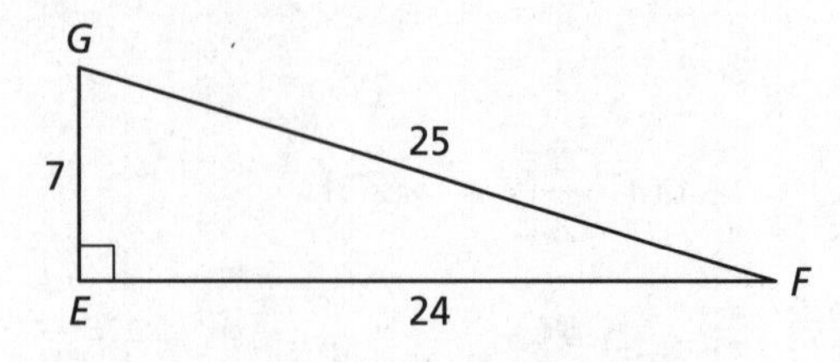

2. The sine of the angle is about 0.38.

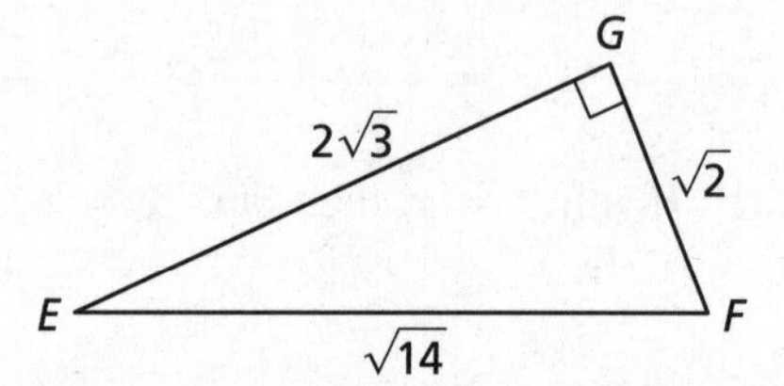

In Exercises 3–6, let $\angle H$ be an acute angle. Use a calculator to approximate the measure of $\angle H$ to the nearest tenth of a degree.

3. $\sin H = 0.2$

4. $\tan H = 1$

5. $\cos H = 0.33$

6. $\sin H = 0.89$

In Exercises 7–10, solve the right triangle. Round decimal answers to the nearest tenth.

7.

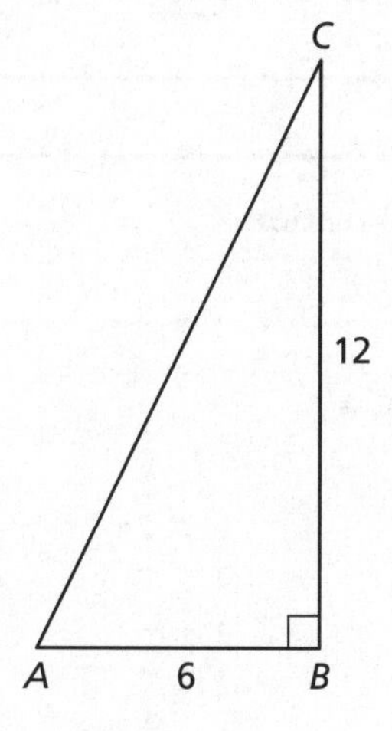

8.

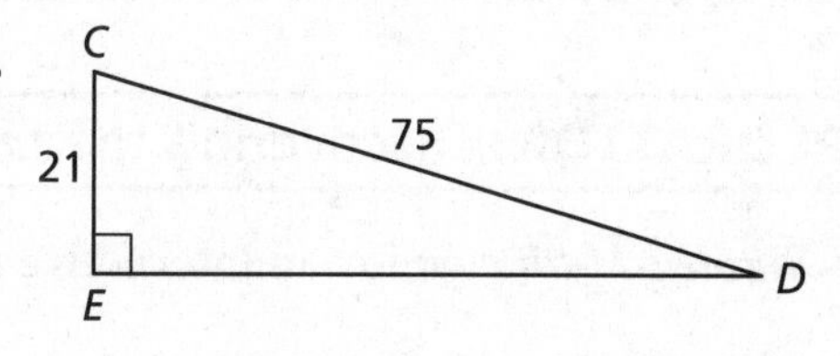

9.

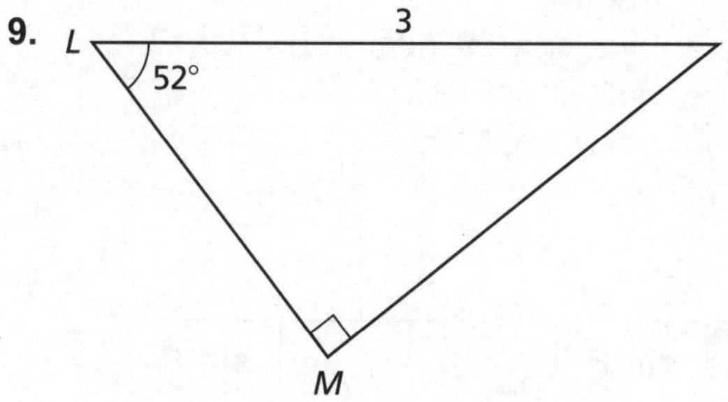

10.

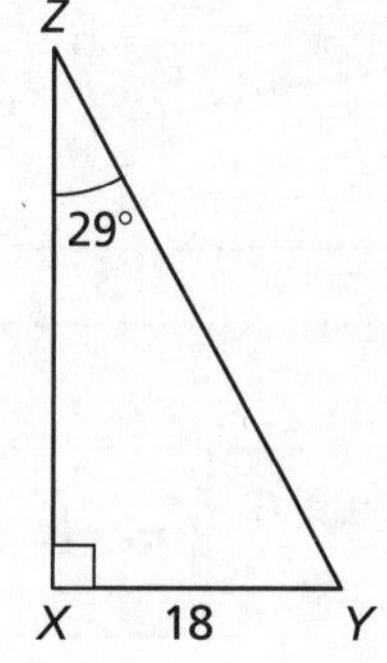

11. A boat is pulled in by a winch on a dock 12 feet above the deck of the boat. When the winch is fully extended to 25 feet, what is the angle of elevation from the boat to the winch?

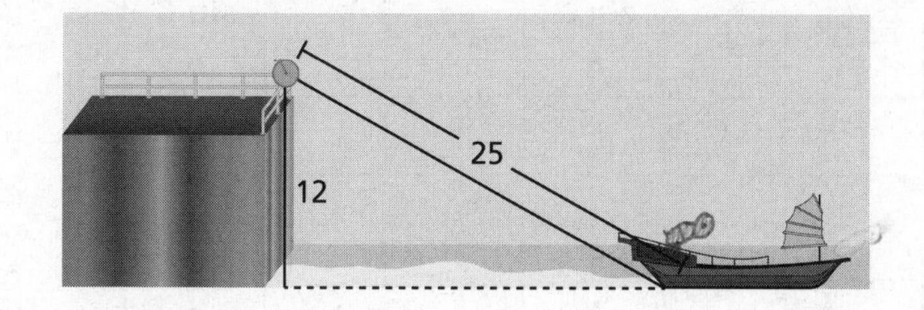

Name ______________________________ Date __________

9.7 Law of Sines and Law of Cosines

For use with Exploration 9.7

Essential Question What are the Law of Sines and the Law of Cosines?

1 EXPLORATION: Discovering the Law of Sines

Go to *BigIdeasMath.com* for an interactive tool to investigate this exploration.

Work with a partner.

a. Complete the table for the triangle shown. What can you conclude?

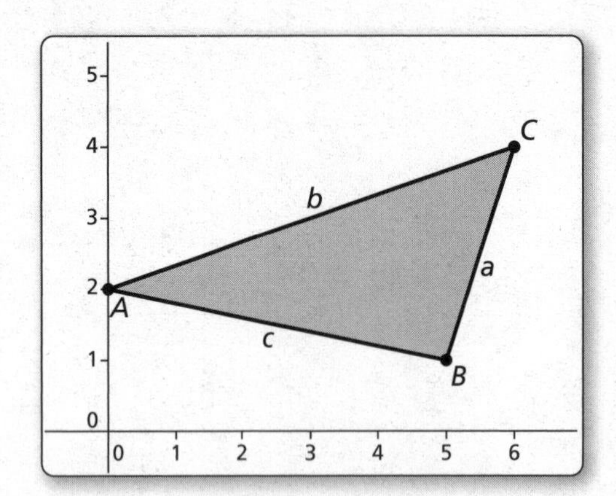

Sample
Segments
$a = 3.16$
$b = 6.32$
$c = 5.10$
Angles
$m\angle A = 29.74°$
$m\angle B = 97.13°$
$m\angle C = 53.13°$

$m\angle A$	a	$\frac{\sin A}{a}$	$m\angle B$	b	$\frac{\sin B}{b}$	$m\angle C$	c	$\frac{\sin C}{c}$

b. Use dynamic geometry software to draw two other triangles. Complete a table for each triangle. Use your results to write a conjecture about the relationship between the sines of the angles and the lengths of the sides of a triangle.

$m\angle A$	a	$\frac{\sin A}{a}$	$m\angle B$	b	$\frac{\sin B}{b}$	$m\angle C$	c	$\frac{\sin C}{c}$

$m\angle A$	a	$\frac{\sin A}{a}$	$m\angle B$	b	$\frac{\sin B}{b}$	$m\angle C$	c	$\frac{\sin C}{c}$

Name______________________________ Date__________

2 EXPLORATION: Discovering the Law of Cosines

Go to *BigIdeasMath.com* for an interactive tool to investigate this exploration.

Work with a partner.

a. Complete the table for the triangle in Exploration 1(a). What can you conclude?

c	c^2	a	a^2	b	b^2	$m\angle C$	$a^2 + b^2 - 2ab \cos C$

b. Use dynamic geometry software to draw two other triangles. Complete a table for each triangle. Use your results to write a conjecture about what you observe in the completed tables.

c	c^2	a	a^2	b	b^2	$m\angle C$	$a^2 + b^2 - 2ab \cos C$

c	c^2	a	a^2	b	b^2	$m\angle C$	$a^2 + b^2 - 2ab \cos C$

Communicate Your Answer

3. What are the Law of Sines and the Law of Cosines?

4. When would you use the Law of Sines to solve a triangle? When would you use the Law of Cosines to solve a triangle?

Name ______________________________ Date __________

9.7 Notetaking with Vocabulary

For use after Lesson 9.7

In your own words, write the meaning of each vocabulary term.

Law of Sines

Law of Cosines

Core Concepts

Area of a Triangle

The area of any triangle is given by one-half the product of the lengths of two sides times the sine of their included angle. For $\triangle ABC$ shown, there are three ways to calculate the area.

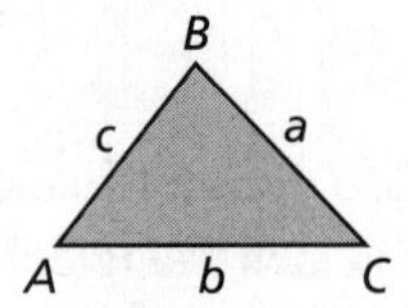

$\text{Area} = \frac{1}{2}bc \sin A$ $\text{Area} = \frac{1}{2}ac \sin B$ $\text{Area} = \frac{1}{2}ab \sin C$

Notes:

9.7 Notetaking with Vocabulary (continued)

Theorems

Theorem 9.9 Law of Sines

The Law of Sines can be written in either of the following forms for $\triangle ABC$ with sides of length a, b, and c.

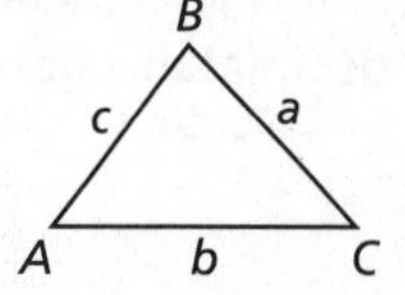

$$\frac{\sin A}{a} = \frac{\sin B}{b} = \frac{\sin C}{c} \qquad \frac{a}{\sin A} = \frac{b}{\sin B} = \frac{c}{\sin C}$$

Notes:

Theorem 9.10 Law of Cosines

If $\triangle ABC$ has sides of length a, b, and c, as shown, then the following are true.

$$a^2 = b^2 + c^2 - 2bc \cos A$$

$$b^2 = a^2 + c^2 - 2ac \cos B$$

$$c^2 = a^2 + b^2 - 2ab \cos C$$

Notes:

Extra Practice

In Exercises 1–3, use a calculator to find the trigonometric ratio. Round your answer to four decimal places.

1. $\sin 225°$ **2.** $\cos 111°$ **3.** $\tan 96°$

In Exercises 4 and 5, find the area of the triangle. Round your answer to the nearest tenth.

4.

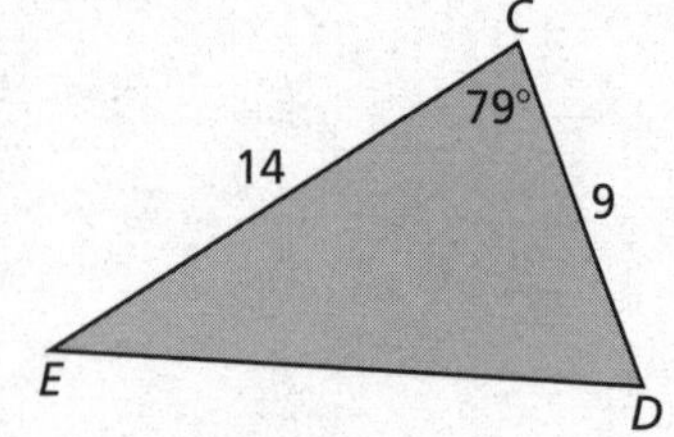

5.

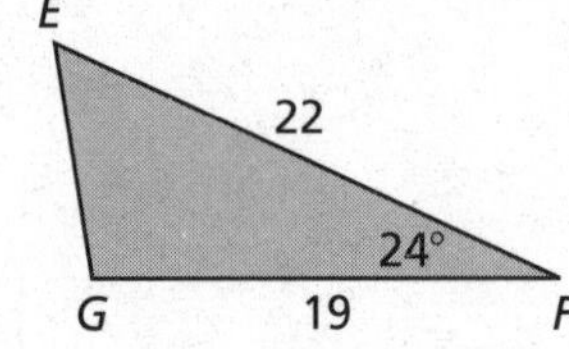

In Exercises 6-8, solve the triangle. Round decimal answers to the nearest tenth.

6.

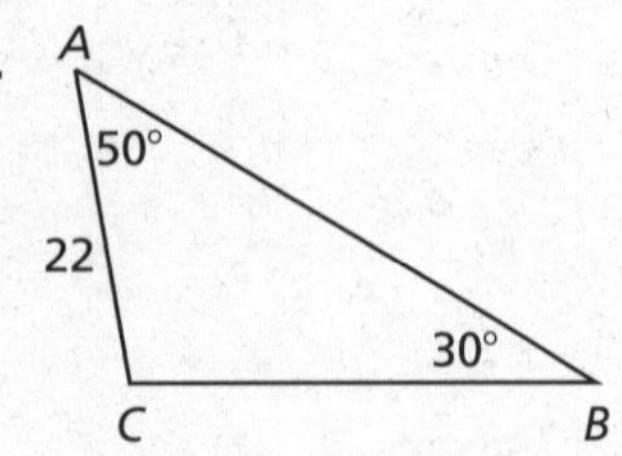

7.

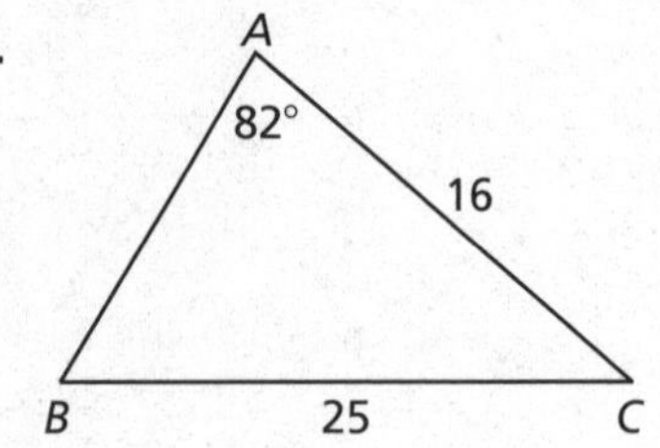

8.

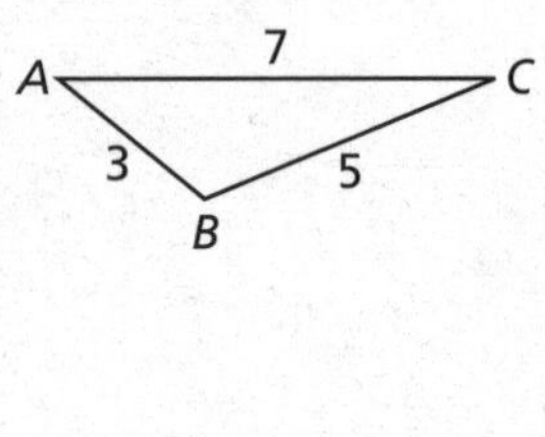

Name_______________________________ Date__________

Chapter 10 Maintaining Mathematical Proficiency

Find the product.

1. $(x - 4)(x - 9)$

2. $(k + 6)(k - 7)$

3. $(y + 5)(y - 13)$

4. $(2r + 3)(3r + 1)$

5. $(4m - 5)(2 - 3m)$

6. $(7w - 1)(6w + 5)$

Solve the equation by completing the square. Round your answer to the nearest hundredth, if necessary.

7. $x^2 + 6x = 10$

8. $p^2 - 14p = 5$

9. $z^2 + 16z + 7 = 0$

10. $z^2 + 5z - 2 = 0$

11. $x^2 + 2x - 5 = 0$

12. $c^2 - c - 1 = 0$

Name ______________________________ Date __________

10.1 Lines and Segments That Intersect Circles

For use with Exploration 10.1

Essential Question What are the definitions of the lines and segments that intersect a circle?

1 EXPLORATION: Lines and Line Segments That Intersect Circles

Work with a partner. The drawing at the right shows five lines or segments that intersect a circle. Use the relationships shown to write a definition for each type of line or segment. Then use the Internet or some other resource to verify your definitions.

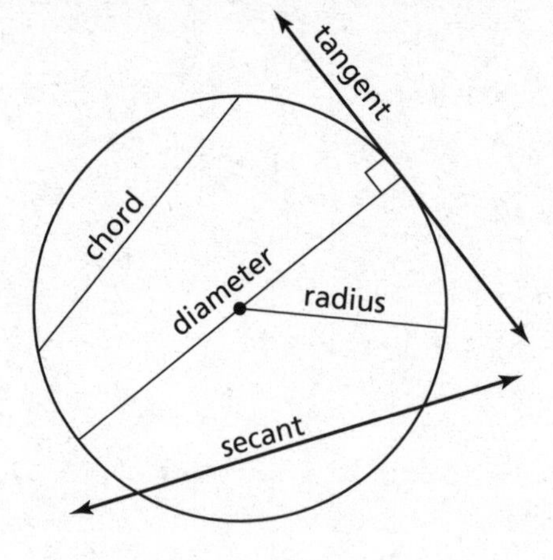

Chord:

Secant:

Tangent:

Radius:

Diameter:

2 EXPLORATION: Using String to Draw a Circle

Work with a partner. Use two pencils, a piece of string, and a piece of paper.

a. Tie the two ends of the piece of string loosely around the two pencils.

b. Anchor one pencil on the paper at the center of the circle. Use the other pencil to draw a circle around the anchor point while using slight pressure to keep the string taut. Do not let the string wind around either pencil.

c. Explain how the distance between the two pencil points as you draw the circle is related to two of the lines or line segments you defined in Exploration 1.

Communicate Your Answer

3. What are the definitions of the lines and segments that intersect a circle?

4. Of the five types of lines and segments in Exploration 1, which one is a subset of another? Explain.

5. Explain how to draw a circle with a diameter of 8 inches.

Name ______________________________ Date __________

10.1 Notetaking with Vocabulary
For use after Lesson 10.1

In your own words, write the meaning of each vocabulary term.

circle

center

radius

chord

diameter

secant

tangent

point of tangency

tangent circles

concentric circles

common tangent

Notes:

Name____ Date__________

10.1 Notetaking with Vocabulary (continued)

Core Concepts

Lines and Segments That Intersect Circles

A segment whose endpoints are the center and any point on a circle is a **radius**.

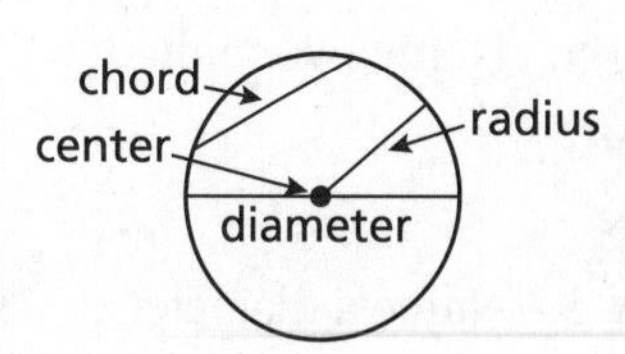

A **chord** is a segment whose endpoints are on a circle. A **diameter** is a chord that contains the center of the circle.

A **secant** is a line that intersects a circle in two points.

A **tangent** is a line in the plane of a circle that intersects the circle in exactly one point, the **point of tangency**. The *tangent ray* $\overrightarrow{AB}$ and the *tangent segment* $\overline{AB}$ are also called tangents.

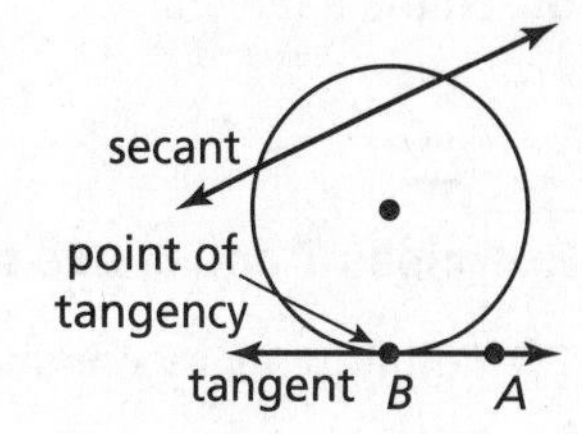

Notes:

Coplanar Circles and Common Tangents

In a plane, two circles can intersect in two points, one point, or no points. Coplanar circles that intersect in one point are called **tangent circles**. Coplanar circles that have a common center are called **concentric circles**.

2 points of intersection

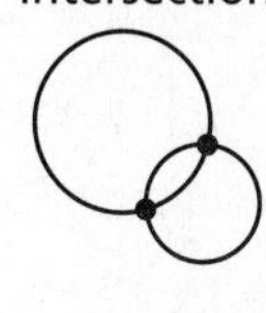

1 point of intersection (tangent circles)

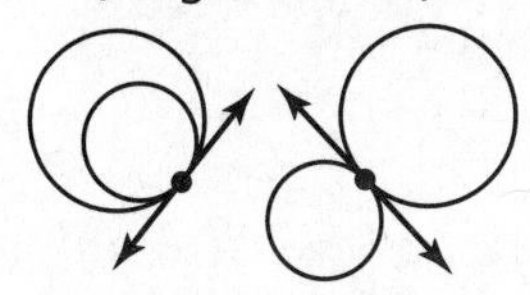

no points of intersection

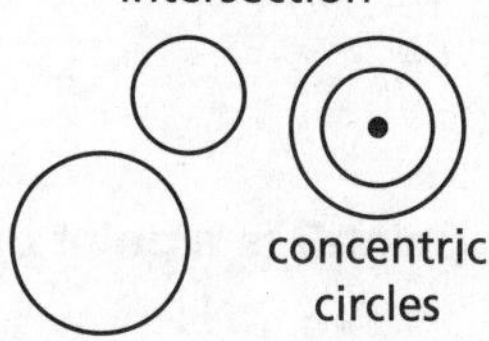

A line or segment that is tangent to two coplanar circles is called a **common tangent**. A *common internal tangent* intersects the segment that joins the centers of the two circles. A *common external tangent* does not intersect the segment that joins the centers of the two circles.

Notes:

Extra Practice

In Exercises 1–6, use the diagram.

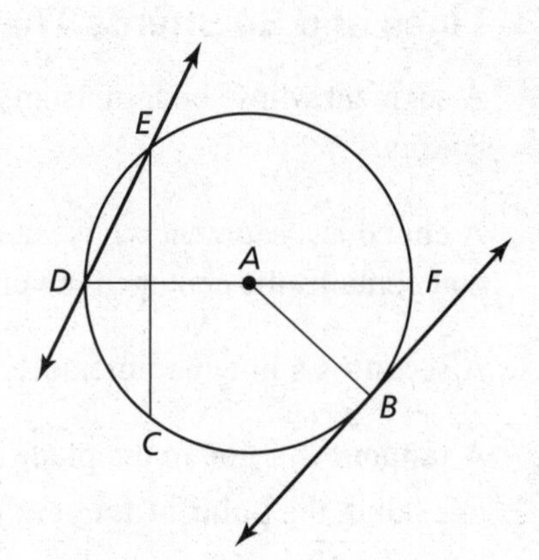

1. Name two radii.

2. Name a chord.

3. Name a diameter.

4. Name a secant.

5. Name a tangent.

6. Name a point of tangency.

In Exercises 7 and 8, use the diagram.

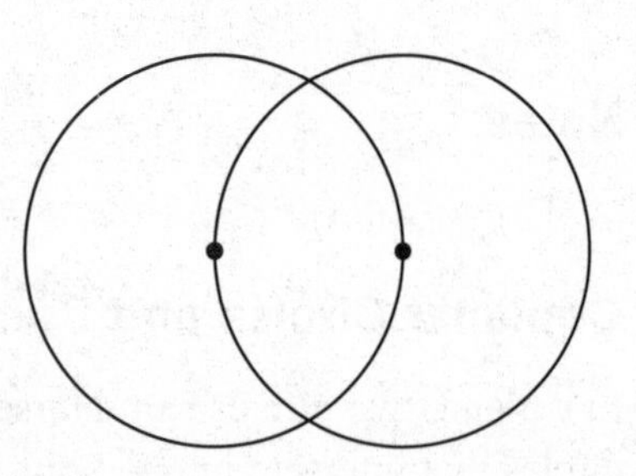

7. Tell how many common tangents the circles have and draw them.

8. Tell whether each common tangent identified in Exercise 7 is internal or external.

In Exercises 9 and 10, point *D* is a point of tangency.

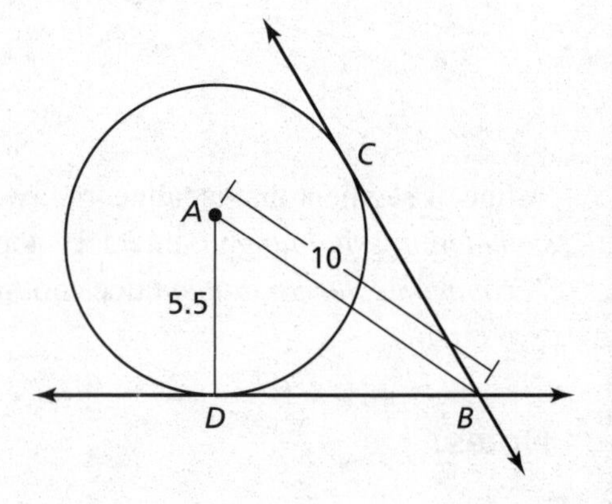

9. Find BD.

10. Point C is also a point of tangency. If $BC = 4x + 6$, find the value of x to the nearest tenth.

Name____ Date__________

10.2 Finding Arc Measures
For use with Exploration 10.2

Essential Question How are circular arcs measured?

A **central angle** of a circle is an angle whose vertex is the center of the circle. A *circular arc* is a portion of a circle, as shown below. The measure of a circular arc is the measure of its central angle.

If $m\angle AOB < 180°$, then the circular arc is called a **minor arc** and is denoted by $\overset{\frown}{AB}$.

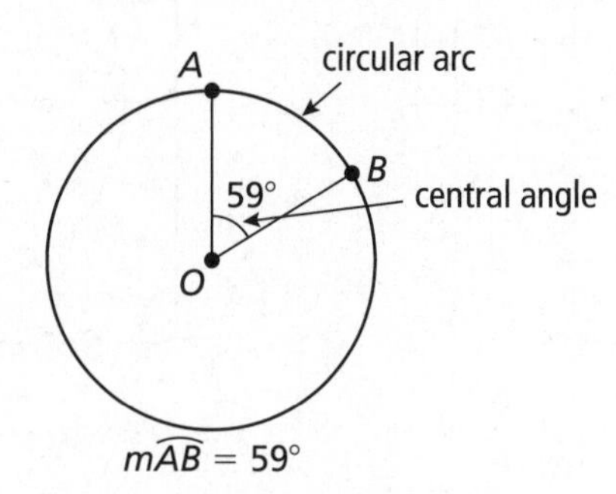

1 EXPLORATION: Measuring Circular Arcs

Go to *BigIdeasMath.com* for an interactive tool to investigate this exploration.

Work with a partner. Use dynamic geometry software to find the measure of $\overset{\frown}{BC}$. Verify your answers using trigonometry.

a.

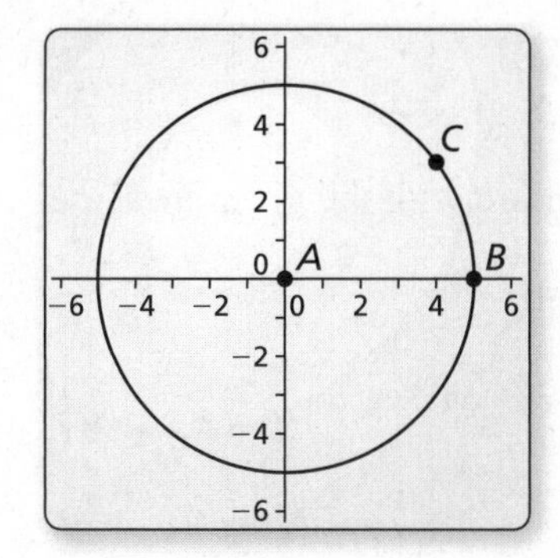

Points
$A(0, 0)$
$B(5, 0)$
$C(4, 3)$

b.

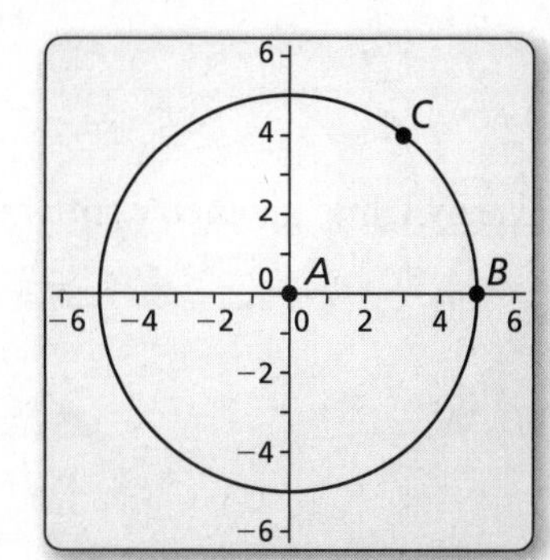

Points
$A(0, 0)$
$B(5, 0)$
$C(3, 4)$

Name ______________________________________ Date ________

1 EXPLORATION: Measuring Circular Arcs (continued)

c.

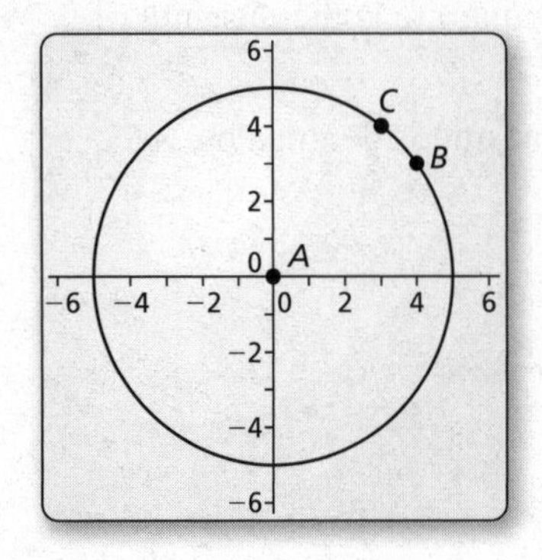

Points
$A(0, 0)$
$B(4, 3)$
$C(3, 4)$

d.

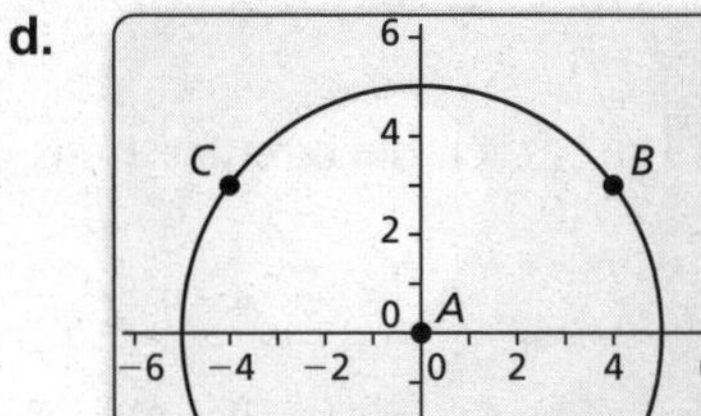

Points
$A(0, 0)$
$B(4, 3)$
$C(-4, 3)$

Communicate Your Answer

2. How are circular arcs measured?

3. Use dynamic geometry software to draw a circular arc with the given measure.

a. 30° **b.** 45°

c. 60° **d.** 90°

Name______________________________ Date__________

10.2 Notetaking with Vocabulary

For use after Lesson 10.2

In your own words, write the meaning of each vocabulary term.

central angle

minor arc

major arc

semicircle

measure of a minor arc

measure of a major arc

adjacent arcs

congruent circles

congruent arcs

similar arcs

Core Concepts

Measuring Arcs

The **measure of a minor arc** is the measure of its central angle. The expression $m\overparen{AB}$ is read as "the measure of arc AB."

The measure of the entire circle is 360°. The **measure of a major arc** is the difference of 360° and the measure of the related minor arc. The measure of a semicircle is 180°.

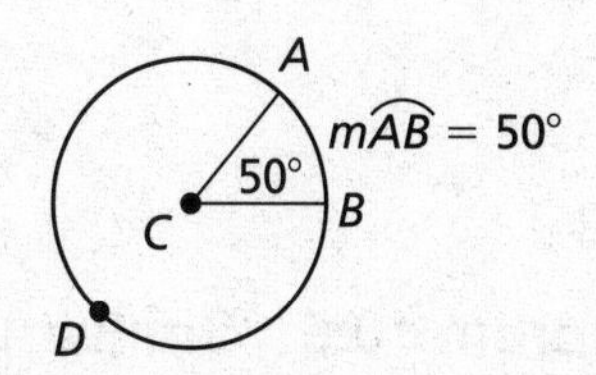

Notes:

Name ____ Date __________

10.2 Notetaking with Vocabulary (continued)

Postulates

Postulate 10.1 Arc Addition Postulate

The measure of an arc formed by two adjacent arcs is the sum of the measures of the two arcs.

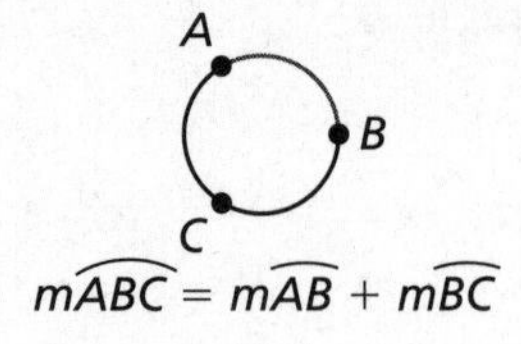

$m\overparen{ABC} = m\overparen{AB} + m\overparen{BC}$

Notes:

Theorems

Theorem 10.3 Congruent Circles Theorem

Two circles are congruent circles if and only if they have the same radius.

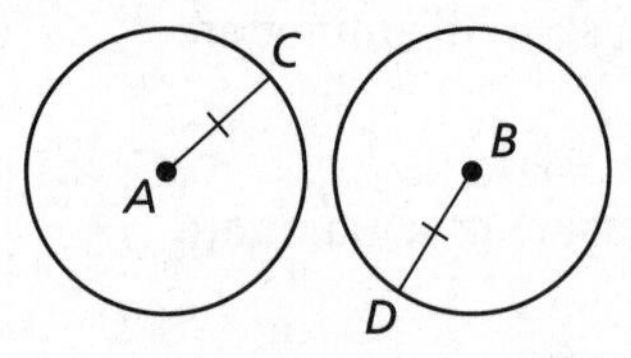

$\odot A \cong \odot B$ if and only if $\overline{AC} \cong \overline{BD}$.

Notes:

Theorem 10.4 Congruent Central Angles Theorem

In the same circle, or in congruent circles, two minor arcs are congruent if and only if their corresponding central angles are congruent.

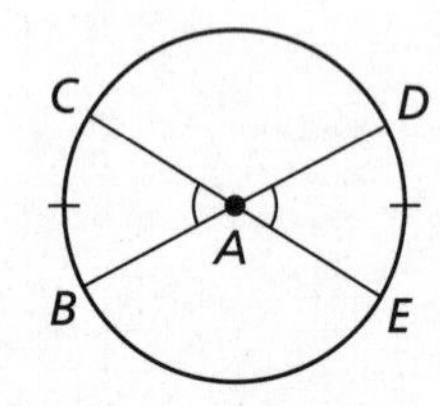

$\overparen{BC} \cong \overparen{DE}$ if and only if $\angle BAC \cong \angle DAE$.

Notes:

Theorem 10.5 Similar Circles Theorem

All circles are similar.

Notes:

Extra Practice

In Exercises 1–8, identify the given arc as a *major arc*, *minor arc*, or *semicircle*. Then find the measure of the arc.

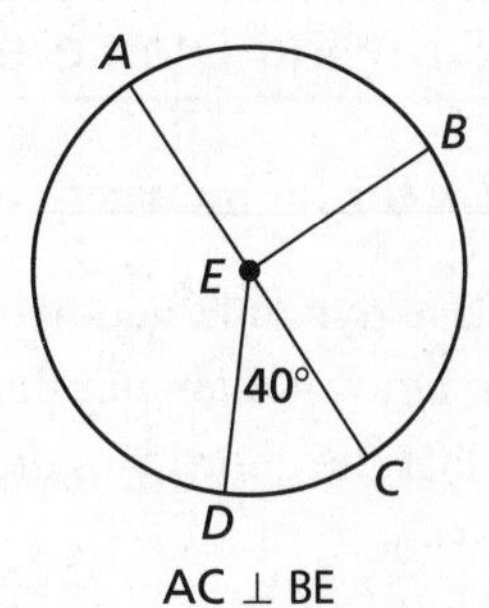

1. $\overset{\frown}{AB}$
2. $\overset{\frown}{ABC}$
3. $\overset{\frown}{ABD}$
4. $\overset{\frown}{BC}$
5. $\overset{\frown}{BAC}$
6. $\overset{\frown}{DAB}$
7. $\overset{\frown}{AD}$
8. $\overset{\frown}{CD}$

9. In $\odot E$ above, tell whether $\overset{\frown}{ABC} \cong \overset{\frown}{ADC}$. Explain why or why not.

10. In $\odot K$, find the measure of $\overset{\frown}{DE}$.

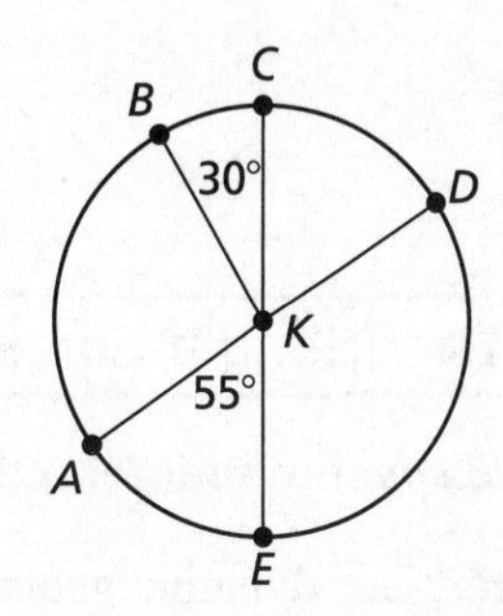

11. Find the value of x. Then find the measure of $\overset{\frown}{AB}$.

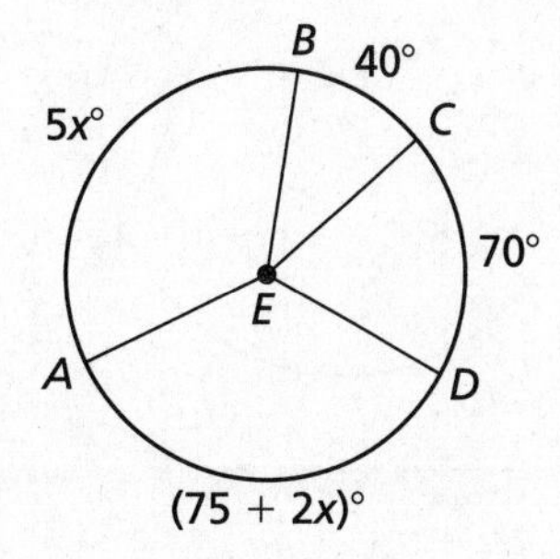

Name ______________________________ Date __________

10.3 Using Chords
For use with Exploration 10.3

Essential Question What are two ways to determine when a chord is a diameter of a circle?

1 EXPLORATION: Drawing Diameters

Go to *BigIdeasMath.com* for an interactive tool to investigate this exploration.

Work with a partner. Use dynamic geometry software to construct a circle of radius 5 with center at the origin. Draw a diameter that has the given point as an endpoint. Explain how you know that the chord you drew is a diameter.

a. $(4, 3)$

b. $(0, 5)$

c. $(-3, 4)$

d. $(-5, 0)$

2 EXPLORATION: Writing a Conjecture about Chords

Go to *BigIdeasMath.com* for an interactive tool to investigate this exploration.

Work with a partner. Use dynamic geometry software to construct a chord $\overline{BC}$ of a circle A. Construct a chord on the perpendicular bisector of $\overline{BC}$. What do you notice? Change the original chord and the circle several times. Are your results always the same? Use your results to write a conjecture.

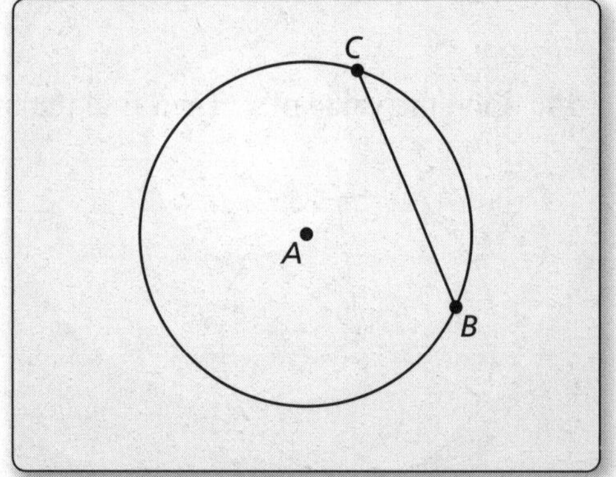

Name____ Date__________

3 EXPLORATION: A Chord Perpendicular to a Diameter

Go to *BigIdeasMath.com* for an interactive tool to investigate this exploration.

Work with a partner. Use dynamic geometry software to construct a diameter $\overline{BC}$ of a circle A. Then construct a chord $\overline{DE}$ perpendicular to $\overline{BC}$ at point F. Find the lengths DF and EF. What do you notice? Change the chord perpendicular to $\overline{BC}$ and the circle several times. Do you always get the same results? Write a conjecture about a chord that is perpendicular to a diameter of a circle.

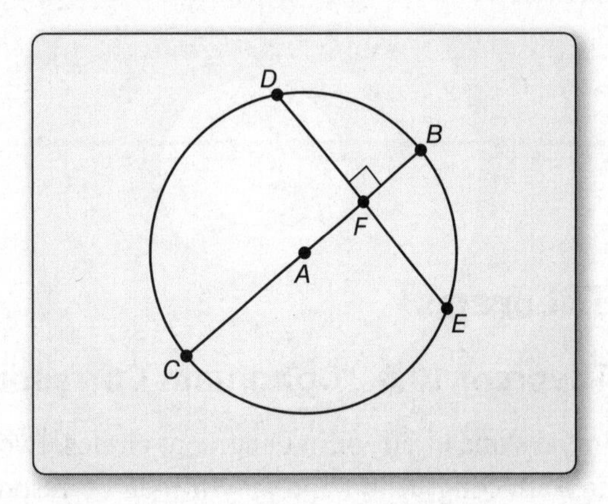

Communicate Your Answer

4. What are two ways to determine when a chord is a diameter of a circle?

Name ____ Date __________

10.3 Notetaking with Vocabulary

For use after Lesson 10.3

In your own words, write the meaning of each vocabulary term.

chord

arc

diameter

Theorems

Theorem 10.6 Congruent Corresponding Chords Theorem

In the same circle, or in congruent circles, two minor arcs are congruent if and only if their corresponding chords are congruent.

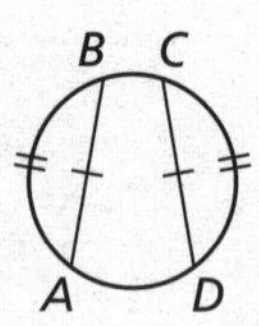

$\overset{\frown}{AB} \cong \overset{\frown}{CD}$ if any only if $\overline{AB} \cong \overline{CD}$.

Notes:

10.3 Notetaking with Vocabulary (continued)

Theorem 10.7 Perpendicular Chord Bisector Theorem

If a diameter of a circle is perpendicular to a chord, then the diameter bisects the chord and its arc.

Notes:

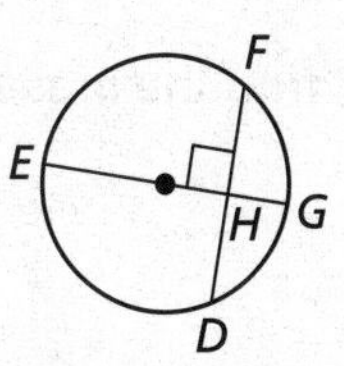

If $\overline{EG}$ is a diameter and $\overline{EG} \perp \overline{DF}$, then $\overline{HD} \cong \overline{HF}$ and $\overset{\frown}{GD} \cong \overset{\frown}{GF}$.

Theorem 10.8 Perpendicular Chord Bisector Converse

If one chord of a circle is a perpendicular bisector of another chord, then the first chord is a diameter.

Notes:

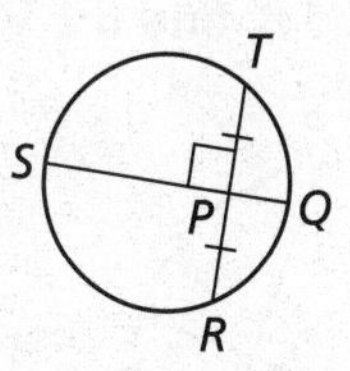

If $\overline{QS}$ is a perpendicular bisector of $\overline{TR}$, then $\overline{QS}$ is a diameter of the circle.

Theorem 10.9 Equidistant Chords Theorem

In the same circle, or in congruent circles, two chords are congruent if and only if they are equidistant from the center.

Notes:

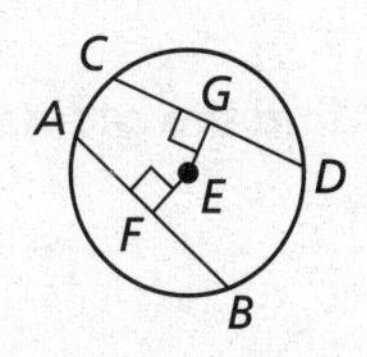

$\overline{AB} \cong \overline{CD}$ if and only if $EF = EG$.

Name ______________________________ Date __________

Extra Practice

In Exercises 1–4, find the measure of the arc or chord in ⊙*Q*.

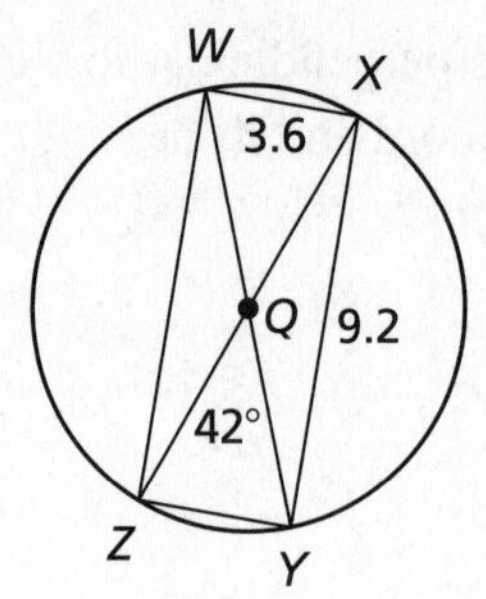

1. $m\overset{\frown}{WX}$

2. YZ

3. WZ

4. $m\overset{\frown}{XY}$

In Exercises 5 and 6, find the value of *x*.

5.

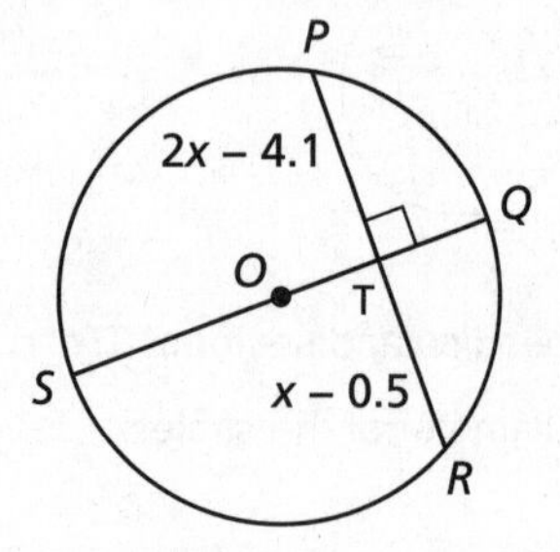

6.

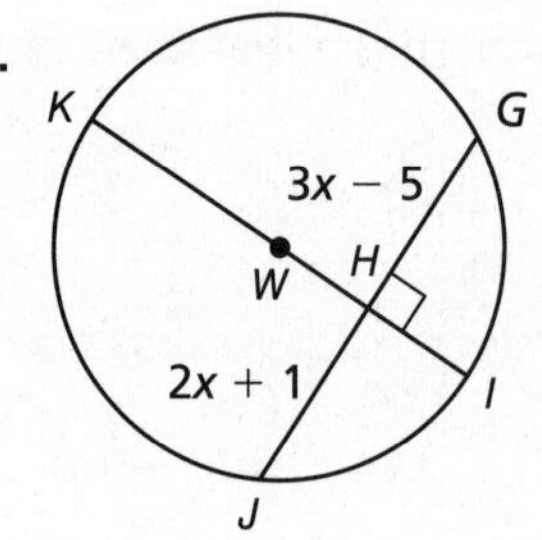

In Exercises 7 and 8, find the radius of the circle.

7.

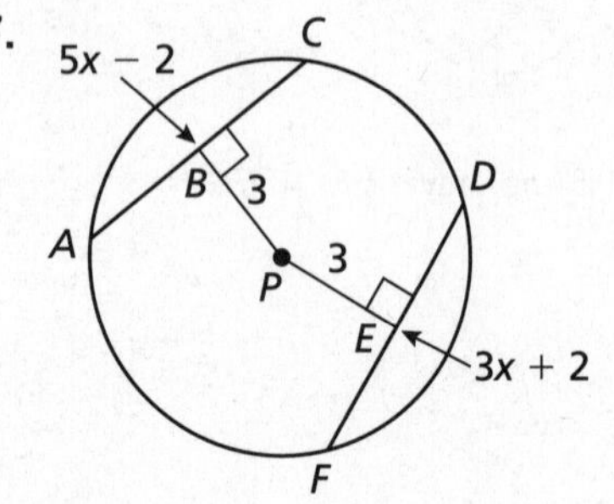

8.

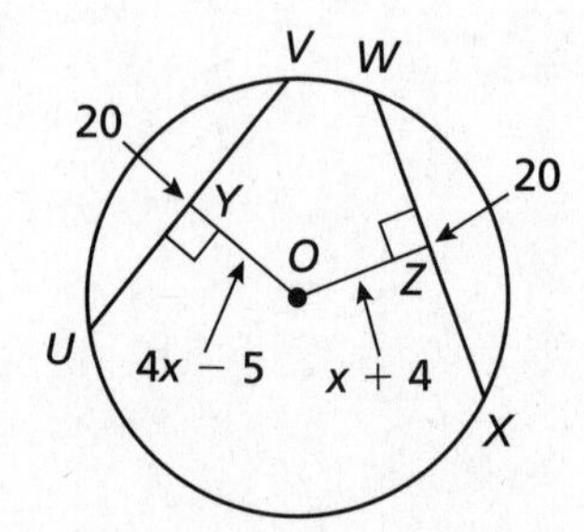

Name ______________________________ Date __________

10.4 Inscribed Angles and Polygons
For use with Exploration 10.4

Essential Question How are inscribed angles related to their intercepted arcs? How are the angles of an inscribed quadrilateral related to each other?

An **inscribed angle** is an angle whose vertex is on a circle and whose sides contain chords of the circle. An arc that lies between two lines, rays, or segments is called an **intercepted arc**. A polygon is an **inscribed polygon** when all its vertices lie on a circle.

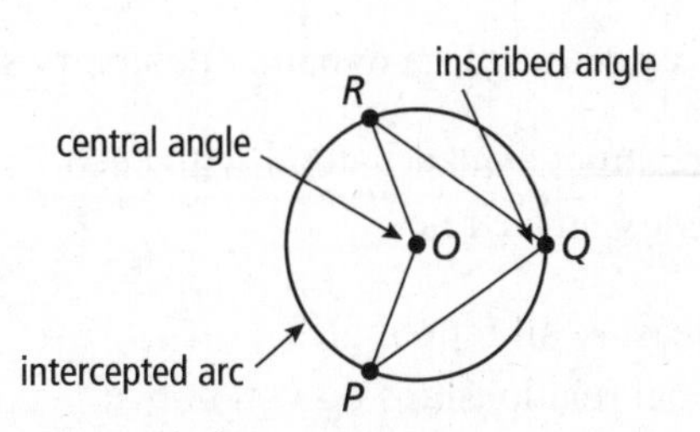

1 EXPLORATION: Inscribed Angles and Central Angles

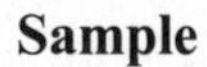

Go to *BigIdeasMath.com* for an interactive tool to investigate this exploration.

Work with a partner. Use dynamic geometry software.

a. Construct an inscribed angle in a circle. Then construct the corresponding central angle.

b. Measure both angles. How is the inscribed angle related to its intercepted arc?

Sample

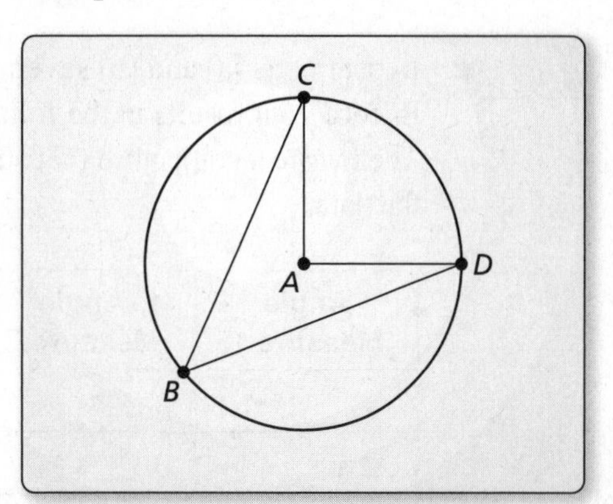

c. Repeat parts (a) and (b) several times. Record your results in the following table. Write a conjecture about how an inscribed angle is related to its intercepted arc.

Measure of Inscribed Angle	Measure of Central Angle	Relationship

2 EXPLORATION: A Quadrilateral with Inscribed Angles

Go to *BigIdeasMath.com* for an interactive tool to investigate this exploration.

Work with a partner. Use dynamic geometry software.

a. Construct a quadrilateral with each vertex on a circle.

Sample

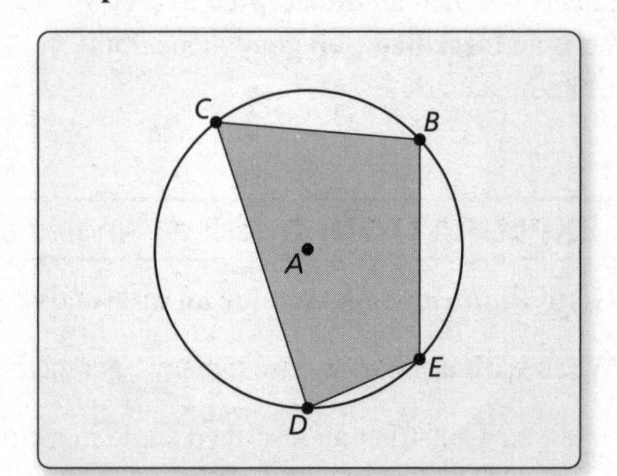

b. Measure all four angles. What relationships do you notice?

c. Repeat parts (a) and (b) several times. Record your results in the following table. Then write a conjecture that summarizes the data.

Angle Measure 1	Angle Measure 2	Angle Measure 3	Angle Measure 4

Communicate Your Answer

3. How are inscribed angles related to their intercepted arcs? How are the angles of an inscribed quadrilateral related to each other?

4. Quadrilateral *EFGH* is inscribed in $\odot C$, and $m\angle E = 80°$. What is $m\angle G$? Explain.

Name____ Date__________

10.4 Notetaking with Vocabulary
For use after Lesson 10.4

In your own words, write the meaning of each vocabulary term.

inscribed angle

intercepted arc

subtend

inscribed polygon

circumscribed circle

Core Concepts

Inscribed Angle and Intercepted Arc

An **inscribed angle** is an angle whose vertex is on a circle and whose sides contain chords of the circle. An arc that lies between two lines, rays, or segments is called an **intercepted arc**. If the endpoints of a chord or arc lie on the sides of an inscribed angle, then the chord or arc is said to **subtend** the angle.

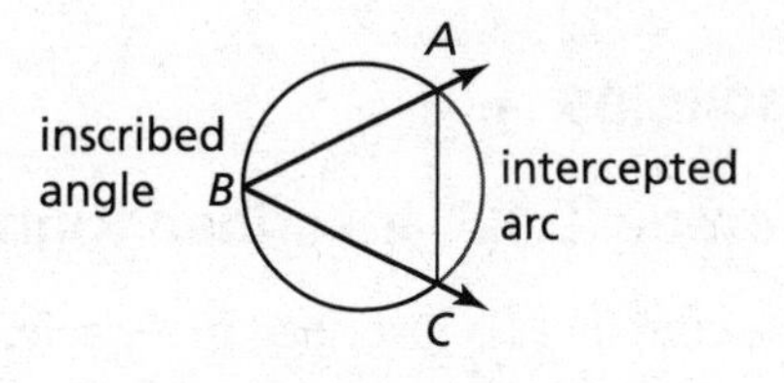

$\angle B$ intercepts $\overset{\frown}{AC}$.
$\overset{\frown}{AC}$ subtends $\angle B$.
$\overline{AC}$ subtends $\angle B$.

Notes:

Theorems

Theorem 10.10 Measure of an Inscribed Angle Theorem

The measure of an inscribed angle is one-half the measure of its intercepted arc.

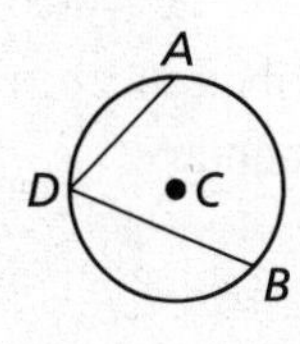

$m\angle ADB = \frac{1}{2}m\overset{\frown}{AB}$

Notes:

Name ____ Date __________

Theorem 10.11 Inscribed Angles of a Circle Theorem

If two inscribed angles of a circle intercept the same arc, then the angles are congruent.

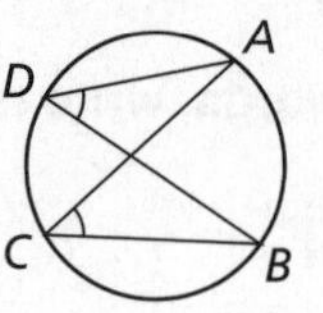

$\angle ADB \cong \angle ACB$

Notes:

Core Concepts

Inscribed Polygon

A polygon is an **inscribed polygon** when all its vertices lie on a circle. The circle that contains the vertices is a **circumscribed circle**.

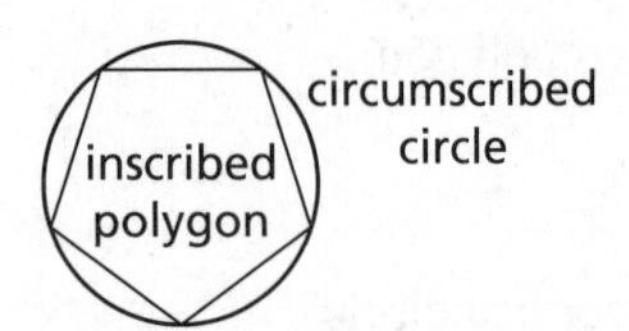

Notes:

Theorems

Theorem 10.12 Inscribed Right Triangle Theorem

If a right triangle is inscribed in a circle, then the hypotenuse is a diameter of the circle. Conversely, if one side of an inscribed triangle is a diameter of the circle, then the triangle is a right triangle and the angle opposite the diameter is the right angle.

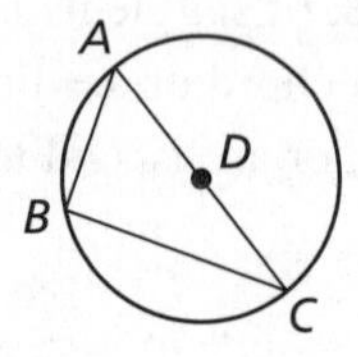

$m\angle ABC = 90°$ if and only if $\overline{AC}$ is a diameter of the circle.

Notes:

Theorem 10.13 Inscribed Quadrilateral Theorem

A quadrilateral can be inscribed in a circle if and only if its opposite angles are supplementary.

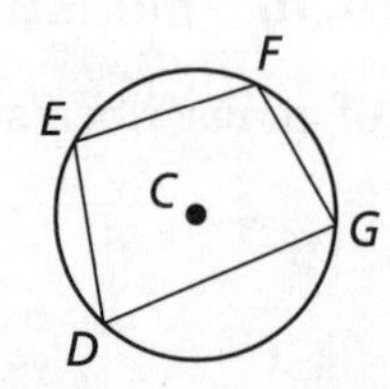

D, E, F, and G lie on $\odot C$ if and only if $m\angle D + m\angle F = m\angle E + m\angle G = 180°$.

Notes:

Name____ Date__________

Extra Practice

In Exercises 1–5, use the diagram to find the indicated measure.

1. $m\angle A$

2. $m\angle C$

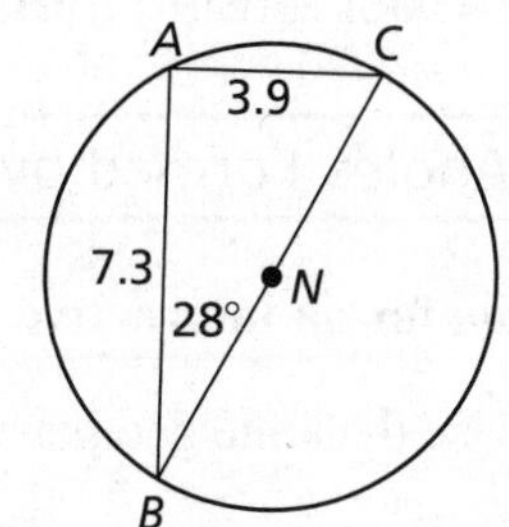

3. BC

4. $m\overparen{AC}$

5. $m\overparen{AB}$

6. Name two pairs of congruent angles.

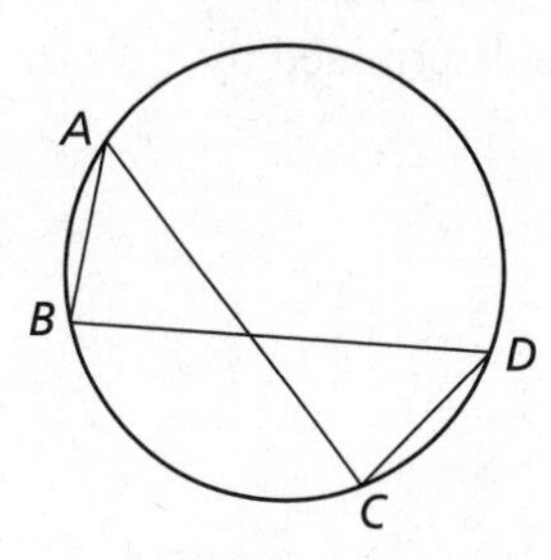

7. Find the value of each variable.

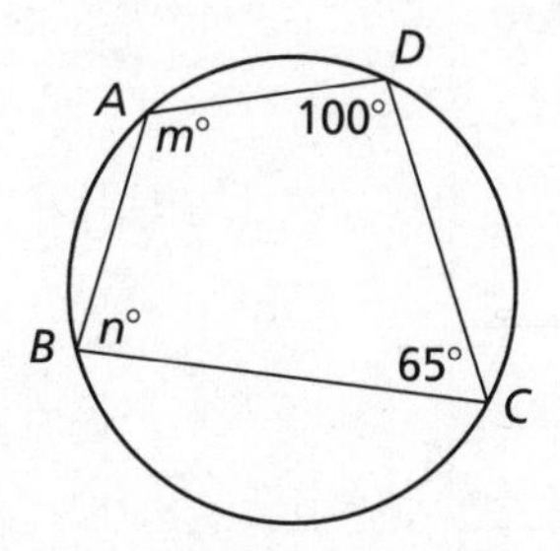

Name ____ Date __________

10.5 Angle Relationships in Circles

For use with Exploration 10.5

Essential Question When a chord intersects a tangent line or another chord, what relationships exist among the angles and arcs formed?

1 EXPLORATION: Angles Formed by a Chord and Tangent Line

Go to *BigIdeasMath.com* for an interactive tool to investigate this exploration.

Work with a partner. Use dynamic geometry software.

Sample

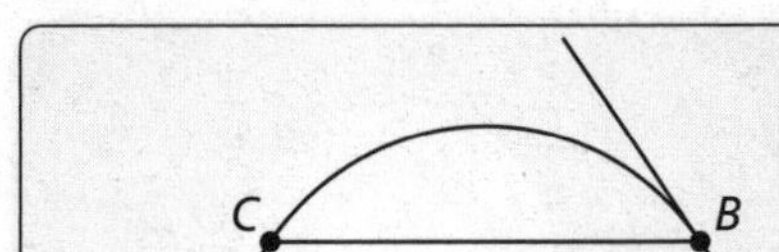
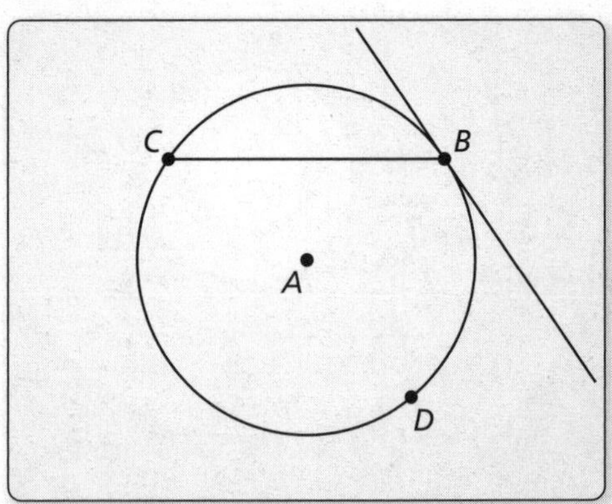

a. Construct a chord in a circle. At one of the endpoints of the chord, construct a tangent line to the circle.

b. Find the measures of the two angles formed by the chord and the tangent line.

c. Find the measures of the two circular arcs determined by the chord.

d. Repeat parts (a)–(c) several times. Record your results in the following table. Then write a conjecture that summarizes the data.

Angle Measure 1	Angle Measure 2	Circular Arc Measure 1	Circular Arc Measure 2

10.5 Angle Relationships in Circles (continued)

2 EXPLORATION: Angles Formed by Intersecting Chords

Go to *BigIdeasMath.com* for an interactive tool to investigate this exploration.

Work with a partner. Use dynamic geometry software.

Sample

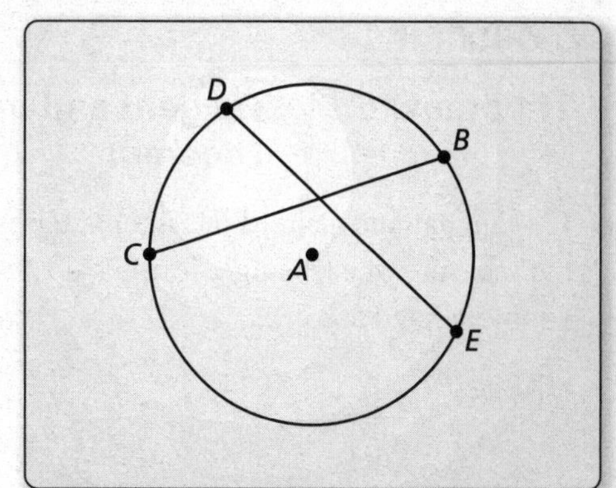

a. Construct two chords that intersect inside a circle.

b. Find the measure of one of the angles formed by the intersecting chords.

c. Find the measures of the arcs intercepted by the angle in part (b) and its vertical angle. What do you observe?

d. Repeat parts (a)–(c) several times. Record your results in the following table. Then write a conjecture that summarizes the data.

Angle Measure	Arc Measures	Observations

Communicate Your Answer

3. When a chord intersects a tangent line or another chord, what relationships exist among the angles and arcs formed?

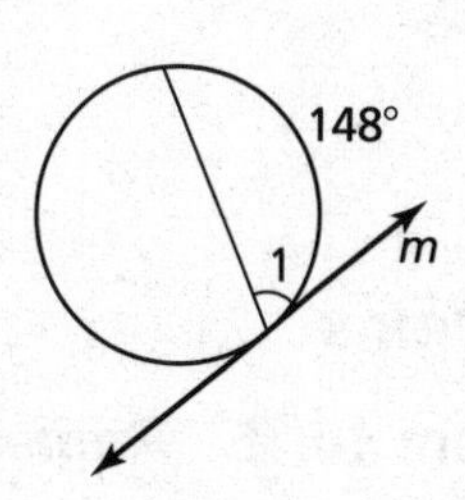

4. Line m is tangent to the circle in the figure at the right. Find the measure of $\angle 1$.

5. Two chords intersect inside a circle to form a pair of vertical angles with measures of 55°. Find the sum of the measures of the arcs intercepted by the two angles.

Name ______________________________ Date __________

10.5 Notetaking with Vocabulary

For use after Lesson 10.5

In your own words, write the meaning of each vocabulary term.

circumscribed angle

Theorems

Theorem 10.14 Tangent and Intersected Chord Theorem

If a tangent and a chord intersect at a point on a circle, then the measure of each angle formed is one-half the measure of its intercepted arc.

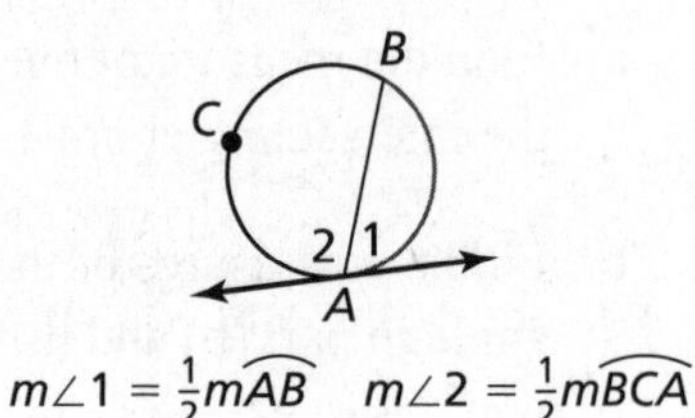

$m\angle 1 = \frac{1}{2} m\overset{\frown}{AB}$ $m\angle 2 = \frac{1}{2} m\overset{\frown}{BCA}$

Notes:

Core Concepts

Intersecting Lines and Circles

If two nonparallel lines intersect a circle, there are three places where the lines can intersect.

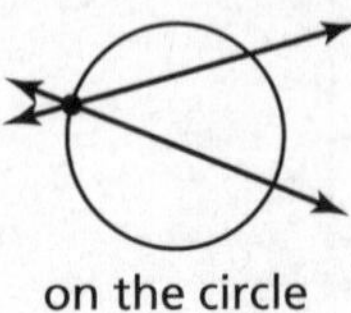
on the circle

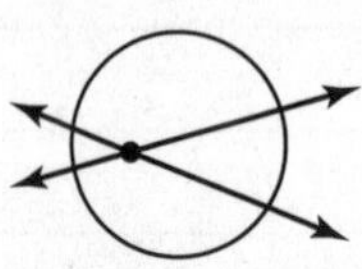
inside the circle

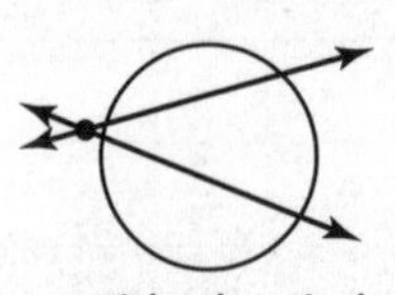
outside the circle

Notes:

Theorems

Theorem 10.15 Angles Inside the Circle Theorem

If two chords intersect *inside* a circle, then the measure of each angle is one-half the *sum* of the measure of the arcs intercepted by the angle and its vertical angle.

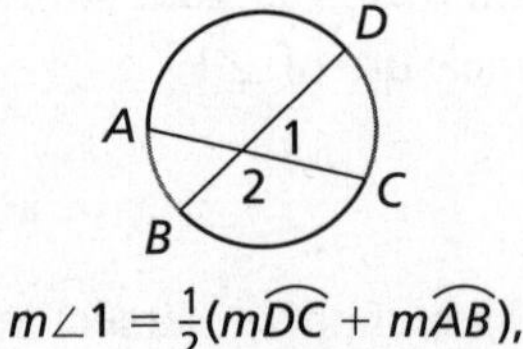

$m\angle 1 = \frac{1}{2}(m\overset{\frown}{DC} + m\overset{\frown}{AB})$,

$m\angle 2 = \frac{1}{2}(m\overset{\frown}{AD} + m\overset{\frown}{BC})$

Notes:

Name____ Date__________

10.5 Notetaking with Vocabulary (continued)

Theorem 10.16 Angles Outside the Circle Theorem

If a tangent and a secant, two tangents, or two secants intersect *outside* a circle, then the measure of the angle formed is one-half the *difference* of the measures of the intercepted arcs.

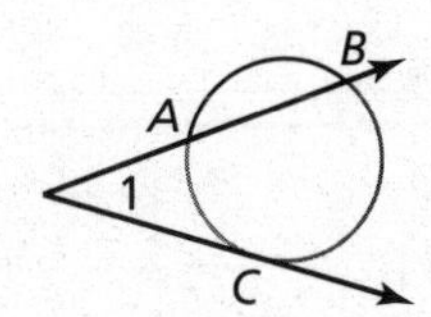

$m\angle 1 = \frac{1}{2}(m\widehat{BC} - m\widehat{AC})$

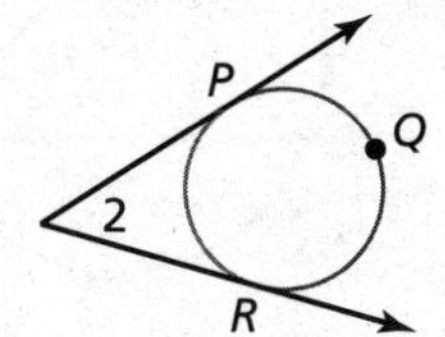

$m\angle 2 = \frac{1}{2}(m\widehat{PQR} - m\widehat{PR})$

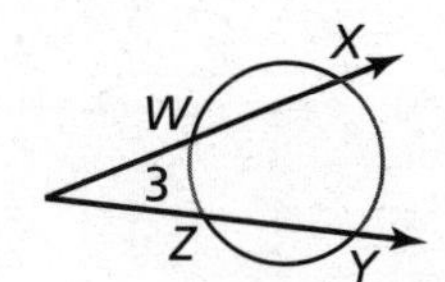

$m\angle 3 = \frac{1}{2}(m\widehat{XY} - m\widehat{WZ})$

Notes:

Core Concepts

Circumscribed Angle

A **circumscribed angle** is an angle whose sides are tangent to a circle.

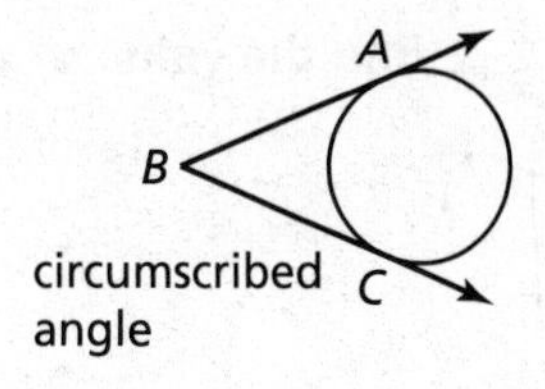

Notes:

Theorems

Theorem 10.17 Circumscribed Angle Theorem

The measure of a circumscribed angle is equal to 180° minus the measure of the central angle that intercepts the same arc.

$m\angle ADB = 180° - m\angle ACB$

Notes:

Extra Practice

In Exercises 1–3, $\overleftrightarrow{CD}$ is tangent to the circle. Find the indicated measure.

1. $m\angle ABC$

2. $m\overparen{AB}$

3. $m\overparen{AEB}$

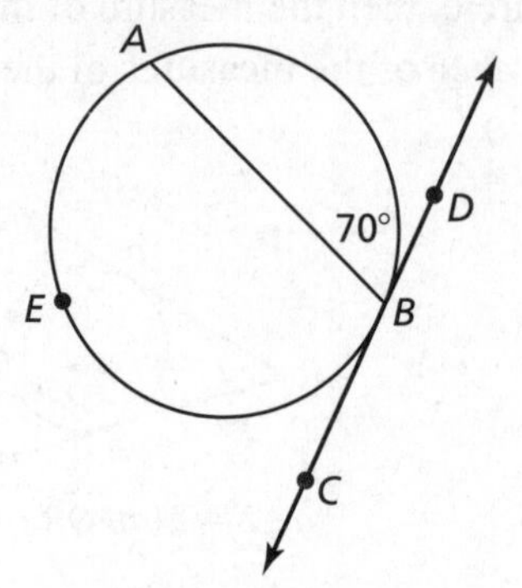

In Exercises 4 and 5, $m\overparen{ADB} = 220°$ and $m\angle B = 21°$. Find the indicated measure.

4. $m\overparen{AB}$

5. $m\angle ACB$

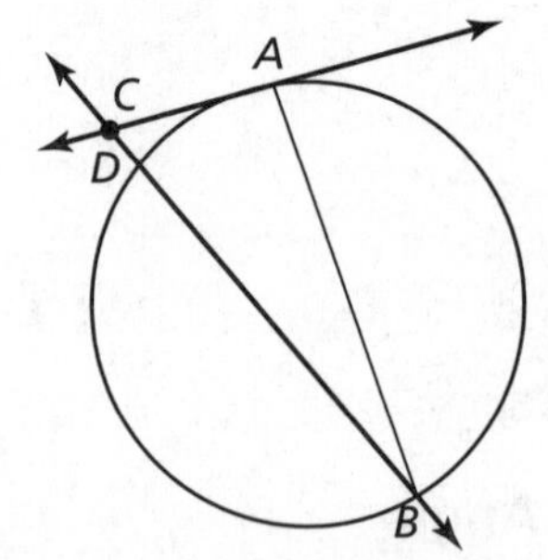

In Exercises 6–9, find the value of *x*.

6.

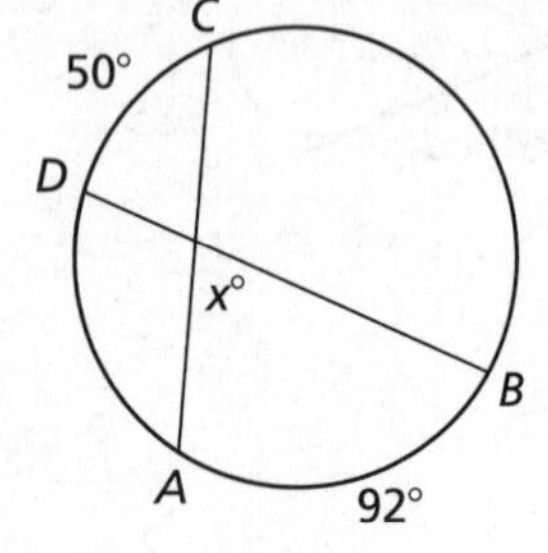

7.

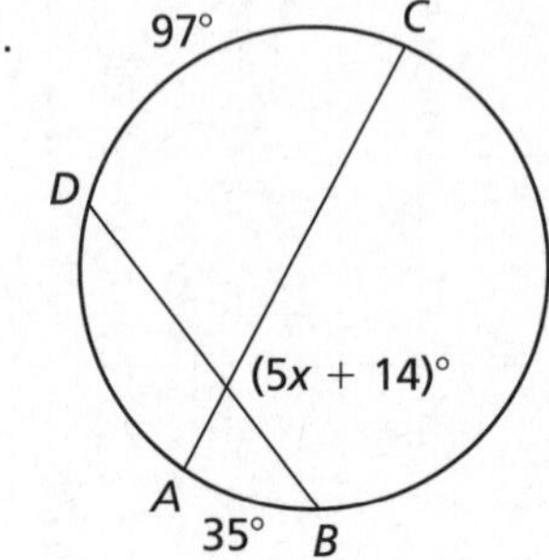

8.

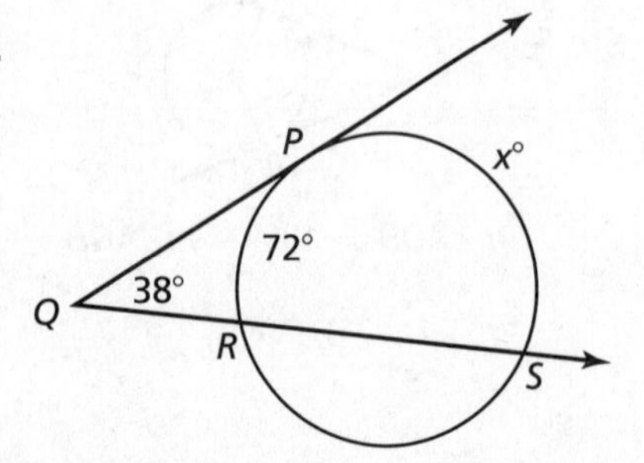

9.

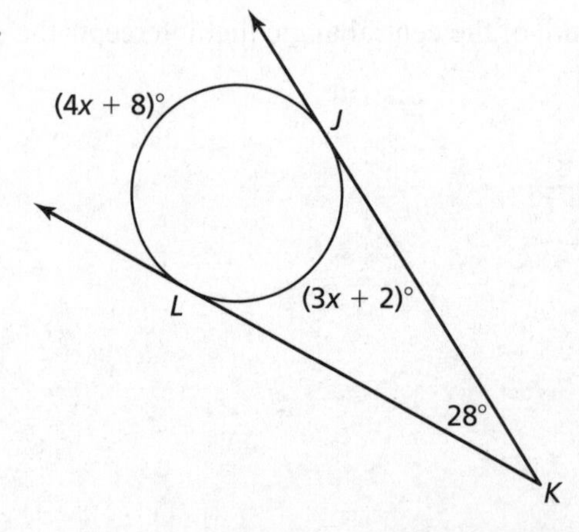

Name____ Date__________

10.6 Segment Relationships in Circles

For use with Exploration 10.6

Essential Question What relationships exist among the segments formed by two intersecting chords or among segments of two secants that intersect outside a circle?

1 EXPLORATION: Segments Formed by Two Intersecting Chords

Go to *BigIdeasMath.com* for an interactive tool to investigate this exploration.

Work with a partner. Use dynamic geometry software.

a. Construct two chords $\overline{BC}$ and $\overline{DE}$ that intersect in the interior of a circle at point F.

Sample

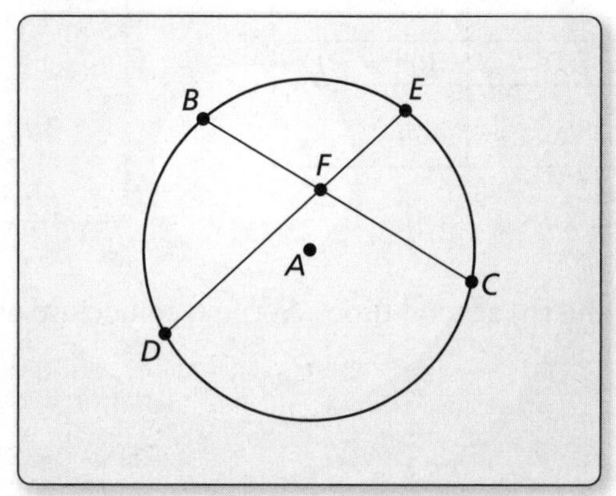

b. Find the segment lengths BF, CF, DF, and EF and complete the table. What do you observe?

BF	CF	$BF \bullet CF$
DF	EF	$DF \bullet EF$

c. Repeat parts (a) and (b) several times. Write a conjecture about your results.

2 EXPLORATION: Secants Intersecting Outside a Circle

Go to *BigIdeasMath.com* for an interactive tool to investigate this exploration.

Work with a partner. Use dynamic geometry software.

a. Construct two secants $\overleftrightarrow{BC}$ and $\overleftrightarrow{BD}$ that intersect at a point B outside a circle, as shown.

Sample

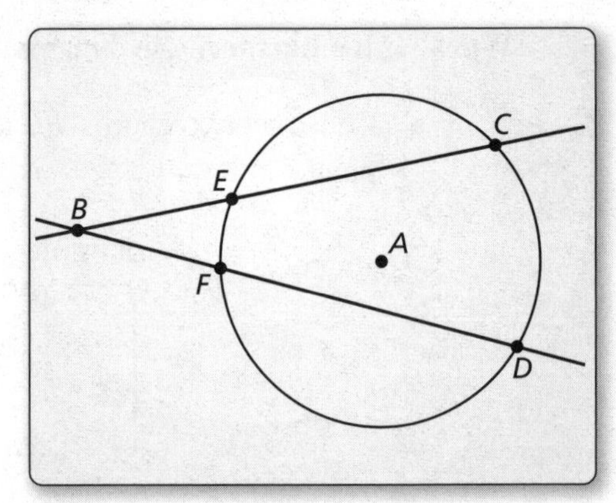

b. Find the segment lengths BE, BC, BF, and BD, and complete the table. What do you observe?

BE	BC	$BE \bullet BC$
BF	BD	$BF \bullet BD$

c. Repeat parts (a) and (b) several times. Write a conjecture about your results.

Communicate Your Answer

3. What relationships exist among the segments formed by two intersecting chords or among segments of two secants that intersect outside a circle?

4. Find the segment length AF in the figure at the right.

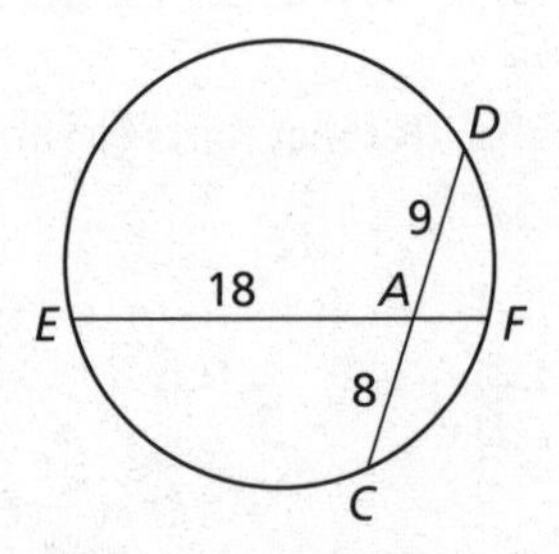

Name____ Date__________

10.6 Notetaking with Vocabulary
For use after Lesson 10.6

In your own words, write the meaning of each vocabulary term.

segments of a chord

tangent segment

secant segment

external segment

Theorems

Theorem 10.18 Segments of Chords Theorem

If two chords intersect in the interior of a circle, then the product of the lengths of the segments of one chord is equal to the product of the lengths of the segments of the other chord.

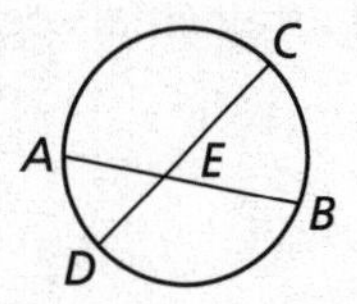

$EA \cdot EB = EC \cdot ED$

Notes:

10.6 Notetaking with Vocabulary (continued)

Core Concepts

Tangent Segment and Secant Segment

A **tangent segment** is a segment that is tangent to a circle at an endpoint. A **secant segment** is a segment that contains a chord of a circle and has exactly one endpoint outside the circle. The part of a secant segment that is outside the circle is called an **external segment**.

external segment Q R
P secant segment
tangent segment S

$\overline{PS}$ is a tangent segment.
$\overline{PR}$ is a secant segment.
$\overline{PQ}$ is the external segment of $\overline{PR}$.

Notes:

Theorems

Theorem 10.19 Segments of Secants Theorem

If two secant segments share the same endpoint outside a circle, then the product of the lengths of one secant segment and its external segment equals the product of the lengths of the other secant segment and its external segment.

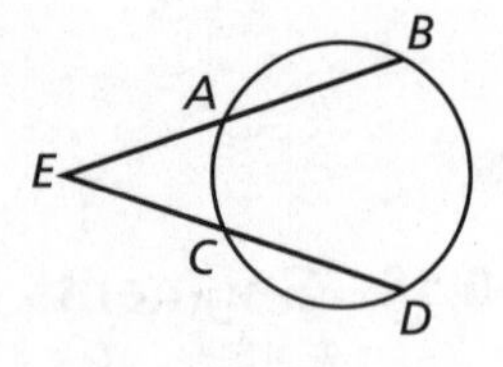

$EA \cdot EB = EC \cdot ED$

Notes:

Theorem 10.20 Segments of Secants and Tangents Theorem

If a secant segment and a tangent segment share an endpoint outside a circle, then the product of the lengths of the secant segment and its external segment equals the square of the length of the tangent segment.

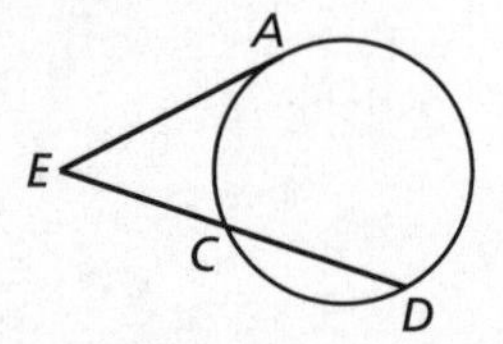

$EA^2 = EC \bullet ED$

Notes:

Extra Practice

In Exercises 1–4, find the value of *x*.

1.

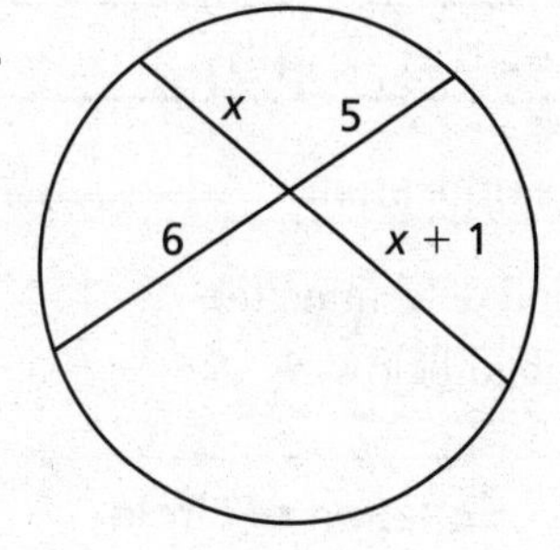

2.

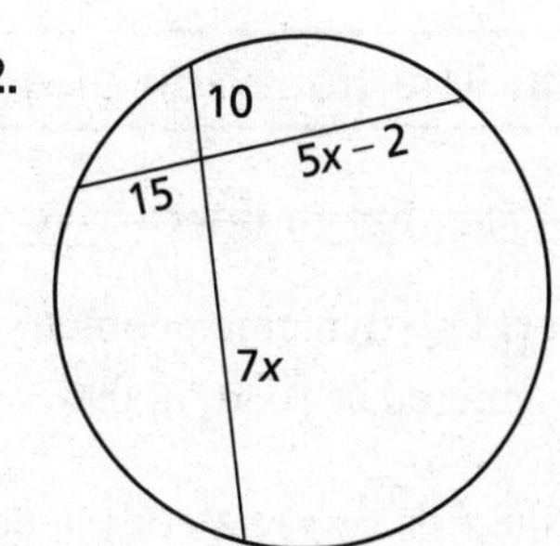

3.

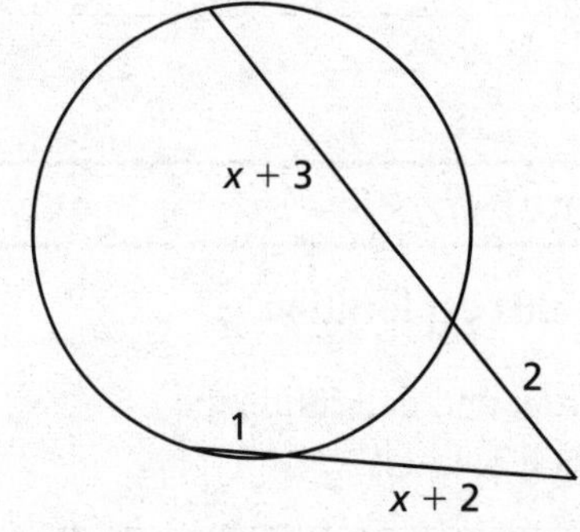

4.

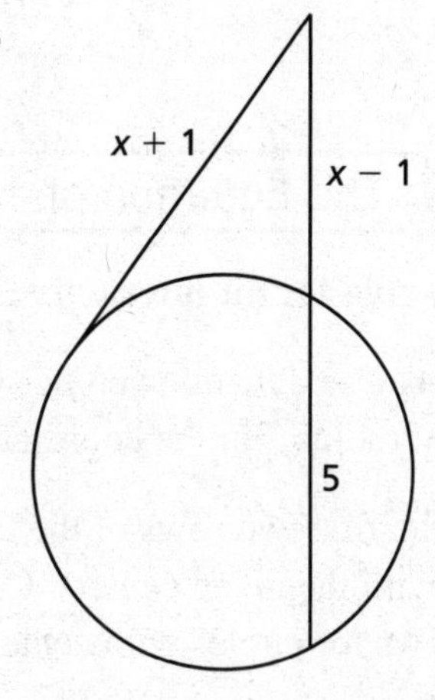

Name ____________________ Date __________

Circles in the Coordinate Plane

For use with Exploration 10.7

Essential Question What is the equation of a circle with center (h, k) and radius r in the coordinate plane?

1 EXPLORATION: The Equation of a Circle with Center at the Origin

Go to *BigIdeasMath.com* for an interactive tool to investigate this exploration.

Work with a partner. Use dynamic geometry software to construct and determine the equations of circles centered at $(0, 0)$ in the coordinate plane, as described below.

a. Complete the first two rows of the table for circles with the given radii. Complete the other rows for circles with radii of your choice.

Radius	Equation of circle
1	
2	

b. Write an equation of a circle with center $(0, 0)$ and radius r.

2 EXPLORATION: The Equation of a Circle with Center (h, k)

Go to *BigIdeasMath.com* for an interactive tool to investigate this exploration.

Work with a partner. Use dynamic geometry software to construct and determine the equations of circles of radius 2 in the coordinate plane, as described below.

a. Complete the first two rows of the table for circles with the given centers. Complete the other rows for circles with centers of your choice.

Center	Equation of circle
$(0, 0)$	
$(2, 0)$	

b. Write an equation of a circle with center (h, k) and radius 2.

c. Write an equation of a circle with center (h, k) and radius r.

3 EXPLORATION: Deriving the Standard Equation of a Circle

Work with a partner. Consider a circle with radius r and center (h, k).

Write the Distance Formula to represent the distance d between a point (x, y) on the circle and the center (h, k) of the circle. Then square each side of the Distance Formula equation.

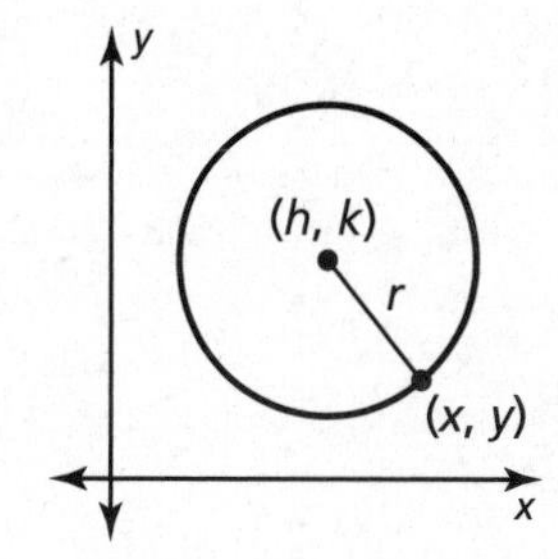

How does your result compare with the equation you wrote in part (c) of Exploration 2?

Communicate Your Answer

4. What is the equation of a circle with center (h, k) and radius r in the coordinate plane?

5. Write an equation of the circle with center $(4, -1)$ and radius 3.

Name ____ Date __________

10.7 Notetaking with Vocabulary

For use after Lesson 10.7

In your own words, write the meaning of each vocabulary term.

standard equation of a circle

Core Concepts

Standard Equation of a Circle

Let (x, y) represent any point on a circle with center (h, k) and radius r. By the Pythagorean Theorem (Theorem 9.1),

$$(x - h)^2 + (y - k)^2 = r^2.$$

This is the **standard equation of a circle** with center (h, k) and radius r.

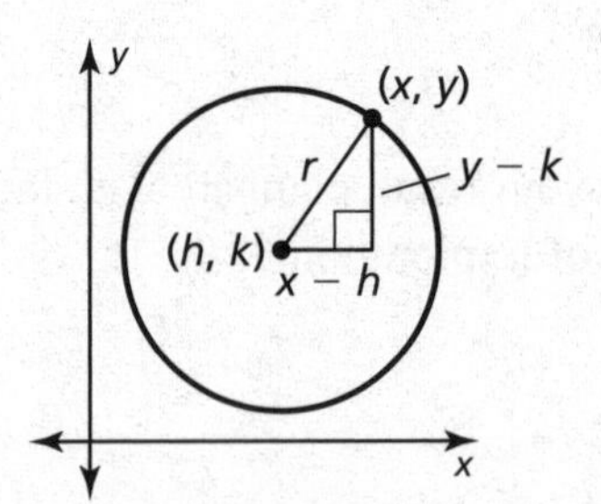

Notes:

Name______________________________ Date__________

10.7 Notetaking with Vocabulary (continued)

Extra Practice

In Exercises 1–4, write the standard equation of the circle.

1.

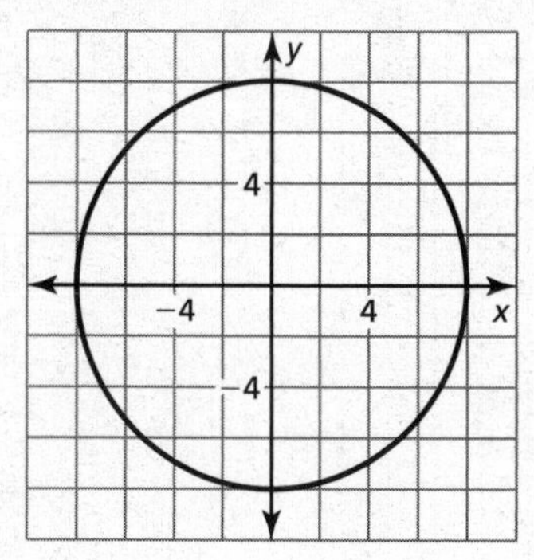

2.

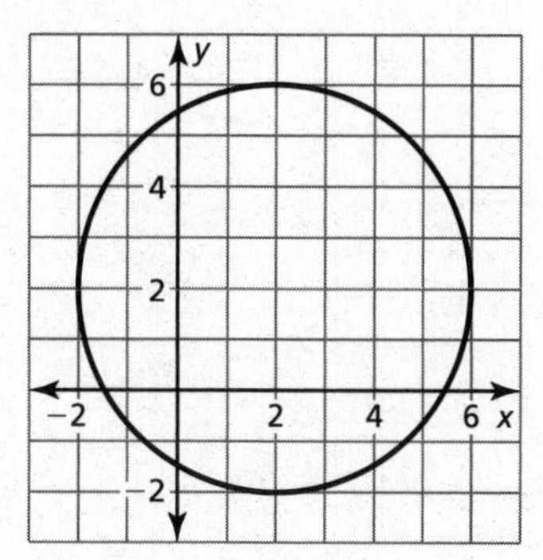

3. a circle with center $(0, 0)$ and radius $\frac{1}{3}$

4. a circle with center $(-3, -5)$ and radius 8

In Exercises 5 and 6, use the given information to write the standard equation of the circle.

5. The center is $(0, 0)$, and a point on the circle is $(4, -3)$.

6. The center is $(4, 5)$, and a point on the circle is $(0, 8)$.

Name ____ Date __________

In Exercises 7–10, find the center and radius of the circle. Then graph the circle.

7. $x^2 + y^2 = 225$

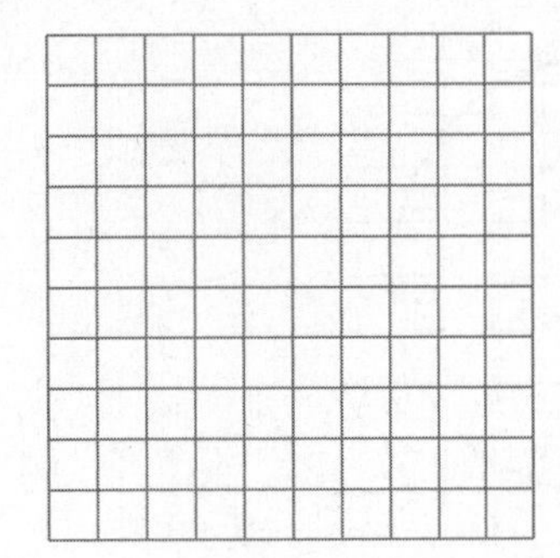

8. $(x - 3)^2 + (y + 2)^2 = 16$

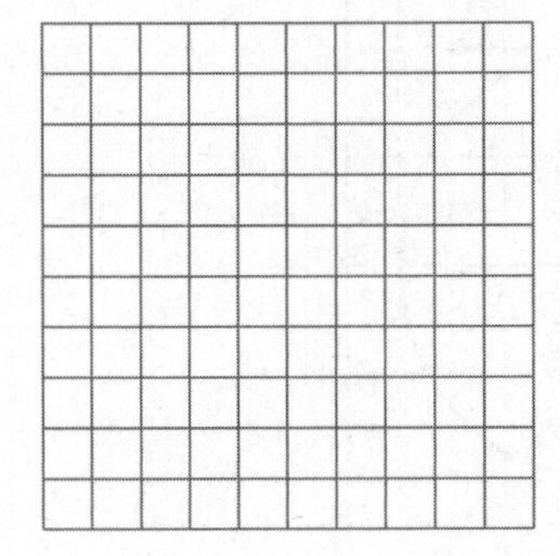

9. $x^2 + y^2 + 2x + 2y = 2$

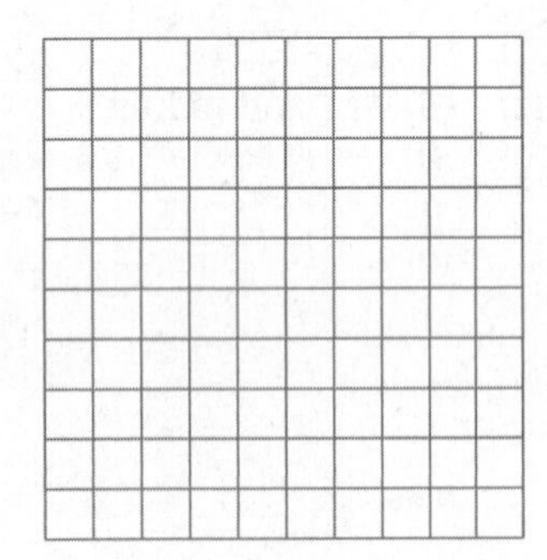

10. $x^2 + y^2 - 3x + y = \frac{5}{2}$

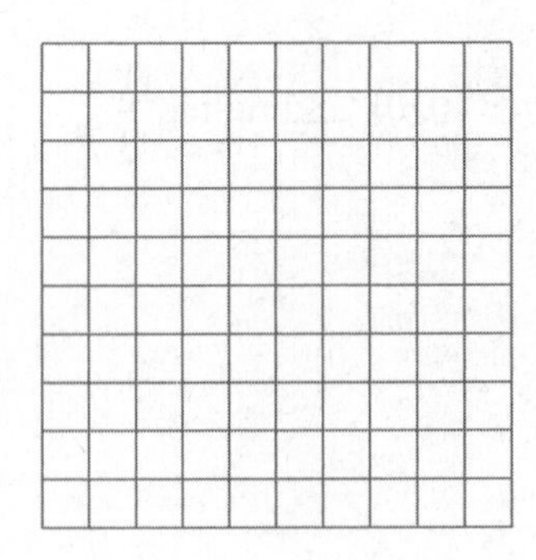

In Exercises 11 and 12, prove or disprove the statement.

11. The point $(-4, 4)$ lies on the circle centered at the origin with radius 6.

12. The point $(-1, 2)$ lies on the circle centered at $(-4, -1)$ with radius $3\sqrt{2}$.

Name______________________________ Date__________

Chapter 11 Maintaining Mathematical Proficiency

Find the surface area of the prism.

1.

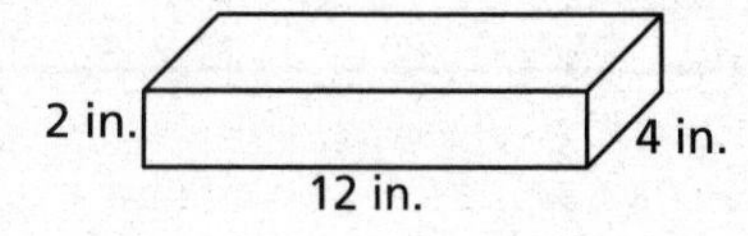

2.

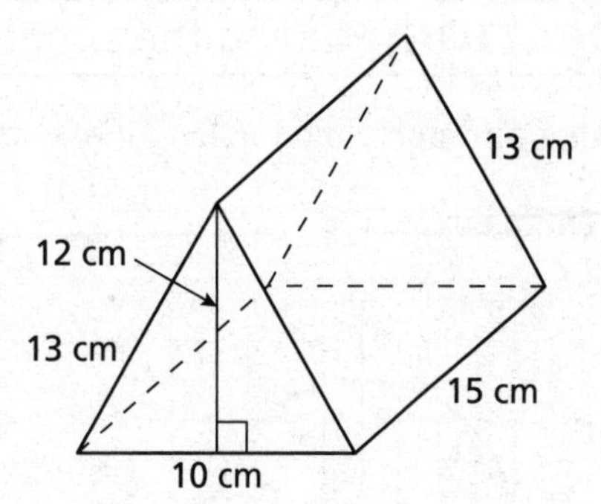

Find the missing dimension.

3. A rectangle has an area of 25 square inches and a length of 10 inches. What is the width of the rectangle?

4. A triangle has an area of 32 square centimeters and a base of 8 centimeters. What is the height of the triangle?

Name ______________________________ Date __________

11.1 Circumference and Arc Length

For use with Exploration 11.1

Essential Question How can you find the length of a circular arc?

1 EXPLORATION: Finding the Length of a Circular Arc

Work with a partner. Find the length of each gray circular arc.

a. entire circle

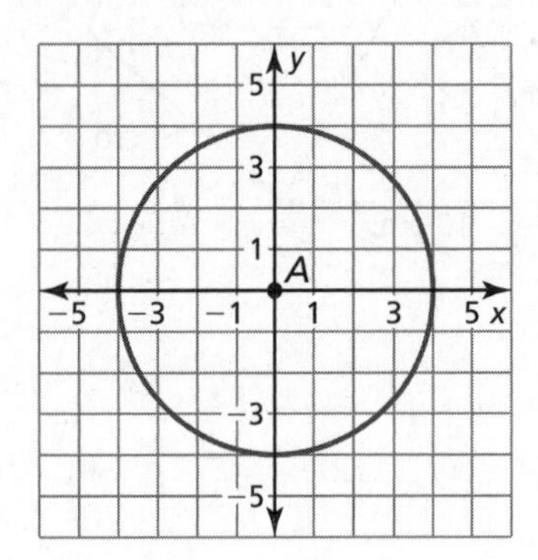

b. one-fourth of a circle

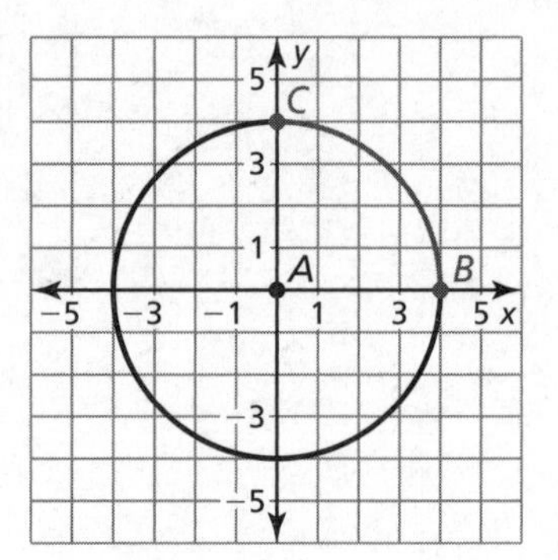

c. one-third of a circle

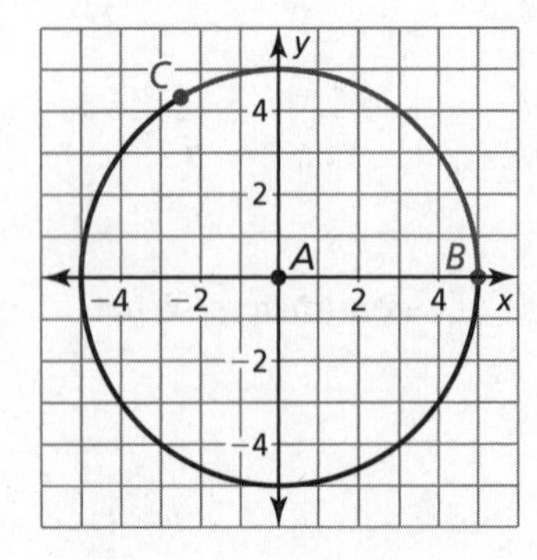

d. five-eighths of a circle

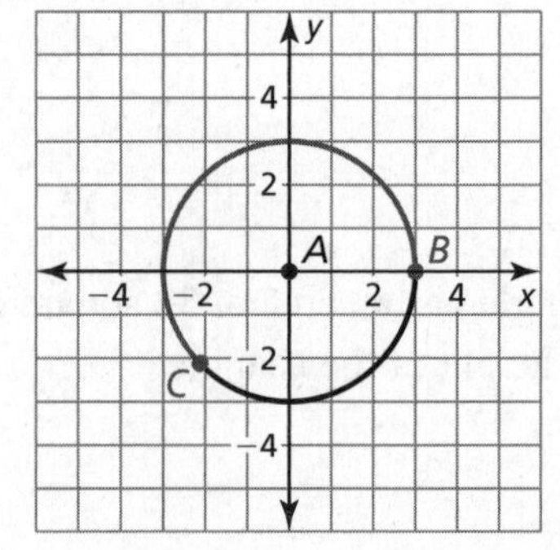

2 EXPLORATION: Writing a Conjecture

Work with a partner. The rider is attempting to stop with the front tire of the motorcycle in the painted rectangular box for a skills test. The front tire makes exactly one-half additional revolution before stopping. The diameter of the tire is 25 inches. Is the front tire still in contact with the painted box? Explain.

Communicate Your Answer

3. How can you find the length of a circular arc?

4. A motorcycle tire has a diameter of 24 inches. Approximately how many inches does the motorcycle travel when its front tire makes three-fourths of a revolution?

Name ______________________________ Date __________

11.1 Notetaking with Vocabulary

For use after Lesson 11.1

In your own words, write the meaning of each vocabulary term.

circumference

arc length

radian

Core Concepts

Circumference of a Circle

The circumference C of a circle is $C = \pi d$ or $C = 2\pi r$, where d is the diameter of the circle and r is the radius of the circle.

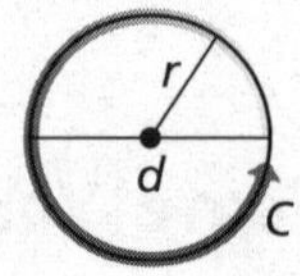

$C = \pi d = 2\pi r$

Notes:

Name____ Date__________

11.1 Notetaking with Vocabulary (continued)

Arc Length

In a circle, the ratio of the length of a given arc to the circumference is equal to the ratio of the measure of the arc to 360°.

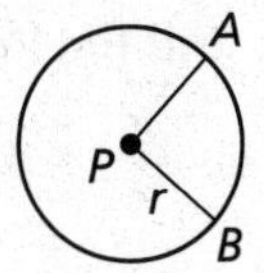

$$\frac{\text{Arc length of } \widehat{AB}}{2\pi r} = \frac{m\widehat{AB}}{360^\circ}, \text{ or}$$

$$\text{Arc length of } \widehat{AB} = \frac{m\widehat{AB}}{360^\circ} \cdot 2\pi r$$

Notes:

Converting Between Degrees and Radians

Degrees to radians

Multiply degree measure by

$$\frac{2\pi \text{ radians}}{360^\circ}, \text{ or } \frac{\pi \text{ radians}}{180^\circ}.$$

Radians to degrees

Multiply radian measure by

$$\frac{360^\circ}{2\pi \text{ radians}}, \text{ or } \frac{180^\circ}{\pi \text{ radians}}.$$

Notes:

Extra Practice

In Exercises 1–5, find the indicated measure.

1. diameter of a circle with a circumference of 10 inches

2. circumference of a circle with a radius of 3 centimeters

3. radius of a circle with a circumference of 8 feet

4. circumference of a circle with a diameter of 2.4 meters

5. arc length of $\widehat{AC}$

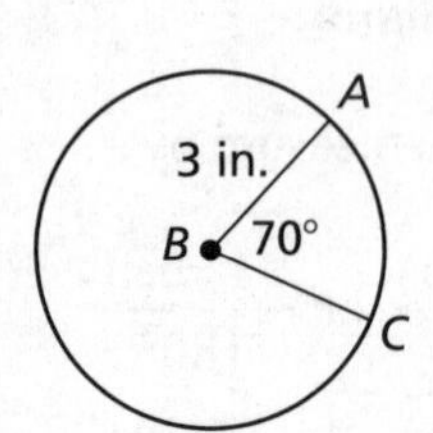

In Exercises 6 and 7, convert the angle measure.

6. Convert 60° to radians.

7. Convert $\frac{5\pi}{6}$ radians to degrees.

Name______________________________ Date__________

11.2 Areas of Circles and Sectors

For use with Exploration 11.2

Essential Question How can you find the area of a sector of a circle?

1 EXPLORATION: Finding the Area of a Sector of a Circle

Work with a partner. A **sector of a circle** is the region bounded by two radii of the circle and their intercepted arc. Find the area of each shaded circle or sector of a circle.

a. entire circle

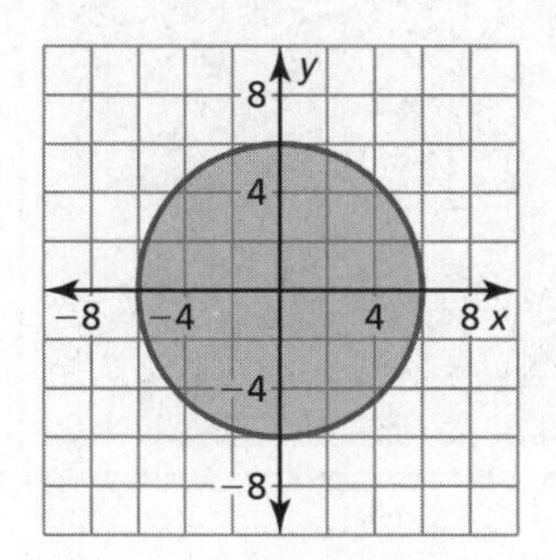

b. one-fourth of a circle

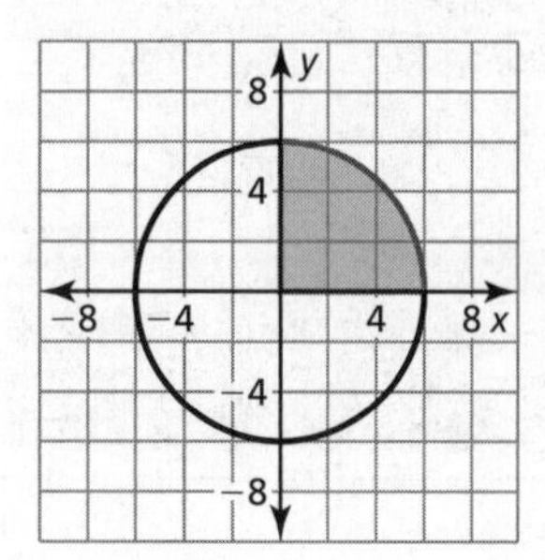

c. seven-eighths of a circle

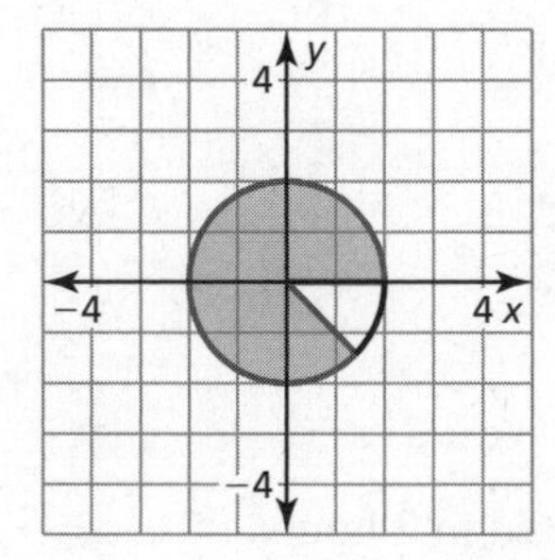

d. two-thirds of a circle

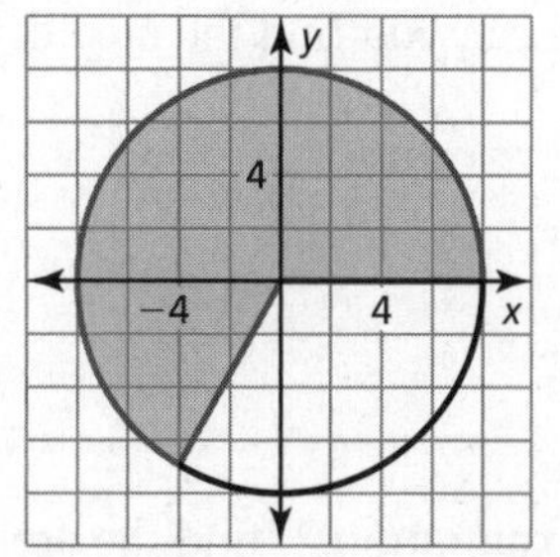

Name ______________________________ Date __________

2 EXPLORATION: Finding the Area of a Circular Sector

Work with a partner. A center pivot irrigation system consists of 400 meters of sprinkler equipment that rotates around a central pivot point at a rate of once every 3 days to irrigate a circular region with a diameter of 800 meters. Find the area of the sector that is irrigated by this system in one day.

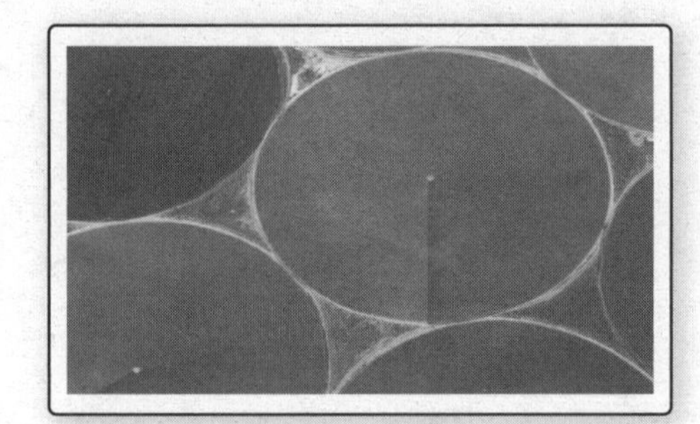

Communicate Your Answer

3. How can you find the area of a sector of a circle?

4. In Exploration 2, find the area of the sector that is irrigated in 2 hours.

Name____ Date__________

11.2 Notetaking with Vocabulary

For use after Lesson 11.2

In your own words, write the meaning of each vocabulary term.

population density

sector of a circle

Core Concepts

Area of a Circle

The area of a circle is

$$A = \pi r^2$$

where r is the radius of the circle.

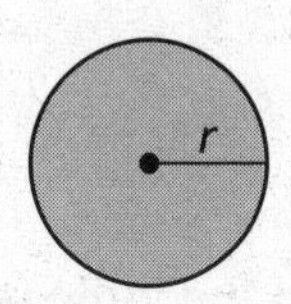

Notes:

Area of a Sector

The ratio of the area of a sector of a circle to the area of the whole circle $\left(\pi r^2\right)$ is equal to the ratio of the measure of the intercepted arc to 360°.

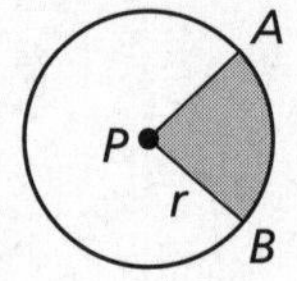

$$\frac{\text{Area of sector } APB}{\pi r^2} = \frac{m\overset{\frown}{AB}}{360^\circ}, \text{ or}$$

$$\text{Area of sector } APB = \frac{m\overset{\frown}{AB}}{360^\circ} \cdot \pi r^2$$

Notes:

Extra Practice

In Exercises 1–2, find the indicated measure.

1. area of $\odot M$

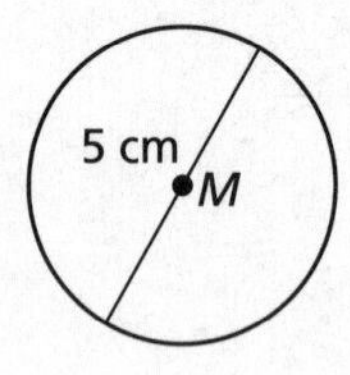

2. area of $\odot R$

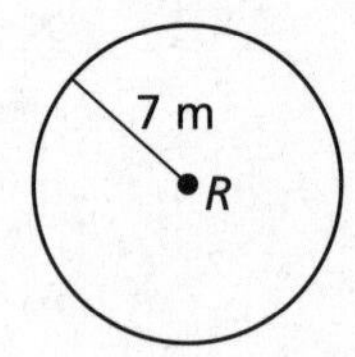

11.2 Notetaking with Vocabulary (continued)

In Exercises 3–8, find the indicated measure.

3. area of a circle with a diameter of 1.8 inches

4. diameter of a circle with an area of 10 square feet

5. radius of a circle with an area of 65 square centimeters

6. area of a circle with a radius of 6.1 yards

7. areas of the sectors formed by $\angle PQR$

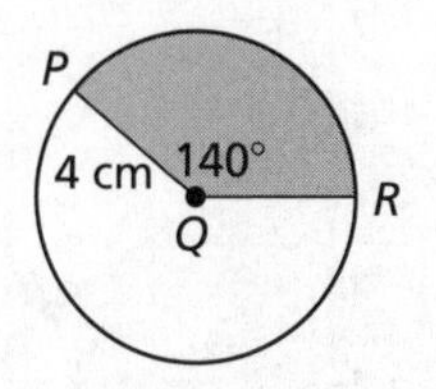

8. area of $\odot Y$

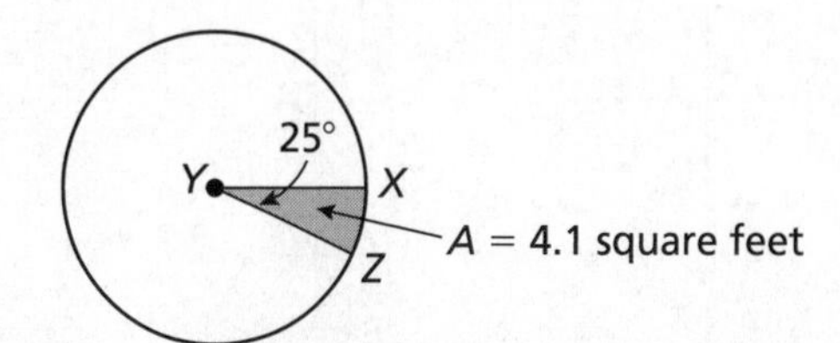

9. About 70,000 people live in a region with a 30-mile radius. Find the population density in people per square mile.

Name ______________________________ Date __________

11.3 Areas of Polygons
For use with Exploration 11.3

Essential Question How can you find the area of a regular polygon?

The **center of a regular polygon** is the center of its circumscribed circle.

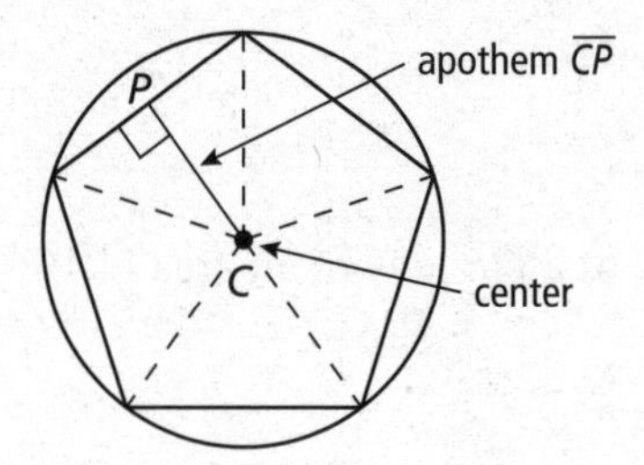

The distance from the center to any side of a regular polygon is called the **apothem of a regular polygon**.

1 EXPLORATION: Finding the Area of a Regular Polygon

Go to *BigIdeasMath.com* for an interactive tool to investigate this exploration.

Work with a partner. Use dynamic geometry software to construct each regular polygon with side lengths of 4, as shown. Find the apothem and use it to find the area of the polygon. Describe the steps that you used.

a.

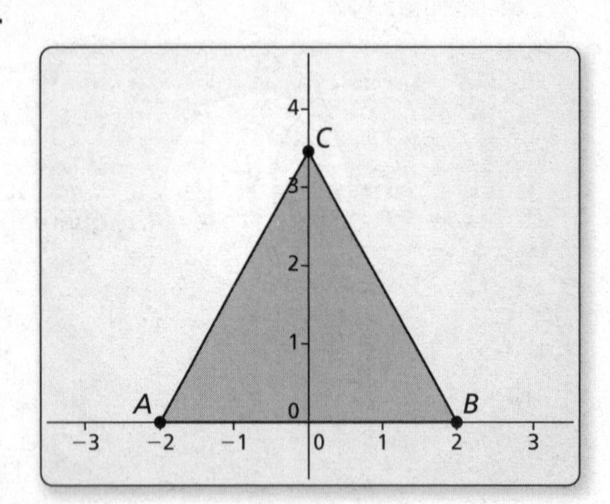

b.

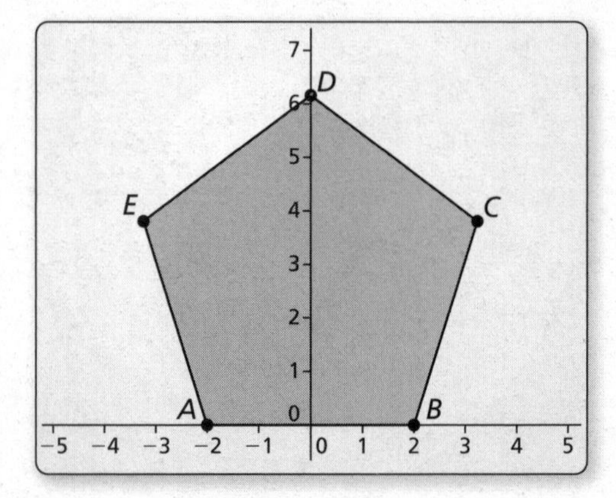

Name____________________ Date__________

1 EXPLORATION: Finding the Area of a Regular Polygon (continued)

c.

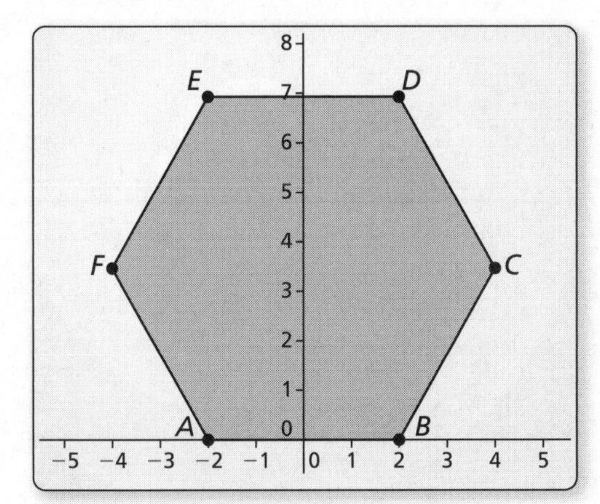

d.

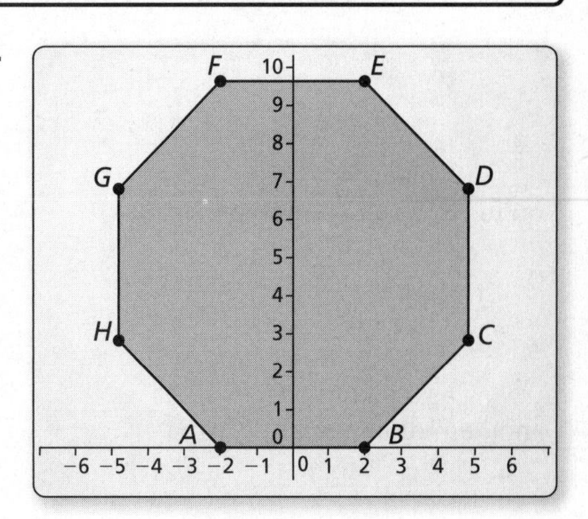

2 EXPLORATION: Writing a Formula for Area

Work with a partner. Generalize the steps you used in Exploration 1 to develop a formula for the area of a regular polygon.

Communicate Your Answer

3. How can you find the area of a regular polygon?

4. Regular pentagon *ABCDE* has side lengths of 6 meters and an apothem of approximately 4.13 meters. Find the area of *ABCDE*.

Name ______________________________ Date __________

11.3 Notetaking with Vocabulary
For use after Lesson 11.3

In your own words, write the meaning of each vocabulary term.

center of a regular polygon

radius of a regular polygon

apothem of a regular polygon

central angle of a regular polygon

Core Concepts

Area of a Rhombus or Kite

The area of a rhombus or kite with diagonals d_1 and d_2 is $\frac{1}{2}d_1d_2$.

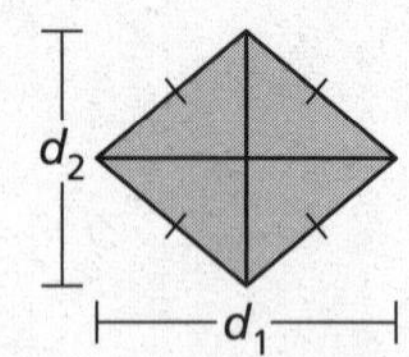

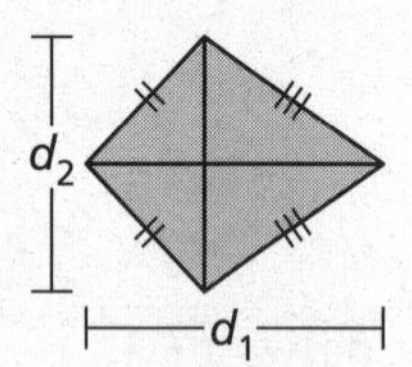

Notes:

Area of a Regular Polygon

The area of a regular n-gon with side length s is one-half the product of the apothem a and the perimeter P.

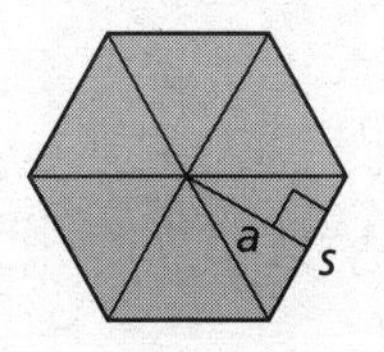

$$A = \frac{1}{2}aP, \text{ or } A = \frac{1}{2}a \cdot ns$$

Notes:

Extra Practice

In Exercises 1 and 2, find the area of the kite or rhombus.

1.

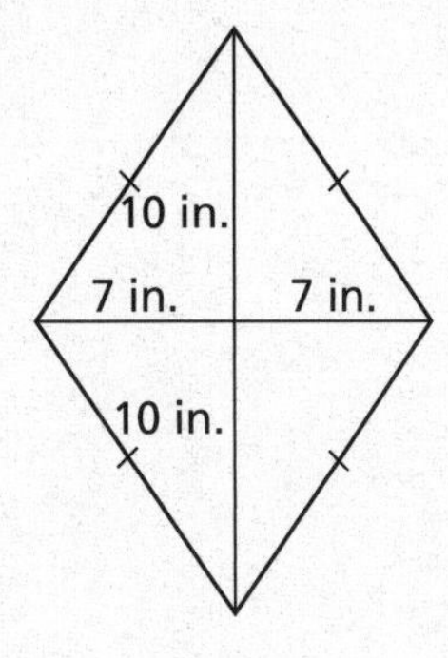

2.

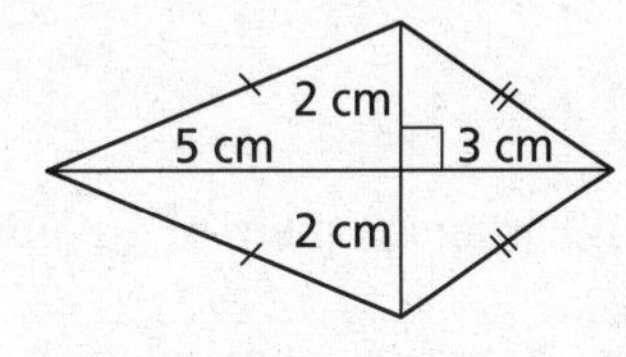

3. Find the measure of a central angle of a regular polygon with 8 sides.

4. The central angles of a regular polygon are 40°. How many sides does the polygon have?

5. A regular pentagon has a radius of 4 inches and a side length of 3 inches.

a. Find the apothem of the pentagon.

b. Find the area of the pentagon.

6. A regular hexagon has an apothem of 10 units.

a. Find the radius of the hexagon and the length of one side.

b. Find the area of the hexagon.

Name____ Date__________

Three-Dimensional Figures

For use with Exploration 11.4

Essential Question What is the relationship between the numbers of vertices V, edges E, and faces F of a polyhedron?

1 EXPLORATION: Analyzing a Property of Polyhedra

Work with a partner. The five *Platonic solids* are shown below. Each of these solids has congruent regular polygons as faces. Complete the table by listing the numbers of vertices, edges, and faces of each Platonic solid.

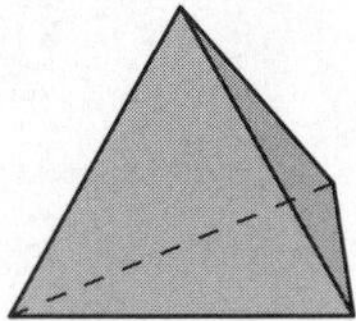
tetrahedron

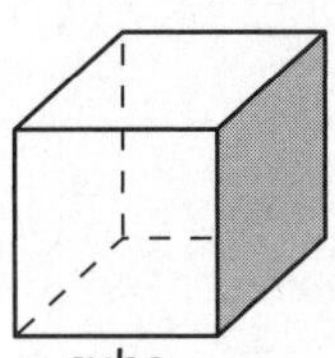
cube

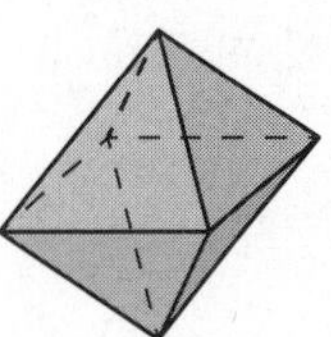
octahedron

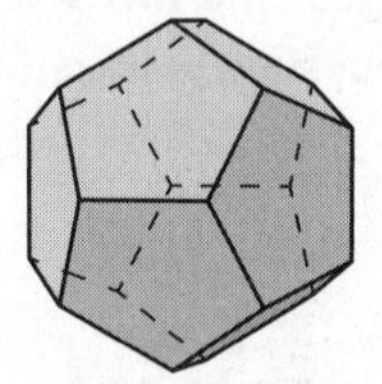
dodecahedron

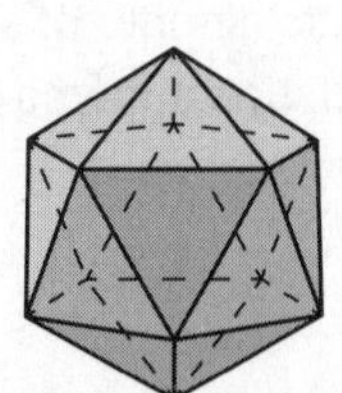
icosahedron

Solid	Vertices, V	Edges, E	Faces, F
tetrahedron			
cube			
octahedron			
dodecahedron			
icosahedron			

Communicate Your Answer

2. What is the relationship between the numbers of vertices V, edges E, and faces F of a polyhedron? (*Note*: Swiss mathematician Leonhard Euler (1707–1783) discovered a formula that relates these quantities.)

3. Draw three polyhedra that are different from the Platonic solids given in Exploration 1. Count the number of vertices, edges, and faces of each polyhedron. Then verify that the relationship you found in Question 2 is valid for each polyhedron.

Name____ Date__________

11.4 Notetaking with Vocabulary

For use after Lesson 11.4

In your own words, write the meaning of each vocabulary term.

polyhedron

face

edge

vertex

cross section

solid of revolution

axis of revolution

Notes:

11.4 Notetaking with Vocabulary (continued)

Core Concepts

Types of Solids

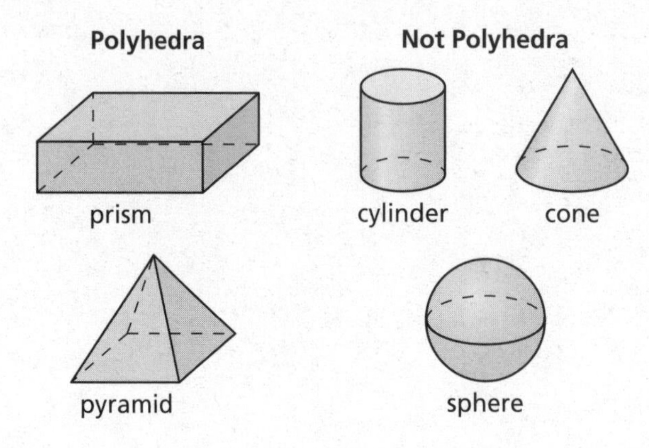

Notes:

Extra Practice

In Exercises 1 and 2, tell whether the solid is a polyhedron. If it is, name the polyhedron.

1.

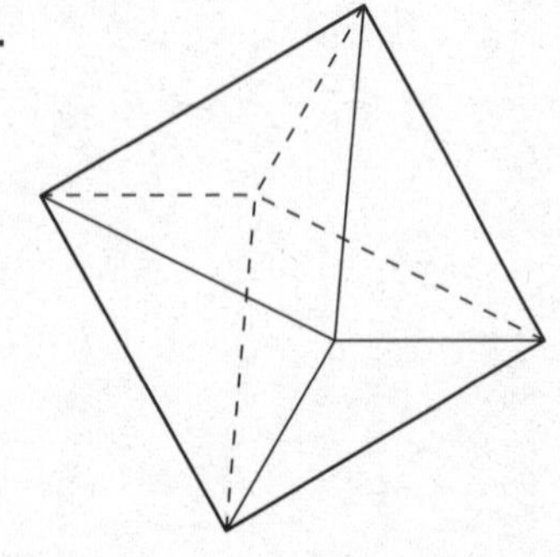

2.

Name____ Date__________

11.4 Notetaking with Vocabulary (continued)

In Exercises 3–6, describe the cross section formed by the intersection of the plane and the solid.

3.

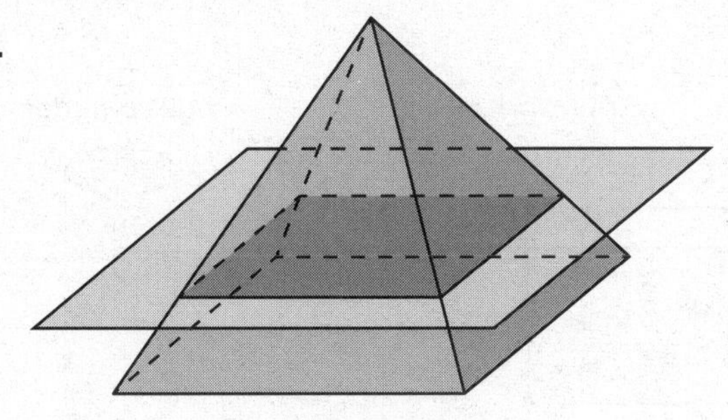

4.

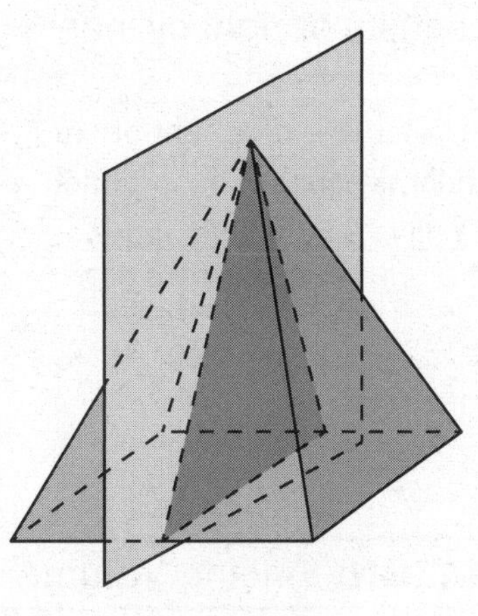

5.

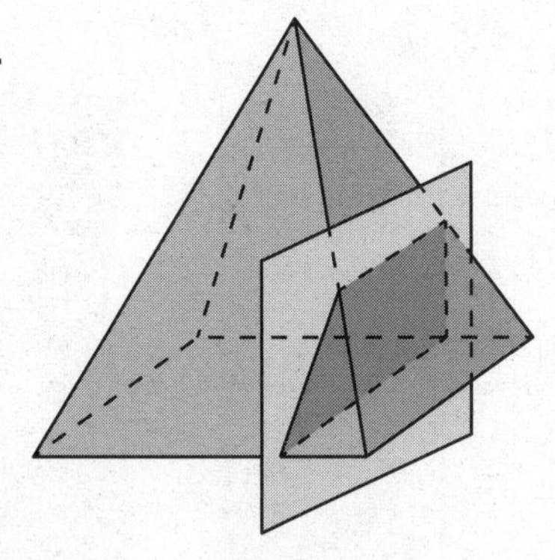

6.

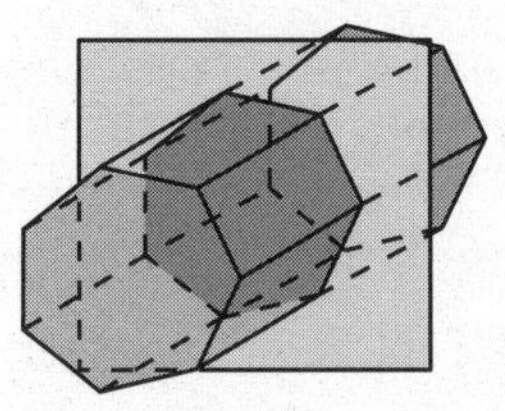

In Exercises 7 and 8, sketch the solid produced by rotating the figure around the given axis. Then identify and describe the solid.

7.

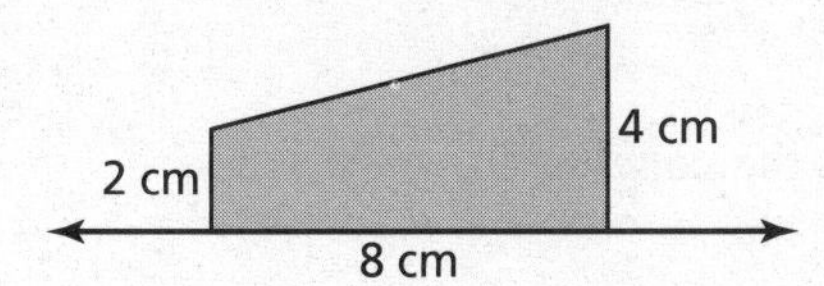

8.

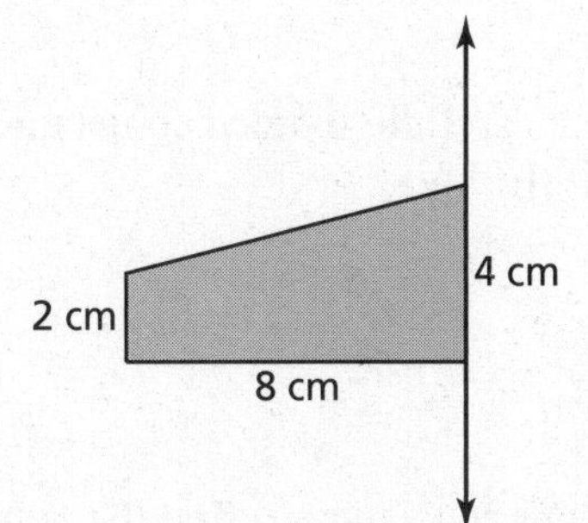

Name ______________________________ Date __________

11.5 Volumes of Prisms and Cylinders

For use with Exploration 11.5

Essential Question How can you find the volume of a prism or cylinder that is not a right prism or right cylinder?

Recall that the volume V of a right prism or a right cylinder is equal to the product of the area of a base B and the height h.

$$V = Bh$$

right prisms

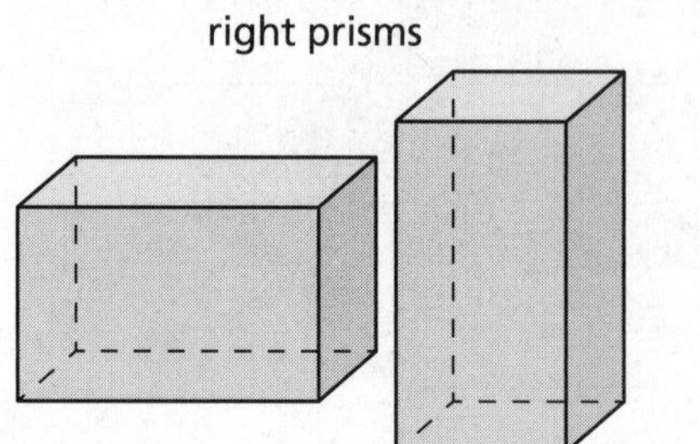

right cylinder

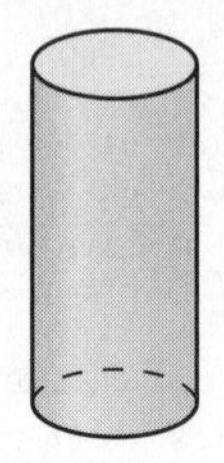

1 EXPLORATION: Finding Volume

Work with a partner. Consider a stack of square papers that is in the form of a right prism.

a. What is the volume of the prism?

b. When you twist the stack of papers, as shown at the right, do you change the volume? Explain your reasoning.

8 in.

2 in. 2 in.

c. Write a carefully worded conjecture that describes the conclusion you reached in part (b).

d. Use your conjecture to find the volume of the twisted stack of papers.

2 EXPLORATION: Finding Volume

Work with a partner. Use the conjecture you wrote in Exploration 1 to find the volume of the cylinder.

a.

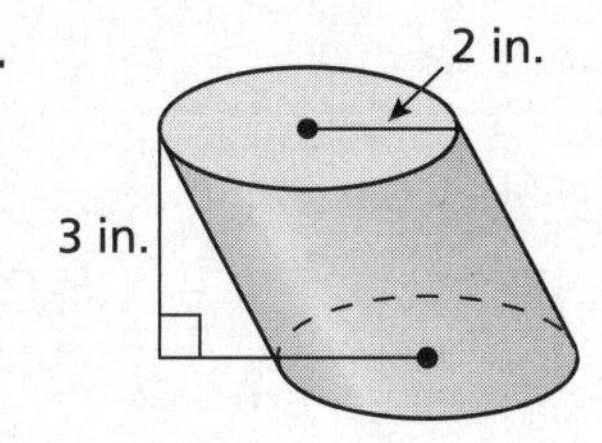

b.

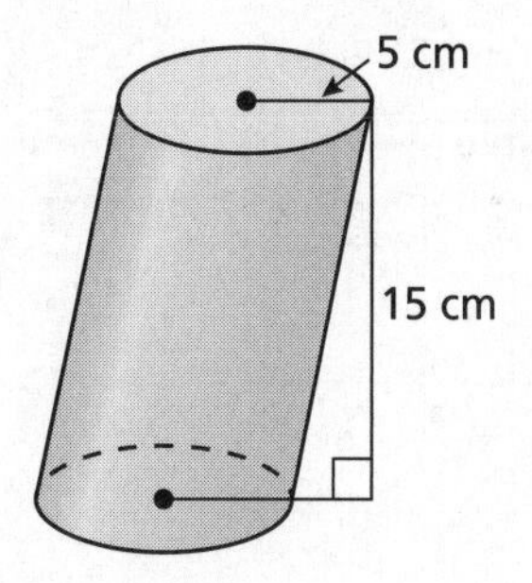

Communicate Your Answer

3. How can you find the volume of a prism or cylinder that is not a right prism or right cylinder?

4. In Exploration 1, would the conjecture you wrote change if the papers in each stack were not squares? Explain your reasoning.

Name ____ Date __________

11.5 Notetaking with Vocabulary
For use after Lesson 11.5

In your own words, write the meaning of each vocabulary term.

volume

Cavalieri's Principle

density

similar solids

Core Concepts

Volume of a Prism

The volume V of a prism is

$$V = Bh$$

where B is the area of a base and h is the height.

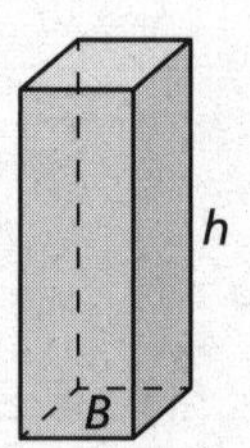

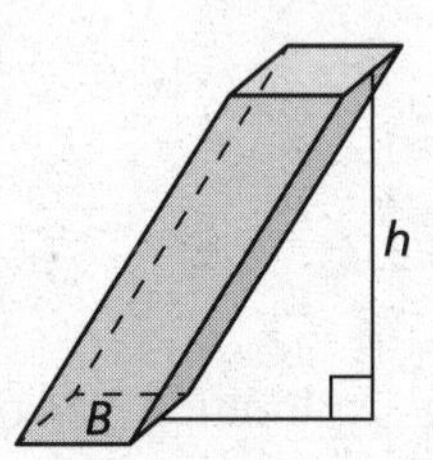

Notes:

Name ______________________________ Date __________

Volume of a Cylinder

The volume V of a cylinder is

$$V = Bh = \pi r^2 h$$

where B is the area of a base, h is the height, and r is the radius of a base.

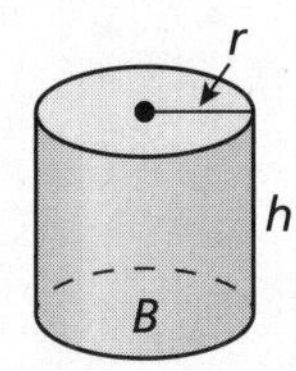

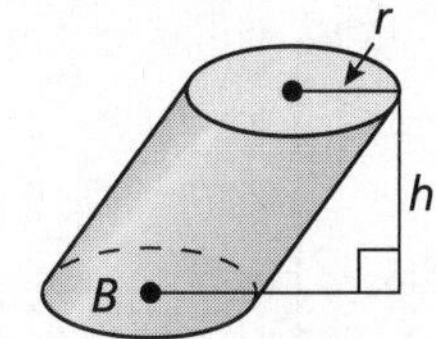

Notes:

Similar Solids

Two solids of the same type with equal ratios of corresponding linear measures, such as heights or radii, are called **similar solids**. The ratio of the corresponding linear measures of two similar solids is called the *scale factor*. If two similar solids have a scale factor of k, then the ratio of their volumes is equal to k^3.

Notes:

Extra Practice

In Exercises 1 and 2, find the volume of the prism.

1.

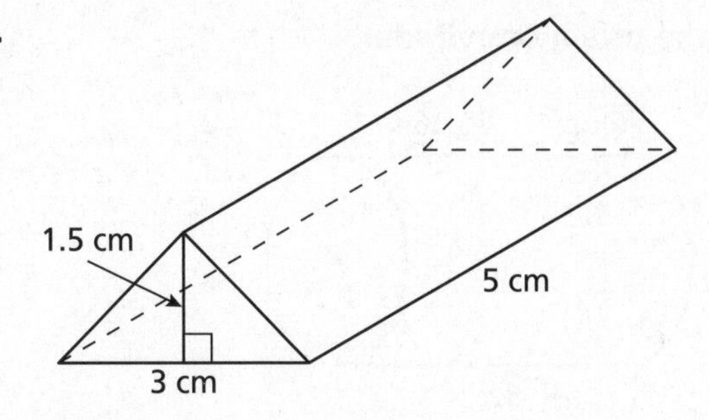

2.

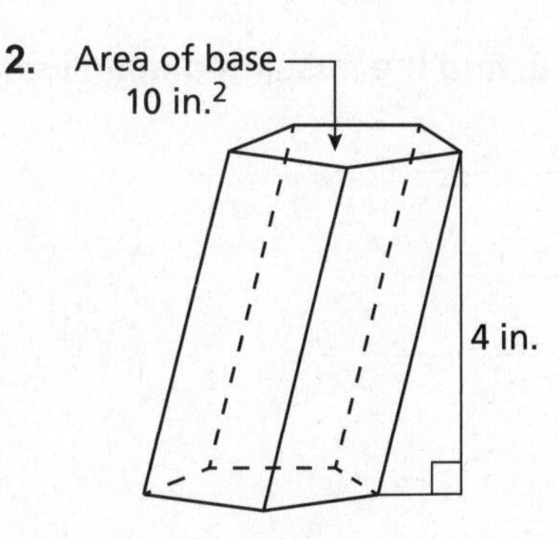

Name ______________________________ Date ________

In Exercises 3 and 4, find the volume of the cylinder.

3.

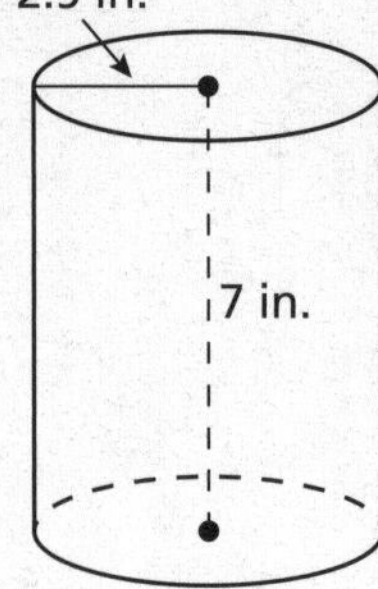

4.

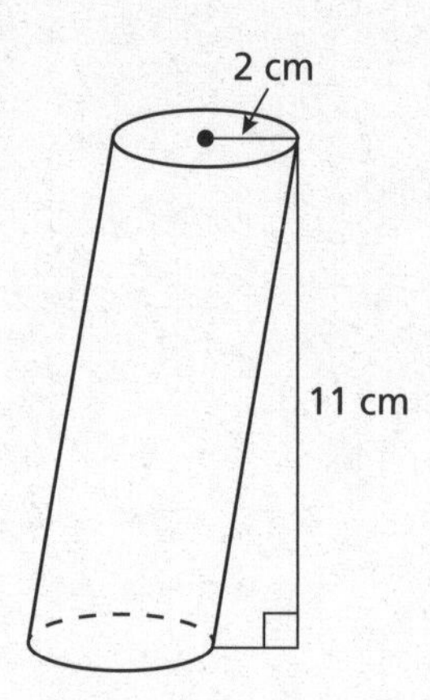

In Exercises 5 and 6, find the indicated measure.

5. height of a cylinder with a base radius of 8 inches and a volume of 2010 cubic inches

6. area of the base of a pentagonal prism with a volume of 50 cubic centimeters and a height of 7.5 centimeters

In Exercises 7 and 8, find the missing dimension of the prism or cylinder.

7.

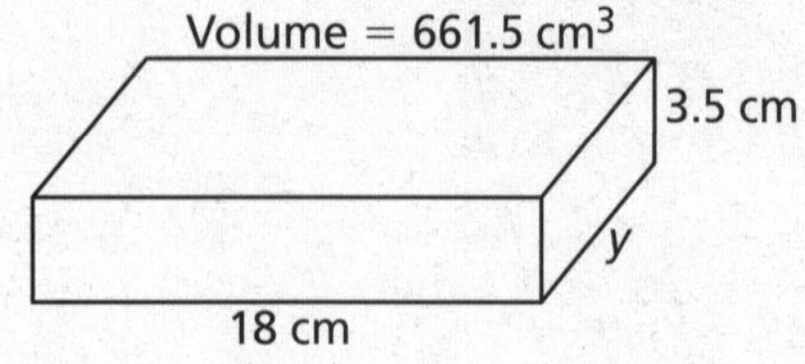

8.

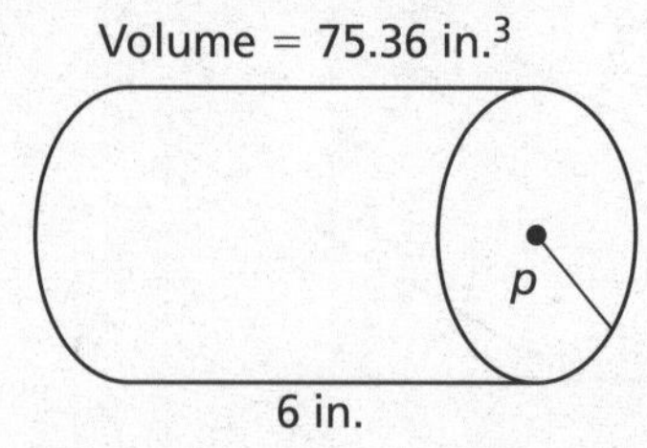

Name_______________________________________ Date__________

11.6 Volumes of Pyramids

For use with Exploration 11.6

Essential Question How can you find the volume of a pyramid?

1 EXPLORATION: Finding the Volume of a Pyramid

Work with a partner. The pyramid and the prism have the same height and the same square base.

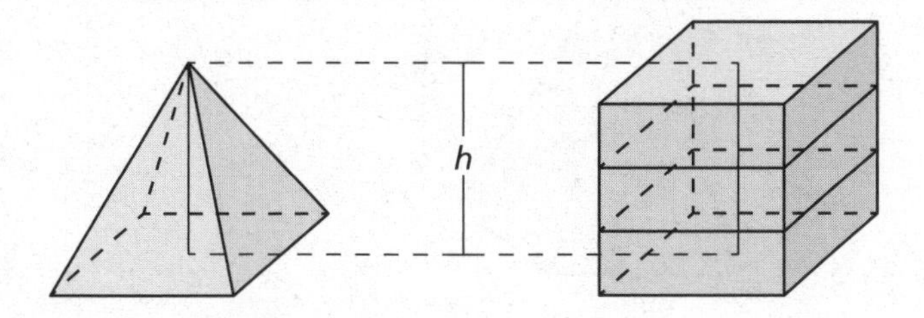

When the pyramid is filled with sand and poured into the prism, it takes three pyramids to fill the prism.

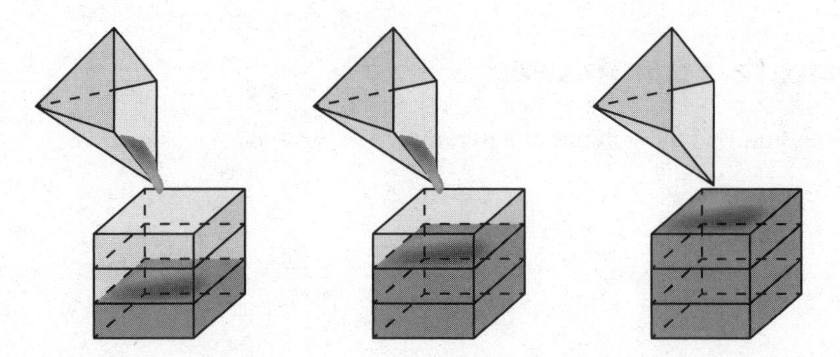

Use this information to write a formula for the volume V of a pyramid.

2 EXPLORATION: Finding the Volume of a Pyramid

Work with a partner. Use the formula you wrote in Exploration 1 to find the volume of the hexagonal pyramid.

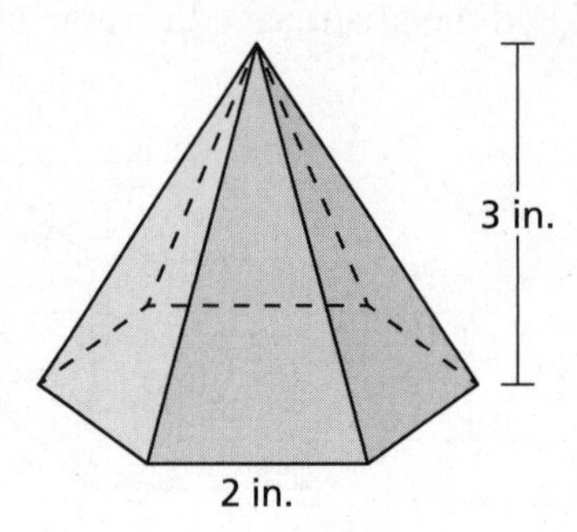

Communicate Your Answer

3. How can you find the volume of a pyramid?

4. In Section 11.7, you will study volumes of cones. How do you think you could use a method similar to the one presented in Exploration 1 to write a formula for the volume of a cone? Explain your reasoning.

Name______________________________ Date__________

11.6 Notetaking with Vocabulary

For use after Lesson 11.6

In your own words, write the meaning of each vocabulary term.

pyramid

composite solid

Core Concepts

Volume of a Pyramid

The volume V of a pyramid is

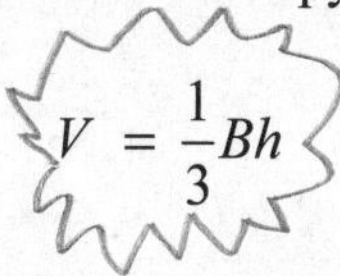

$$V = \frac{1}{3}Bh$$

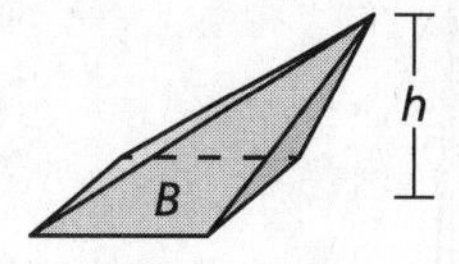

where B is the area of a base and h is the height.

Notes:

Name ______________________ Date __________

Extra Practice

In Exercises 1–6, find the volume of the pyramid.

1.

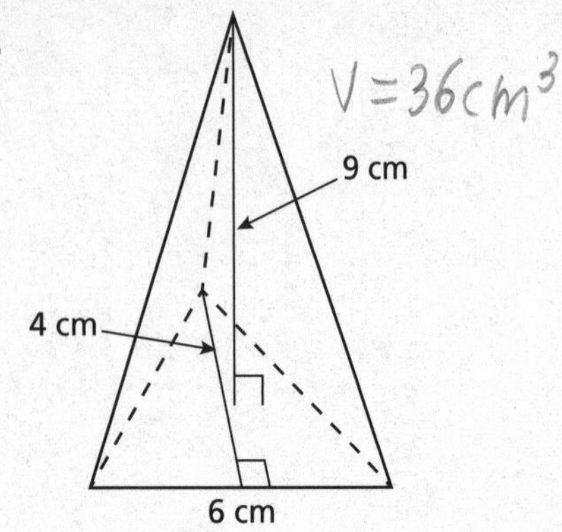

2.

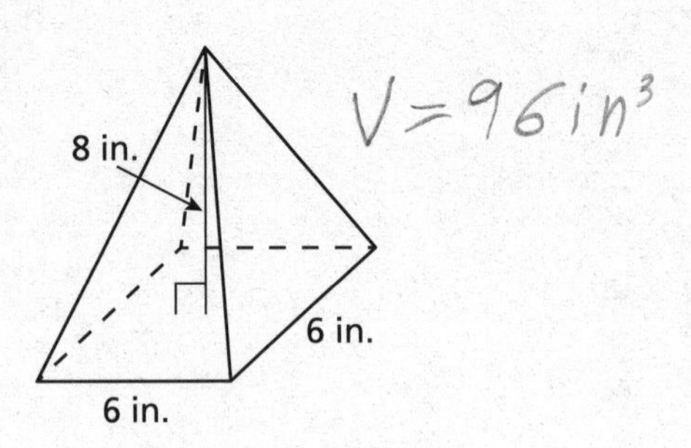

3.

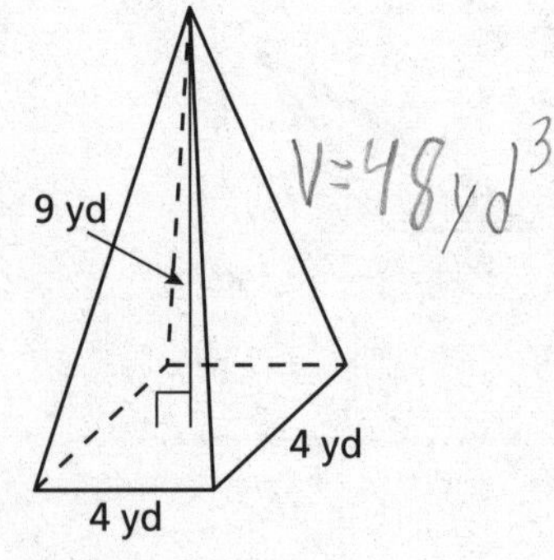

4.

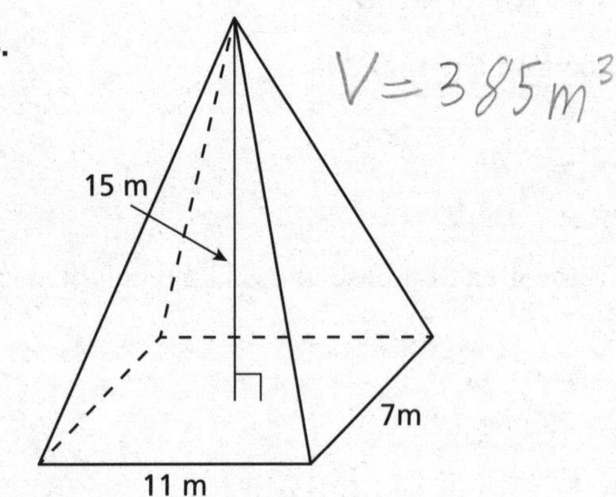

5.

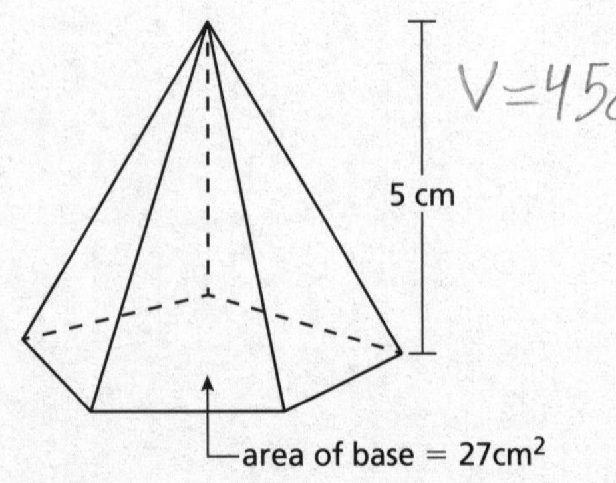

6.

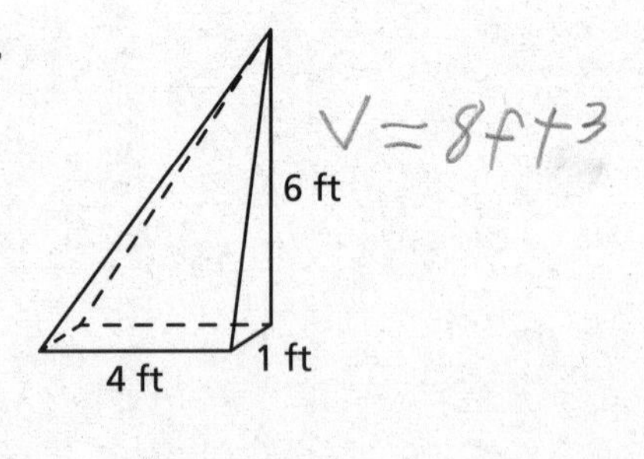

Name____ Date__________

11.6 Notetaking with Vocabulary (continued)

In Exercises 7–9, find the indicated measure.

7. A pyramid with a square base has a volume of 128 cubic inches and a height of 6 inches. Find the side length of the square base.

8. A pyramid with a rectangular base has a volume of 6 cubic feet. The length of the rectangular base is 3 feet and the width of the base is 1.5 feet. Find the height of the pyramid.

9. A pyramid with a triangular base has a volume of 18 cubic centimeters. The height of the pyramid is 9 centimeters and the height of the triangular base is 3 centimeters. Find the width of the base.

10. The pyramids are similar. Find the volume of pyramid B.

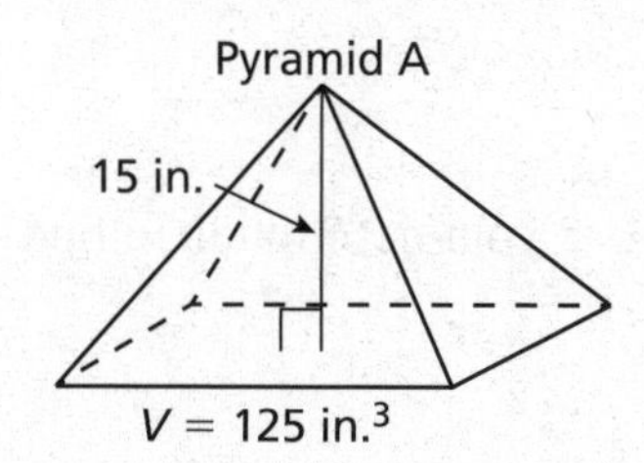

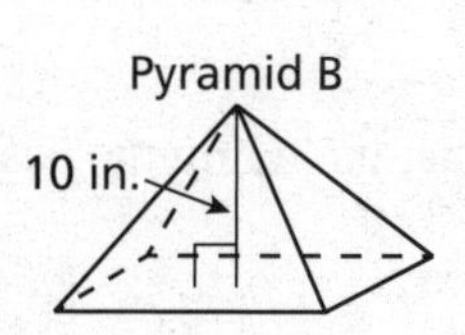

Name ______________________________ Date __________

11.7 Surface Areas and Volumes of Cones

For use with Exploration 11.7

Essential Question How can you find the surface area and the volume of a cone?

1 EXPLORATION: Finding the Surface Area of a Cone

Work with a partner. Construct a circle with a radius of 3 inches. Mark the circumference of the circle into six equal parts, and label the length of each part. Then cut out one sector of the circle and make a cone.

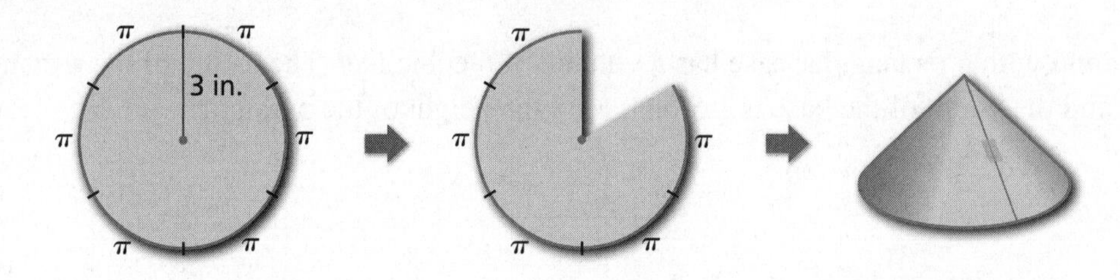

a. Explain why the base of the cone is a circle. What are the circumference and radius of the base?

b. What is the area of the original circle? What is the area with one sector missing?

c. Describe the surface area of the cone, including the base. Use your description to find the surface area.

2 EXPLORATION: Finding the Volume of a Cone

Work with a partner. The cone and the cylinder have the same height and the same circular base.

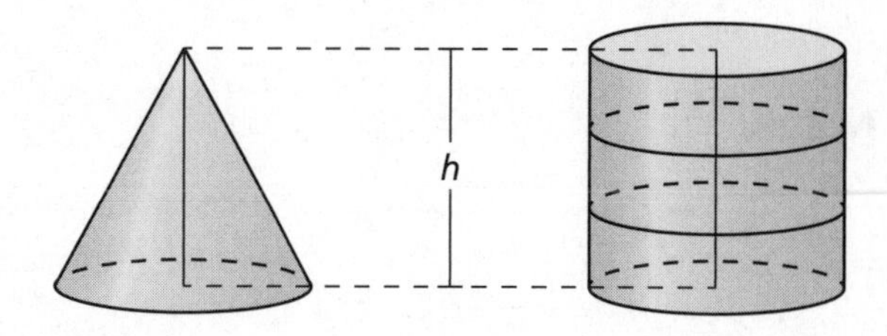

When the cone is filled with sand and poured into the cylinder, it takes three cones to fill the cylinder.

Use this information to write a formula for the volume V of a cone.

Communicate Your Answer

3. How can you find the surface area and the volume of a cone?

4. In Exploration 1, cut another sector from the circle and make a cone. Find the radius of the base and the surface area of the cone. Repeat this three times, recording your results in a table. Describe the pattern.

Radius of Base	Surface Area of Cone

Name ______________________________ Date __________

11.7 Notetaking with Vocabulary
For use after Lesson 11.7

In your own words, write the meaning of each vocabulary term.

lateral surface of a cone

Notes:

Core Concepts

Surface Area of a Right Cone

The surface area S of a right cone is

$$S = \pi r^2 + \pi r\ell$$

where r is the radius of the base and ℓ is the slant height.

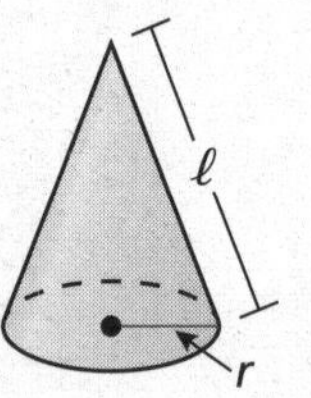

Notes:

Name______________________________ Date__________

Volume of a Cone

The volume V of a cone is

$$V = \frac{1}{3}Bh = \frac{1}{3}\pi r^2 h$$

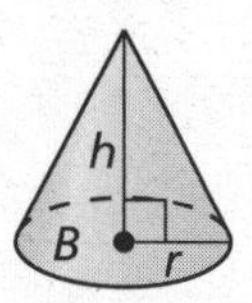

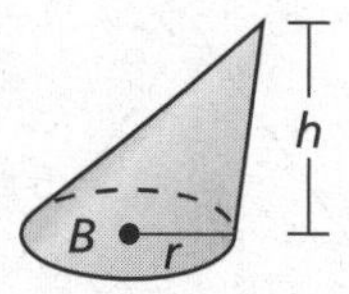

where B is the area of a base, h is the height, and r is the radius of the base.

Notes:

Extra Practice

In Exercises 1 and 2, find the surface area of the right cone.

1.

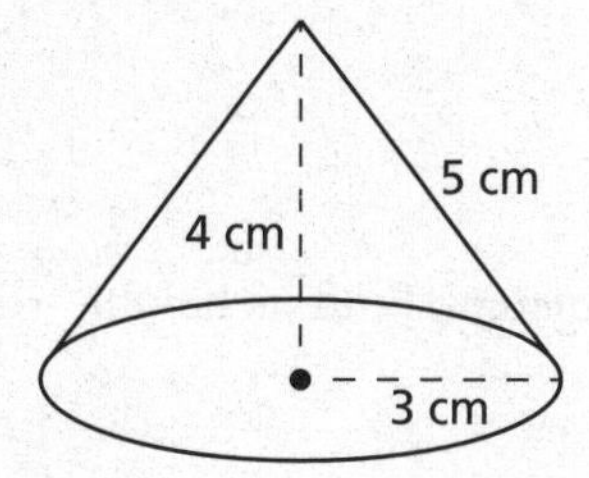

2. A right cone has a diameter of 1.8 inches and a height of 3 inches.

Name ______________________ Date ________

11.7 Notetaking with Vocabulary (continued)

In Exercises 3 and 4, find the volume of the cone.

3.

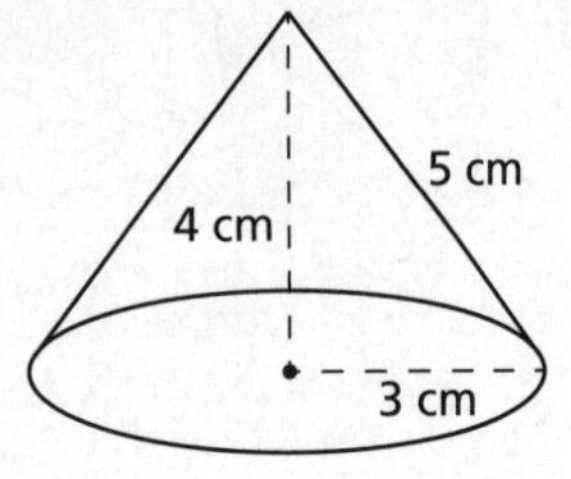

4. A right cone has a radius of 5 feet and a slant height of 13 feet.

In Exercises 5–7, find the indicated measure.

5. A right cone has a surface area of 440 square inches and a radius of 7 inches. Find its slant height.

6. A right cone has a volume of 528 cubic meters and a diameter of 12 meters. Find its height.

7. Cone A and cone B are similar. The radius of cone A is 4 cm and the radius of cone B is 10 cm. The volume of cone A is 134 cm^3, find the volume of cone B.

8. Find the volume of the composite solid.

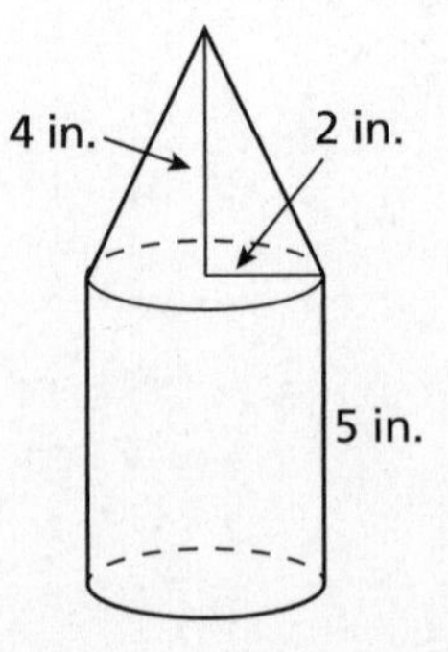

Name____ Date__________

11.8 Surface Areas and Volumes of Spheres

For use with Exploration 11.8

Essential Question How can you find the surface area and the volume of a sphere?

1 EXPLORATION: Finding the Surface Area of a Sphere

Work with a partner. Remove the covering from a baseball or softball.

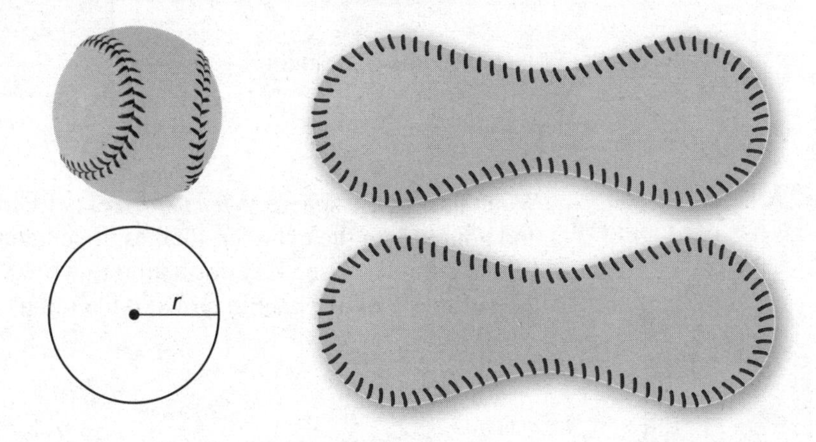

You will end up with two "figure 8" pieces of material, as shown above. From the amount of material it takes to cover the ball, what would you estimate the surface area S of the ball to be? Express your answer in terms of the radius r of the ball.

S = ______________________ Surface area of a sphere

Use the Internet or some other resource to confirm that the formula you wrote for the surface area of a sphere is correct.

11.8 Surface Areas and Volumes of Spheres (continued)

2 EXPLORATION: Finding the Volume of a Sphere

Work with a partner. A cylinder is circumscribed about a sphere, as shown. Write a formula for the volume V of the cylinder in terms of the radius r.

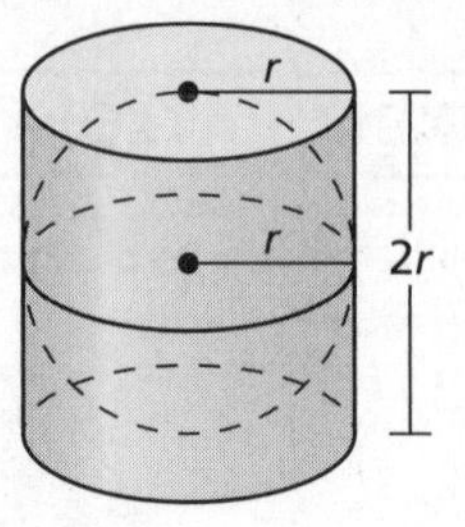

$V =$ ____________________ Volume of cylinder

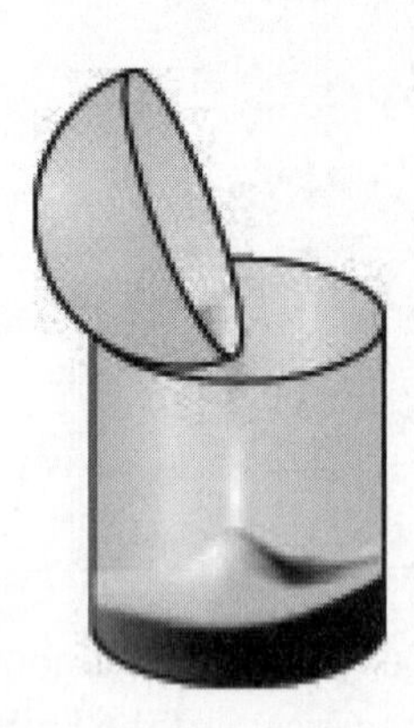

When half of the sphere (a *hemisphere*) is filled with sand and poured into the cylinder, it takes three hemispheres to fill the cylinder. Use this information to write a formula for the volume V of a sphere in terms of the radius r

$V =$ ____________________ Volume of a sphere

Communicate Your Answer

3. How can you find the surface area and the volume of a sphere?

4. Use the results of Explorations 1 and 2 to find the surface area and the volume of a sphere with a radius of (a) 3 inches and (b) 2 centimeters.

Name______________________________ Date__________

11.8 Notetaking with Vocabulary
For use after Lesson 11.8

In your own words, write the meaning of each vocabulary term.

chord of a sphere

great circle

Core Concepts

Surface Area of a Sphere

The surface area S of a sphere is

$$S = 4\pi r^2$$

where r is the radius of the sphere.

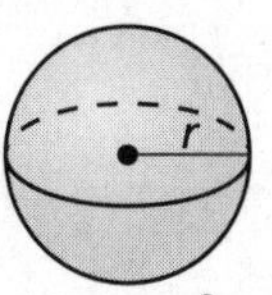

Notes:

Name ______________________________ Date __________

11.8 Notetaking with Vocabulary (continued)

Volume of a Sphere

The volume V of a sphere is

$$V = \frac{4}{3}\pi r^3$$

where r is the radius of the sphere.

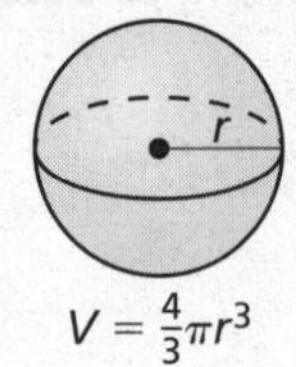

Notes:

Extra Practice

In Exercises 1–4, find the surface area of the solid.

1.

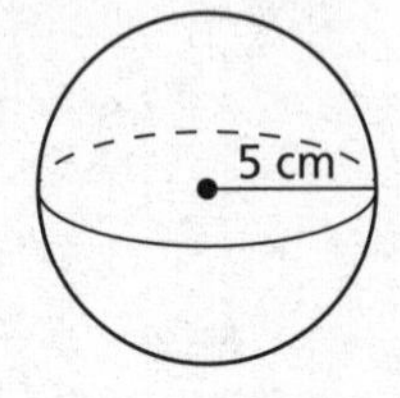

2.

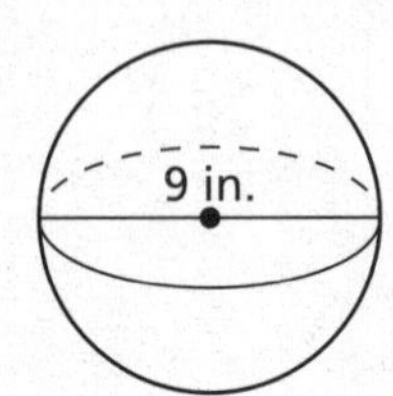

3.

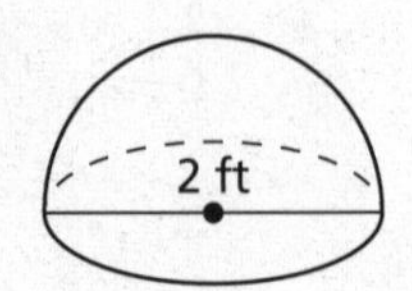

4.

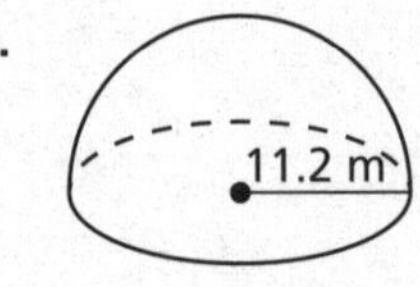

Name________________________________ Date__________

11.8 Notetaking with Vocabulary (continued)

In Exercises 5–8, find the volume of the sphere.

5.

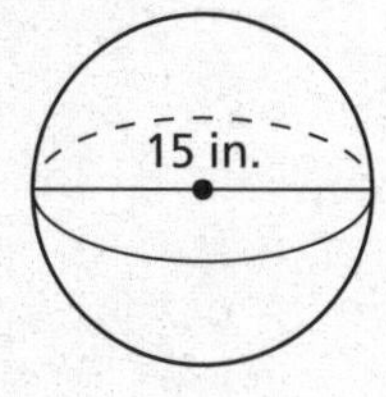

6.

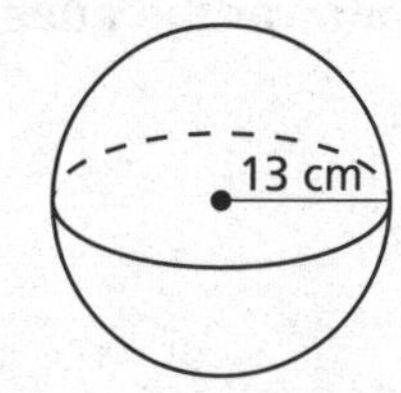

7.

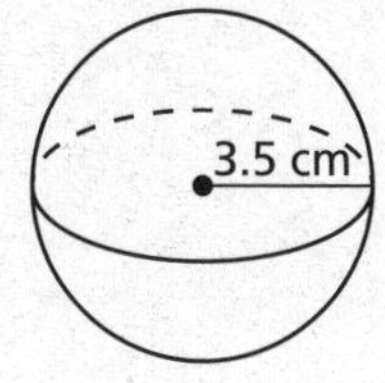

8.

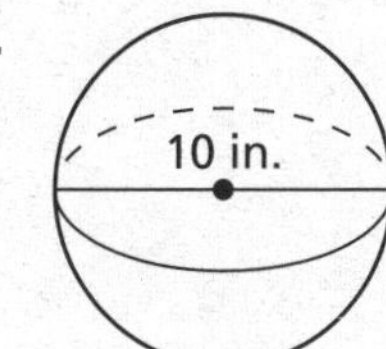

In Exercises 9–11, find the indicated measure.

9. Find the diameter of a sphere with a surface area of 144π square centimeters.

10. Find the volume of a sphere with a surface area of 256π square inches.

11. Find the volume of a sphere with a surface area of 400π square feet.

Name ____ Date __________

Chapter 12 Maintaining Mathematical Proficiency

Write and solve a proportion to answer the question.

1. What percent of 260 is 65?

2. What number is 32% of 75?

3. 15.01 is what percent of 19?

Display the data in a histogram.

4.

	Number of Strikeouts in One Game		
Strikeouts	0–3	4–7	8–11
Frequency	34	20	8

5.

	Number of Days of Exercise in One Week			
Days of Exercise	0–1	2–3	4–5	6–7
Frequency	4	26	22	6

Name____ Date__________

12.1 Sample Spaces and Probability

For use with Exploration 12.1

Essential Question How can you list the possible outcomes in the sample space of an experiment?

The **sample space** of an experiment is the set of all possible outcomes for that experiment.

1 EXPLORATION: Finding the Sample Space of an Experiment

Go to *BigIdeasMath.com* for an interactive tool to investigate this exploration.

Work with a partner. In an experiment, three coins are flipped. List the possible outcomes in the sample space of the experiment.

2 EXPLORATION: Finding the Sample Space of an Experiment

Go to *BigIdeasMath.com* for an interactive tool to investigate this exploration.

Work with a partner. List the possible outcomes in the sample space of the experiment.

a. One six-sided die is rolled.

b. Two six-sided dice are rolled.

3 EXPLORATION: Finding the Sample Space of an Experiment

Go to *BigIdeasMath.com* for an interactive tool to investigate this exploration.

Work with a partner. In an experiment, a spinner is spun.

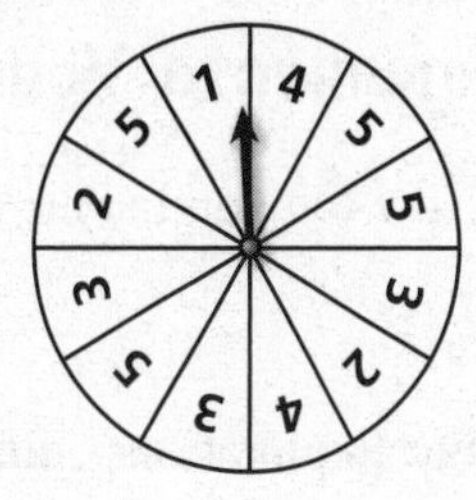

a. How many ways can you spin a 1? 2? 3? 4? 5?

Name ______________________________ Date ________

3 EXPLORATION: Finding the Sample Space of an Experiment (continued)

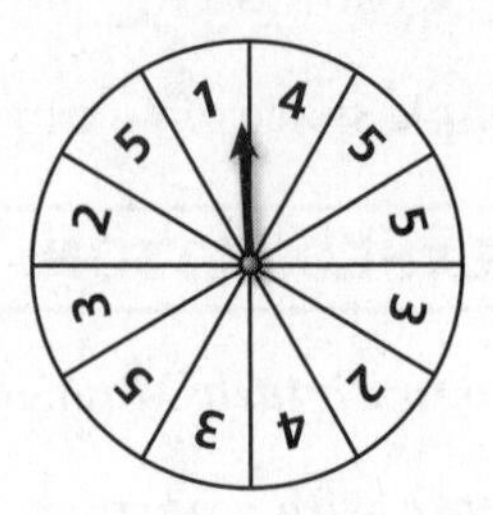

b. List the sample space.

c. What is the total number of outcomes?

4 EXPLORATION: Finding the Sample Space of an Experiment

Go to *BigIdeasMath.com* for an interactive tool to investigate this exploration.

Work with a partner. In an experiment, a bag contains 2 blue marbles and 5 red marbles. Two marbles are drawn from the bag.

a. How many ways can you choose two blue? a red then blue? a blue then red? two red?

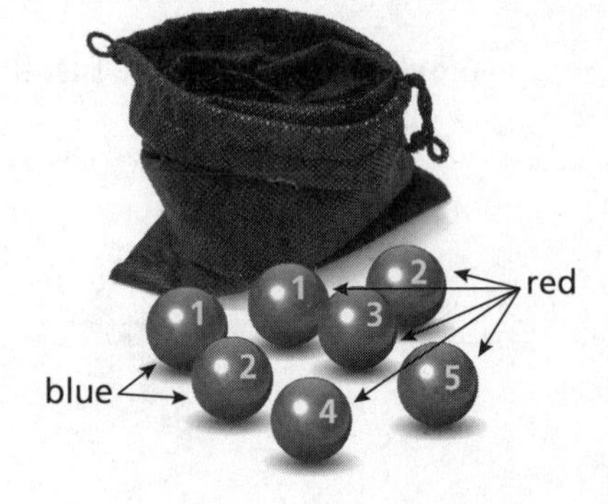

b. List the sample space.

c. What is the total number of outcomes?

Communicate Your Answer

5. How can you list the possible outcomes in the sample space of an experiment?

6. For Exploration 3, find the ratio of the number of each possible outcome to the total number of outcomes. Then find the sum of these ratios. Repeat for Exploration 4. What do you observe?

Name____ Date__________

12.1 Notetaking with Vocabulary
For use after Lesson 12.1

In your own words, write the meaning of each vocabulary term.

probability experiment

outcome

event

sample space

probability of an event

theoretical probability

geometric probability

experimental probability

Core Concepts

Probability of the Complement of an Event

The probability of the complement of event A is

$$P(\overline{A}) = 1 - P(A).$$

Notes:

Extra Practice

In Exercises 1 and 2, find the number of possible outcomes in the sample space. Then list the possible outcomes.

1. A stack of cards contains the thirteen clubs from a standard deck of cards. You pick a card from the stack and flip two coins.

2. You spin a spinner with the numbers 1–5 on it and roll a die.

3. When two tiles with numbers between 1 and 10 are chosen from two different bags, there are 100 possible outcomes. Find the probability that (a) the sum of the two numbers is not 10 and (b) the product of the numbers is greater than 10.

4. At a school dance, the parents sell pizza slices. The table shows the number of pizza slices that are available. A student chooses a slice at random. What is the probability that the student chooses a thin crust slice with pepperoni?

	Pepperoni	**Plain Cheese**
Thin Crust	34	36
Thick Crust	8	12

Name____ Date__________

12.1 Notetaking with Vocabulary (continued)

5. Find the probability that the polynomial $x^2 - x - c$ can be factored if c is a randomly chosen integer from 1 to 12.

6. You throw a dart at the board shown. Your dart is equally likely to hit any point inside the square board.

a. What is the probability your dart lands in the smallest triangle?

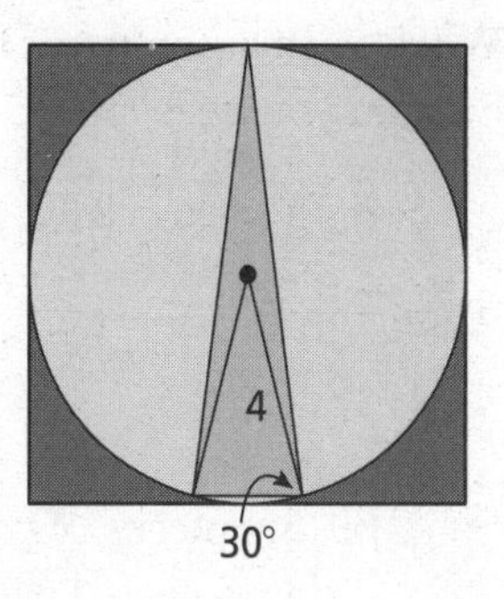

b. What is the probability your dart does not land anywhere in the circle?

7. The sections of a spinner are numbered 1 through 12. Each section of the spinner has the same area. You spin the spinner 180 times. The table shows the results. For which number is the experimental probability of stopping on the number the same as the theoretical probability?

Spinner Results											
1	**2**	**3**	**4**	**5**	**6**	**7**	**8**	**9**	**10**	**11**	**12**
13	21	22	20	11	8	14	9	15	12	18	17

Name ______________________________ Date __________

12.2 Independent and Dependent Events

For use with Exploration 12.2

Essential Question How can you determine whether two events are independent or dependent?

Two events are **independent events** when the occurrence of one event does not affect the occurrence of the other event. Two events are **dependent events** when the occurrence of one event *does* affect the occurrence of the other event.

1 EXPLORATION: Identifying Independent and Dependent Events

Work with a partner. Determine whether the events are independent or dependent. Explain your reasoning.

a. Two six-sided dice are rolled.

b. Six pieces of paper, numbered 1 through 6, are in a bag. Two pieces of paper are selected one at a time without replacement.

2 EXPLORATION: Finding Experimental Probabilities

Go to *BigIdeasMath.com* for an interactive tool to investigate this exploration.

Work with a partner.

a. In Exploration 1(a), experimentally estimate the probability that the sum of the two numbers rolled is 7. Describe your experiment.

b. In Exploration 1(b), experimentally estimate the probability that the sum of the two numbers selected is 7. Describe your experiment.

3 EXPLORATION: Finding Theoretical Probabilities

Work with a partner.

a. In Exploration 1(a), find the theoretical probability that the sum of the two numbers rolled is 7. Then compare your answer with the experimental probability you found in Exploration 2(a).

b. In Exploration 1(b), find the theoretical probability that the sum of the two numbers selected is 7. Then compare your answer with the experimental probability you found in Exploration 2(b).

c. Compare the probabilities you obtained in parts (a) and (b).

Communicate Your Answer

4. How can you determine whether two events are independent or dependent?

5. Determine whether the events are independent or dependent. Explain your reasoning.

a. You roll a 4 on a six-sided die and spin red on a spinner.

b. Your teacher chooses a student to lead a group, chooses another student to lead a second group, and chooses a third student to lead a third group.

Name ____ Date __________

12.2 Notetaking with Vocabulary
For use after Lesson 12.2

In your own words, write the meaning of each vocabulary term.

independent events

dependent events

conditional probability

Core Concepts

Probability of Independent Events

Words Two events A and B are independent events if and only if the probability that both events occur is the product of the probabilities of the events.

Symbols $P(A \text{ and } B) = P(A) \bullet P(B)$

Notes:

Probability of Dependent Events

Words If two events A and B are dependent events, then the probability that both events occur is the product of the probability of the first event and the conditional probability of the second event given the first event.

Symbols $P(A \text{ and } B) = P(A) \bullet P(B|A)$

Example Using the information in Example 2:

$$P(\text{girl first and girl second}) = P(\text{girl first}) \bullet P(\text{girl second}|\text{girl first})$$

$$= \frac{9}{12} \bullet \frac{6}{9} = \frac{1}{2}$$

Notes:

Extra Practice

In Exercises 1 and 2, determine whether the events are independent. Explain your reasoning.

1. You have three white golf balls and two yellow golf balls in a bag. You randomly select one golf ball to hit now and another golf ball to place in your pocket. Use a sample space to determine whether randomly selecting a white golf ball first and then a white golf ball second are independent events.

2. Your friend writes a phone number down on a piece of paper but the last three numbers get smudged after being in your pocket all day long. You decide to randomly choose numbers for each of the three digits. Use a sample space to determine whether guessing the first digit correctly and the second digit correctly are independent events.

12.2 Notetaking with Vocabulary (continued)

3. You are trying to guess a three-letter password that uses only the letters A, E, I, O, U, and Y. Letters can be used more than once. Find the probability that you pick the correct password "YOU."

4. You are trying to guess a three-letter password that uses only the letters A, E, I, O, U, and Y. Letters *cannot* be used more than once. Find the probability that you pick the correct password "AIE."

5. The table shows the number of male and female college students who played collegiate basketball and collegiate soccer in the United States in a recent year.

	Collegiate Soccer	Collegiate Basketball
Male	37,240	31,863
Female	36,523	28,002

a. Find the probability that a randomly selected collegiate soccer player is female.

b. Find the probability that a randomly selected male student is a collegiate basketball player.

Name____ Date__________

12.3 Two-Way Tables and Probability

For use with Exploration 12.3

Essential Question How can you construct and interpret a two-way table?

1 EXPLORATION: Completing and Using a Two-Way Table

Work with a partner. A *two-way table* displays the same information as a Venn diagram. In a two-way table, one category is represented by the rows and the other category is represented by the columns.

The Venn diagram shows the results of a survey in which 80 students were asked whether they play a musical instrument and whether they speak a foreign language. Use the Venn diagram to complete the two-way table. Then use the two-way table to answer each question.

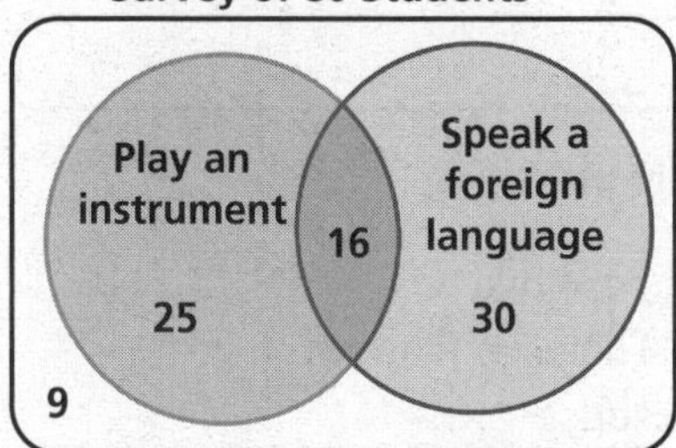

	Play an Instrument	Do Not Play an Instrument	Total
Speak a Foreign Language			
Do Not Speak a Foreign Language			
Total			

a. How many students play an instrument?

b. How many students speak a foreign language?

c. How many students play an instrument and speak a foreign language?

d. How many students do not play an instrument and do not speak a foreign language?

e. How many students play an instrument and do not speak a foreign language?

2 EXPLORATION: Two-Way Tables and Probability

Work with a partner. In Exploration 1, one student is selected at random from the 80 students who took the survey. Find the probability that the student

a. plays an instrument.

2 EXPLORATION: Two-Way Tables and Probability (continued)

b. speaks a foreign language.

c. plays an instrument and speaks a foreign language.

d. does not play an instrument and does not speak a foreign language.

e. plays an instrument and does not speak a foreign language.

3 EXPLORATION: Conducting a Survey

Go to *BigIdeasMath.com* for an interactive tool to investigate this exploration.

Work with your class. Conduct a survey of students in your class. Choose two categories that are different from those given in Explorations 1 and 2. Then summarize the results in both a Venn diagram and a two-way table. Discuss the results.

Communicate Your Answer

4. How can you construct and interpret a two-way table?

5. How can you use a two-way table to determine probabilities?

Name____ Date__________

12.3 Notetaking with Vocabulary
For use after Lesson 12.3

In your own words, write the meaning of each vocabulary term.

two-way table

joint frequency

marginal frequency

joint relative frequency

marginal relative frequency

conditional relative frequency

Core Concepts

Relative and Conditional Relative Frequencies

A **joint relative frequency** is the ratio of a frequency that is not in the total row or the total column to the total number of values or observations.

A **marginal relative frequency** is the sum of the joint relative frequencies in a row or a column.

A **conditional relative frequency** is the ratio of a joint relative frequency to the marginal relative frequency. You can find a conditional relative frequency using a row total or a column total of a two-way table.

Notes:

Extra Practice

In Exercises 1 and 2, complete the two-way table.

1.

		Arrival		
		Tardy	On Time	Total
Method	Walk	22		
	City Bus			60
	Total		58	130

2.

		Response		
		Yes	No	Total
Age	Under 21		24	25
	Over 21	29		
	Total	30		75

3. A survey was taken of 100 families with one child and 86 families with two or more children to determine whether they were saving for college. Of those, 94 of the families with one child and 60 of the families with two or more children were saving for college. Organize these results in a two-way table. Then find and interpret the marginal frequencies.

12.3 Notetaking with Vocabulary (continued)

4. In a survey, 214 ninth graders played video games every day of the week and 22 ninth graders did not play video games every day of the week. Of those that played every day of the week, 36 had trouble sleeping at night. Of those that did not play every day of the week, 7 had trouble sleeping at night. Make a two-way table that shows the joint and marginal relative frequencies.

5. For financial reasons, a school district is debating about eliminating a Computer Programming class at the high school. The district surveyed parents, students, and teachers. The results, given as joint relative frequencies, are shown in the two-way table.

		Population		
		Parents	**Students**	**Teachers**
Response	**Yes**	0.58	0.08	0.10
	No	0.06	0.15	0.03

a. What is the probability that a randomly selected parent voted to eliminate the class?

b. What is the probability that a randomly selected student did not want to eliminate the class?

c. Determine whether voting to eliminate the class and being a teacher are independent events.

Name ______________________________ Date __________

12.4 Probability of Disjoint and Overlapping Events

For use with Exploration 12.4

Essential Question How can you find probabilities of disjoint and overlapping events?

Two events are **disjoint**, or **mutually exclusive**, when they have no outcomes in common. Two events are **overlapping** when they have one or more outcomes in common.

1 EXPLORATION: Disjoint Events and Overlapping Events

Go to *BigIdeasMath.com* for an interactive tool to investigate this exploration.

Work with a partner. A six-sided die is rolled. Draw a Venn diagram that relates the two events. Then decide whether the events are disjoint or overlapping.

a. Event A: The result is an even number.
Event B: The result is a prime number.

b. Event A: The result is 2 or 4.
Event B: The result is an odd number.

2 EXPLORATION: Finding the Probability that Two Events Occur

Work with a partner. A six-sided die is rolled. For each pair of events, find (a) $P(A)$, (b) $P(B)$, (c) $P(A \text{ and } B)$, and (d) $P(A \text{ or } B)$.

a. Event A: The result is an even number.
Event B: The result is a prime number.

b. Event A: The result is a 2 or 4.
Event B: The result is an odd number.

Name____ Date__________

3 EXPLORATION: Discovering Probability Formulas

Go to *BigIdeasMath.com* for an interactive tool to investigate this exploration.

Work with a partner.

a. In general, if event A and event B are disjoint, then what is the probability that event A or event B will occur? Use a Venn diagram to justify your conclusion.

b. In general, if event A and event B are overlapping, then what is the probability that event A or event B will occur? Use a Venn diagram to justify your conclusion.

c. Conduct an experiment using a six-sided die. Roll the die 50 times and record the results. Then use the results to find the probabilities described in Exploration 2. How closely do your experimental probabilities compare to the theoretical probabilities you found in Exploration 2?

Communicate Your Answer

4. How can you find probabilities of disjoint and overlapping events?

5. Give examples of disjoint events and overlapping events that do not involve dice.

Name ______________________________ Date __________

12.4 Notetaking with Vocabulary
For use after Lesson 12.4

In your own words, write the meaning of each vocabulary term.

compound event

overlapping events

disjoint or mutually exclusive events

Core Concepts

Probability of Compound Events

If A and B are any two events, then the probability of A or B is

$$P(A \text{ or } B) = P(A) + P(B) - P(A \text{ and } B).$$

If A and B are disjoint events, then the probability of A or B is

$$P(A \text{ or } B) = P(A) + P(B).$$

Notes:

12.4 Notetaking with Vocabulary (continued)

Extra Practice

1. Events A and B are disjoint. $P(A) = \frac{2}{3}$ and $P(B) = \frac{1}{6}$. Find $P(A \text{ or } B)$.

2. $P(A) = 0.8$, $P(B) = 0.05$, and $P(A \text{ or } B) = 0.6$. Find $P(A \text{ and } B)$.

In Exercises 3–6, a vehicle is randomly chosen from a parking lot. The parking lot contains three red minivans, two blue minivans, three blue convertibles, one black pickup truck, three black motorcycles, one red motorcycle and two blue scooters. Find the probability of selecting the type of vehicle.

3. A red vehicle or a minivan

4. A scooter or a black vehicle

5. A black vehicle or a motorcycle

6. A four-wheeled vehicle or a blue vehicle

7. During a basketball game, the coach needs to select a player to make the free throw after a technical foul on the other team. There is a 68% chance that the coach will select you and a 26% chance that the coach will select your friend. What is the probability that you or your friend is selected to make the free throw?

8. Two six-sided dice are rolled. Find the probability of rolling the same number twice.

9. Out of 120 student parents, 90 of them can chaperone the Homecoming dance or the Prom. There are 40 parents who can chaperone the Homecoming dance and 65 parents who can chaperone the Prom. What is the probability that a randomly selected parent can chaperone both the Homecoming dance and the Prom?

10. A football team scores a touchdown first 75% of the time when they start with the ball. The team does not score first 51% of the time when their opponent starts with the ball. The team who gets the ball first is determined by a coin toss. What is the probability that the team scores a touchdown first?

Name______________________________ Date__________

12.5 Permutations and Combinations

For use with Exploration 12.5

Essential Question How can a tree diagram help you visualize the number of ways in which two or more events can occur?

1 EXPLORATION: Reading a Tree Diagram

Work with a partner. Two coins are flipped and the spinner is spun. The tree diagram shows the possible outcomes.

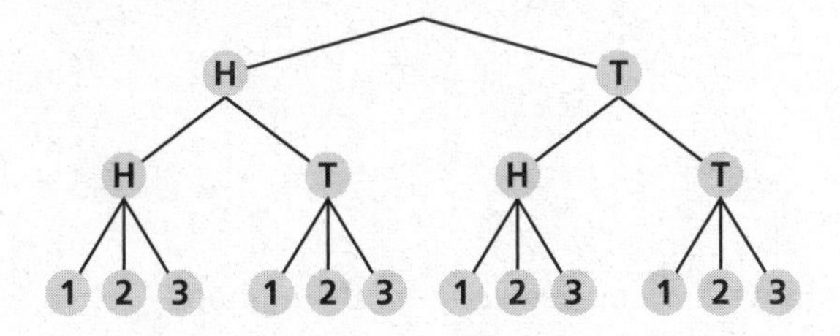

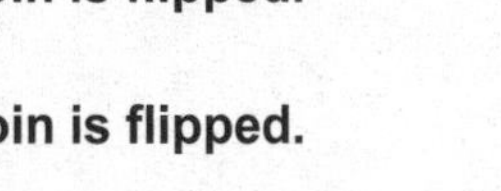

Spinner is spun.

a. How many outcomes are possible?

b. List the possible outcomes.

2 EXPLORATION: Reading a Tree Diagram

Work with a partner. Consider the tree diagram below.

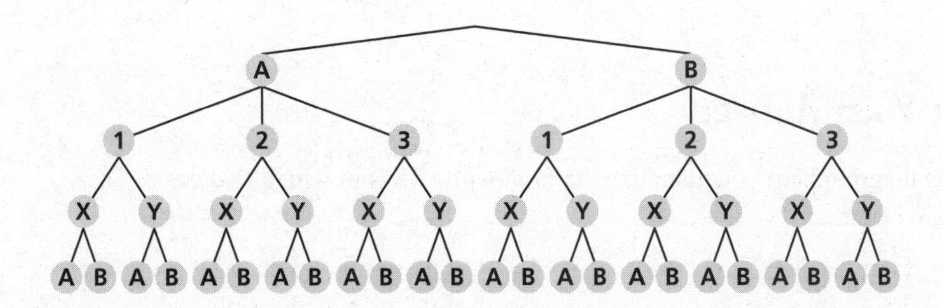

a. How many events are shown?

b. What outcomes are possible for each event?

c. How many outcomes are possible?

d. List the possible outcomes.

Name ______________________________ Date __________

3 EXPLORATION: Writing a Conjecture

Work with a partner.

a. Consider the following general problem: Event 1 can occur in m ways and event 2 can occur in n ways. Write a conjecture about the number of ways the two events can occur. Explain your reasoning.

b. Use the conjecture you wrote in part (a) to write a conjecture about the number of ways *more than* two events can occur. Explain your reasoning.

c. Use the results of Explorations 1(a) and 2(c) to verify your conjectures.

Communicate Your Answer

4. How can a tree diagram help you visualize the number of ways in which two or more events can occur?

5. In Exploration 1, the spinner is spun a second time. How many outcomes are possible?

Name____ Date______________

12.5 Notetaking with Vocabulary

For use after Lesson 12.5

In your own words, write the meaning of each vocabulary term.

permutation

n factorial

combination

Core Concepts

Permutations

Formulas	**Examples**
The number of permutations of n objects is given by $_nP_n = n!.$	The number of permutations of 4 objects is $_4P_4 = 4! = 4 \bullet 3 \bullet 2 \bullet 1 = 24.$
The number of permutations of n objects taken r at a time, where $r \le n$, is given by $_nP_r = \frac{n!}{(n-r)!}.$	The number of permutations of 4 objects taken 2 at a time is $_4P_2 = \frac{4!}{(4-2)!} = \frac{4 \bullet 3 \bullet \cancel{2!}}{\cancel{2!}} = 12.$

Notes:

12.5 Notetaking with Vocabulary (continued)

Combinations

Formula The number of combinations of n objects taken r at a time, where $r \leq n$, is given by

$${}_nC_r = \frac{n!}{(n-r)! \bullet r!}.$$

Example The number of combinations of 4 objects taken 2 at a time is

$${}_4C_2 = \frac{4!}{(4-2)! \bullet 2!}$$

$$= \frac{4 \bullet 3 \bullet \cancel{2!}}{\cancel{2!} \bullet (2 \bullet 1)}$$

$$= 6.$$

Notes:

Extra Practice

In Exercises 1 and 2, find the number of ways you can arrange (a) all of the numbers and (b) 3 of the numbers in the given amount.

1. $2,564,783

2. $4,128,675,309

3. Your rock band has nine songs recorded but you only want to put five of them on your demo CD to hand out to local radio stations. How many possible ways could the five songs be ordered on your demo CD?

4. A witness at the scene of a hit-and-run accident saw that the car that caused the accident had a license plate with only the letters I, R, L, T, O, and A. Find the probability that the license plate starts with a T and ends with an R.

5. How many possible combinations of three colors can be chosen from the seven colors of the rainbow?

Name ______________________________ Date __________

12.6 Binomial Distributions

For use with Exploration 12.6

Essential Question How can you determine the frequency of each outcome of an event?

1 EXPLORATION: Analyzing Histograms

Go to *BigIdeasMath.com* for an interactive tool to investigate this exploration.

Work with a partner. The histograms show the results when n coins are flipped.

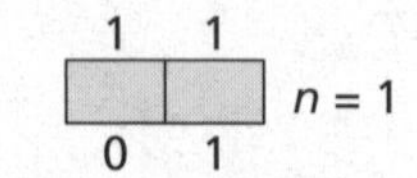

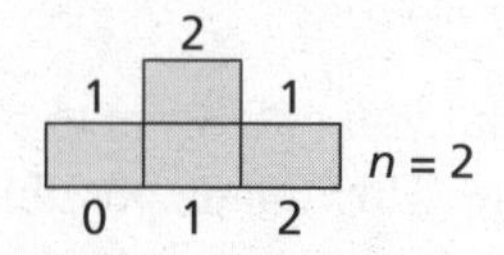

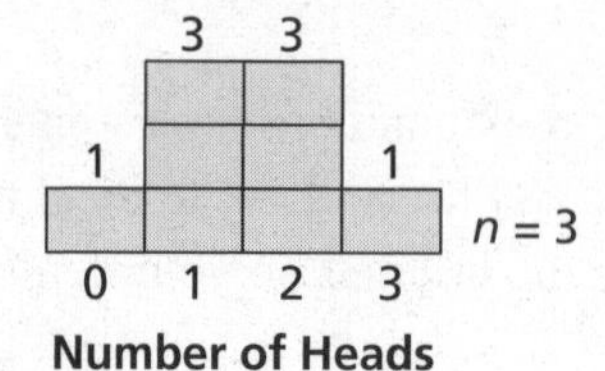

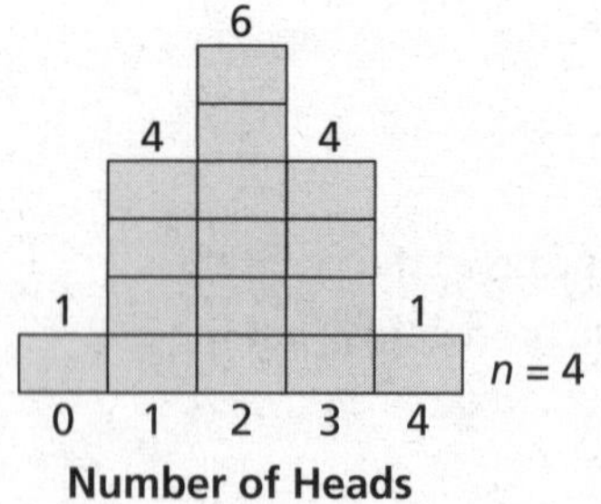

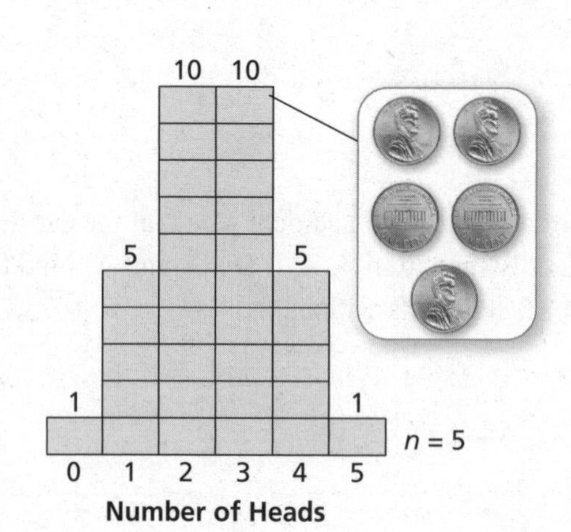

a. In how many ways can 3 heads occur when 5 coins are flipped?

b. Draw a histogram that shows the numbers of heads that can occur when 6 coins are flipped.

c. In how many ways can 3 heads occur when 6 coins are flipped?

2 EXPLORATION: Determining the Number of Occurrences

Work with a partner.

a. Complete the table showing the numbers of ways in which 2 heads can occur when n coins are flipped.

n	3	4	5	6	7
Occurrences of 2 heads					

b. Determine the pattern shown in the table. Use your result to find the number of ways in which 2 heads can occur when 8 coins are flipped.

Communicate Your Answer

3. How can you determine the frequency of each outcome of an event?

4. How can you use a histogram to find the probability of an event?

Name ______________________________ Date __________

12.6 Notetaking with Vocabulary
For use after Lesson 12.6

In your own words, write the meaning of each vocabulary term.

random variable

probability distribution

binomial distribution

binomial experiment

Core Concepts

Probability Distributions

A **probability distribution** is a function that gives the probability of each possible value of a random variable. The sum of all the probabilities in a probability distribution must equal 1.

Probability Distribution for Rolling a Six-Sided Die						
x	1	2	3	4	5	6
$P(x)$	$\frac{1}{6}$	$\frac{1}{6}$	$\frac{1}{6}$	$\frac{1}{6}$	$\frac{1}{6}$	$\frac{1}{6}$

Notes:

Binomial Experiments

A **binomial experiment** meets the following conditions.

- There are n independent trials.
- Each trial has only two possible outcomes: success and failure.
- The probability of success is the same for each trial. This probability is denoted by p. The probability of failure is $1 - p$.

For a binomial experiment, the probability of exactly k successes in n trials is

$$P(k \text{ successes}) = {}_nC_k p^k (1 - p)^{n-k}.$$

Notes:

Extra Practice

1. Make a table and draw a histogram showing the probability distribution for the random variable P if P = the product when two six-sided dice are rolled.

2. Use the probability distribution to determine (a) the number that is most likely to be spun on a spinner, and (b) the probability of spinning a perfect square.

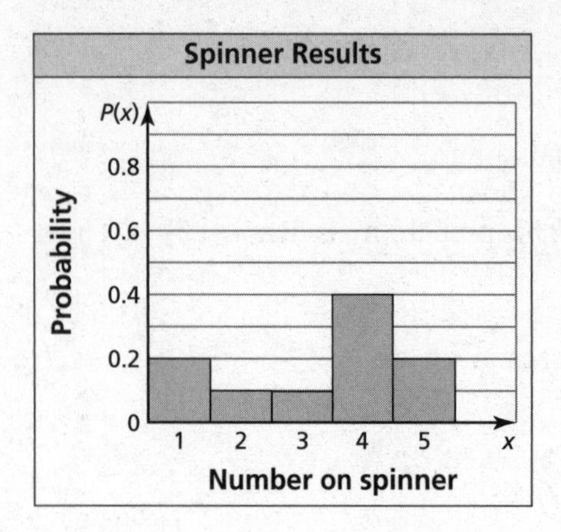

3. Calculate the probability of flipping a coin twenty times and getting nineteen heads.

4. According to a survey, 78% of women in a city watch professional football. You ask four randomly chosen women from the city whether they watch professional football.

   a. Draw a histogram of the binomial distribution for your survey.

   b. What is the most likely outcome of your survey?

   c. What is the probability that at most one woman watches professional football?